The Elementary Process of Bremsstrahlung

World Scientific Lecture Notes in Physics

ISSN: 1793-1436

*Published titles**

Vol. 65: Universal Fluctuations: The Phenomenology of Hadronic Matter
R Botet and M Ploszajczak

Vol. 66: Microcanonical Thermodynamics: Phase Transitions in "Small" Systems
D H E Gross

Vol. 67: Quantum Scaling in Many-Body Systems
M A Continentino

Vol. 69: Deparametrization and Path Integral Quantization of Cosmological Models
C Simeone

Vol. 70: Noise Sustained Patterns: Fluctuations and Nonlinearities
Markus Loecher

Vol. 71: The QCD Vacuum, Hadrons and Superdense Matter (2nd ed.)
Edward V Shuryak

Vol. 72: Massive Neutrinos in Physics and Astrophysics (3rd ed.)
R Mohapatra and P B Pal

Vol. 73: The Elementary Process of Bremsstrahlung
W Nakel and E Haug

Vol. 74: Lattice Gauge Theories: An Introduction (3rd ed.)
H J Rothe

Vol. 75: Field Theory: A Path Integral Approach (2nd ed.)
A Das

Vol. 76: Effective Field Approach to Phase Transitions and Some Applications to Ferroelectrics (2nd ed.)
J A Gonzalo

Vol. 77: Principles of Phase Structures in Particle Physics
H Meyer-Ortmanns and T Reisz

Vol. 78: Foundations of Quantum Chromodynamics: An Introduction to Perturbation Methods in Gauge Theories (3rd ed.)
T Muta

Vol. 79: Geometry and Phase Transitions in Colloids and Polymers
W Kung

Vol. 80: Introduction to Supersymmetry (2nd ed.)
H J W Müller-Kirsten and A Wiedemann

Vol. 81: Classical and Quantum Dynamics of Constrained Hamiltonian Systems
H J Rothe and K D Rothe

Vol. 82: Lattice Gauge Theories: An Introduction (4th ed.)
H J Rothe

*For the complete list of published titles, please visit
http://www.worldscientific.com/series/wslnp

World Scientific Lecture Notes in Physics – Vol. 73

The Elementary Process of Bremsstrahlung

Eberhard Haug
Institut für Astronomie und Astrophysik
Universität Tübingen, Germany

Werner Nakel
Physikalisches Institut
Universität Tübingen, Germany

World Scientific

Published by

World Scientific Publishing Co. Pte. Ltd.
5 Toh Tuck Link, Singapore 596224
USA office: 27 Warren Street, Suite 401-402, Hackensack, NJ 07601
UK office: 57 Shelton Street, Covent Garden, London WC2H 9HE

British Library Cataloguing-in-Publication Data
A catalogue record for this book is available from the British Library.

World Scientific Lecture Notes in Physics — Vol. 73
THE ELEMENTARY PROCESS OF BREMSSTRAHLUNG

Copyright © 2004 by World Scientific Publishing Co. Pte. Ltd.

All rights reserved. This book, or parts thereof, may not be reproduced in any form or by any means, electronic or mechanical, including photocopying, recording or any information storage and retrieval system now known or to be invented, without written permission from the Publisher.

For photocopying of material in this volume, please pay a copying fee through the Copyright Clearance Center, Inc., 222 Rosewood Drive, Danvers, MA 01923, USA. In this case permission to photocopy is not required from the publisher.

ISBN 978-981-238-578-9

Printed in Singapore

Preface

This book is based on a review article about the elementary process of bremsstrahlung [W. Nakel, *Phys. Rep.* **243** (1994) 317]. When the author was encouraged by the editors of World Scientific to expand the review article to a lecture notes volume, he asked a colleague (E.H.) who has been working on the theory of bremsstrahlung, to be the co-author.

The aim of the book is to represent the theoretical and experimental developments in the elementary bremsstrahlung process, i.e., the triply differential cross section, and the status of the comparison and extent of agreement between them. It will be shown how far theory proceeded in the effort to describe the coupling of the radiation field to the electron-atom system. Although the main importance is attached to electron-atom bremsstrahlung, we also discuss electron-electron bremsstrahlung, electron-positron bremsstrahlung, and further bremsstrahlung processes.

The book is mainly addressed to graduate students and is therefore endowed with classical and semiclassical considerations on the bremsstrahlung process and with problems to be solved. But it will also be of interest to physicists who are actively engaged in research since it summarizes and discusses the extensive theoretical work in the original literature. Last but not least, the book is addressed to those readers who may want to get a more general view on this subject since bremsstrahlung is related to the fundamentals of theory and appears in many branches of physics and in technical applications.

The authors are indebted to Prof. H. Ruder, Prof. G.J. Wagner, and Dr. H. Lindel for their support of the work, and to Prof. P. Grabmayr for his valuable help in the preparation of this book. We are very grateful to

Prof. L.C. Maximon for helpful discussions. We wish to thank Dipl.-Phys. E. Gaertig, F. Engeser and U. Vogl for their extensive technical help in getting the manuscript into its final form.

We acknowledge the permission by Elsevier Science to reproduce the following figures from the article which appeared in *Physics Reports* [86]: Figs. 4.1, 4.2, 4.4, 4.5, 4.7 to 4.14, 4.16, 5.6, and 6.1 to 6.6.

Eberhard Haug
Werner Nakel

Universität Tübingen, *September 2003*

Note from the authors

The authors would be grateful if the readers would inform them about any errors which they may find. Corrections for this book which come to the authors' attention will be posted on the World Wide Web at
http://www.pit.physik.uni-tuebingen.de/nakel/nakel_ee.html
but can also be obtained by writing to the authors. The contact e-mail address is nakel@pit.physik.uni-tuebingen.de.

Contents

Preface   v

Chapter 1. Introduction   1
1.1 General introduction 1
1.2 Short historical note 4
1.3 Notations and definitions 5

Chapter 2. Classical and semiclassical considerations on the bremsstrahlung process   9
2.1 Electron-nucleus bremsstrahlung 9
    2.1.1 Screening effects 16
    2.1.2 Bremsstrahlung linear polarization 18
    2.1.3 Bremsstrahlung from transversely polarized electrons . 22
2.2 Electron-electron bremsstrahlung 23
2.3 Weizsäcker-Williams method of virtual quanta 26

Chapter 3. Theory of the elementary process of electron-nucleus bremsstrahlung   27
3.1 Introduction 27
3.2 Bremsstrahlung cross section 28
3.3 Born approximation (Bethe-Heitler formula) 32
    3.3.1 Atomic screening 47
    3.3.2 Nuclear recoil effects 50
    3.3.3 Effects of nuclear structure 51
3.4 Approximations with the Sommerfeld-Maue wave function . . . 55

	3.4.1	Sommerfeld-Maue wave function	55
	3.4.2	Spin formalism	61
	3.4.3	Cross section	63
	3.4.4	Nonrelativistic approximation	71
	3.4.5	The short-wavelength limit	72
	3.4.6	Approximation for the long-wavelength limit and for high electron energies	74
	3.4.7	Cross section in second Born approximation	76
3.5	Calculation using relativistic partial-wave expansions		79
	3.5.1	Partial wave expansion	79
	3.5.2	Calculations with a relativistic self-consistent-field potential	89
	3.5.3	Calculation for a pure Coulomb potential	93
3.6	Spin-dependent cross section and bremsstrahlung asymmetry		97
3.7	Bremsstrahlung polarization		104
	3.7.1	General polarization correlation	104
	3.7.2	Linear polarization in Born approximation	108
	3.7.3	Circular polarization from polarized electrons in Born approximation	112
3.8	Radiative corrections to bremsstrahlung		115

Chapter 4. Experiments on the elementary process of electron-nucleus bremsstrahlung — 119

4.1	Survey of experimental devices		119
4.2	Electron-photon coincidence experiments without regard to polarization variables		121
	4.2.1	Angular distributions of photons for fixed directions of outgoing electrons	121
	4.2.2	Angular distributions of electrons for fixed photon directions	123
	4.2.3	Energy distributions for fixed electron and photon directions	125
	4.2.4	Wide-angle bremsstrahlung experiments at very high energies	127
4.3	Electron-photon coincidence experiments including polarization variables		129
	4.3.1	Linear polarization of bremsstrahlung emitted by unpolarized electrons	129

	4.3.2	Photon emission asymmetry of bremsstrahlung from transversely polarized electrons	136
	4.3.3	Further polarization correlations	143
4.4	Tagged photons		143

Chapter 5. Theory of the elementary process of electron-electron bremsstrahlung   **149**

5.1	Introduction		149
5.2	Kinematics		150
	5.2.1	Center-of-mass system	153
	5.2.2	Laboratory system	154
5.3	Cross section		159
	5.3.1	Laboratory system	163
	5.3.2	Center-of-mass system	164
	5.3.3	Center-of-mass system of the outgoing electrons	164
5.4	Bremsstrahlung in the field of bound electrons		168

Chapter 6. Experiments on the elementary process of electron-electron bremsstrahlung   **177**

6.1	Electron-photon coincidence experiments without regard to polarization variables		177
	6.1.1	Angular distribution of photons for fixed electron direction	178
	6.1.2	Energy distributions	180
6.2	Electron-photon coincidence experiments including polarization variables		184
	6.2.1	Linear polarization of electron-electron bremsstrahlung emitted by unpolarized electrons	184
	6.2.2	Further polarization correlations	187

Chapter 7. Integrated cross sections and further bremsstrahlung processes   **189**

7.1	Integrated cross sections	189
7.2	Positron-nucleus bremsstrahlung	192
7.3	Electron-positron bremsstrahlung	194
7.4	Two-photon bremsstrahlung	198
7.5	Polarization bremsstrahlung	205
7.6	Crystalline targets: coherent bremsstrahlung	208

7.7	Bremsstrahlung from heavy particles	208
7.8	Bremsstrahlung in nuclear decays	209
	7.8.1 Bremsstrahlung in β decay	209
	7.8.2 Bremsstrahlung in orbital-electron capture	209
	7.8.3 Bremsstrahlung in α decay	209
7.9	Bremsstrahlung in magnetic fields	210
	7.9.1 Electron-nucleus bremsstrahlung in strong magnetic fields	210
	7.9.2 Synchrotron radiation (magnetobremsstrahlung)	212
7.10	Stimulated bremsstrahlung	212

Conclusion	213
Appendix A Problems	**215**
Appendix B Squared matrix element of electron-electron bremsstrahlung	**243**
Bibliography	251
Index	257

Chapter 1
Introduction

1.1 General introduction

The emission of a photon in the scattering of an electron from an atom is called bremsstrahlung (braking radiation). Apart from the interest in the nature of the process itself, there are a variety of reasons why the bremsstrahlung process occupies such an important place in physics. Firstly, the process is related to the fundamentals of the theory since it is a consequence of the general coupling of the electromagnetic field and matter fields. Therefore bremsstrahlung appears in nearly all branches of physics: atomic and nuclear, solid-state and elementary-particle physics. Moreover, bremsstrahlung is an important tool in many areas of experimental research, in the field of astrophysics, and it has a wide range of technical applications.

The bremsstrahlung process is generally considered to be well understood. However, the comparisons between experiment and theory have for the most part been made for cases where only the emitted photons are considered, disregarding the decelerated outgoing electrons. Thus the results are necessarily integrated over all electron scattering angles, whereby important features are lost and the check of the theory is not as strong as it could be. When, on the other hand, the bremsstrahlung photons are detected in coincidence with the decelerated electrons scattered into a fixed direction, information on the elementary process of bremsstrahlung can be obtained and a more stringent check of the theoretical work becomes possible.

The quantity measured in the coincidence experiments and calculated by the pertinent theory is the triply differential cross section, differential in photon energy and angle, and in the scattering angle of the outgoing electron. However, the most detailed independently observable quantity is the triply differential cross section including all polarization correlations, corresponding to an electron-photon coincidence experiment with a spin-polarized primary beam and detectors sensitive to polarization of the photons and of the outgoing electrons. Such a complete scattering experiment has not yet been done, but several experiments including single polarization variables were performed, providing benchmarks for comparison with theoretical predictions.

The main force acting by the atom on the incident electron, leading to bremsstrahlung emission, is due to the Coulomb field of the nuclear charge. The effect of the atomic electrons is two-fold: on the one hand, the atomic electrons screen the Coulomb field of the nucleus as a static charge distribution and reduce the cross section for bremsstrahlung emission. The recoil momentum is taken up by the atom as a whole. On the other hand, the atomic electrons act as individual particles and the bremsstrahlung process may also take place in the collision with an atomic electron which then absorbs the recoil momentum and is ejected. This process is called electron-electron bremsstrahlung. With the aid of coincidence experiments it is possible to differentiate exactly between these two components.

The book is organized as follows: Chapter 2 is devoted to classical and semiclassical considerations on the bremsstrahlung process. In four main sections (Chapters 3 to 6) we will deal with the elementary processes of (screened) electron-nucleus bremsstrahlung and of electron-electron bremsstrahlung. We start in Chapter 3 with the discussion of theoretical predictions for relativistic electron-nucleus bremsstrahlung. Electron-photon coincidence experiments on electron-nucleus bremsstrahlung are discussed in Chapter 4. In all the experiments it turned out that there is a strong angular correlation with the photons emitted on the same side relative to the primary beam as the decelerated outgoing electrons (Section 4.2.1). A measured angular distribution of decelerated electrons for fixed photon direction shows a sharp forward peaking (Section 4.2.2).

Only two coincidence experiments were done for electron-nucleus bremsstrahlung including measurements of the polarizational variables. We first discuss measurements of linear polarization of bremsstrahlung emitted by unpolarized electrons (Section 4.3.1). Here the radiation was found to be almost completely polarized with the electric vector in the emission plane. Decreases of the degree of polarization were attributed to spin-flip radiation processes. In Section 4.3.2 we discuss the measurements of the photon emission asymmetry of bremsstrahlung from transversely polarized electrons, where spin-orbit effects are observed which are usually masked in non-coincidence experiments. Even in the case of zero deflection of the decelerated outgoing electrons a non-zero photon emission asymmetry was observed.

Further polarization correlations for which the elementary process has not yet been measured, are discussed in Section 4.3.3. In all experiments no spin analysis of the outgoing electrons is performed to date.

Section 4.4 deals with the application of the coincidence technique in producing quasimonochromatic photon beams, so-called tagged photons.

In Chapters 5 and 6 we investigate the process of electron-electron bremsstrahlung. In contrast to the electron-nucleus system the electron-electron system has no electric dipole moment and the electron-electron bremsstrahlung shows features of quadrupole radiation. The theoretical part starts with the kinematics of the process (Section 5.2). The calculation of the elementary radiation process in free electron-electron collisions is a straightforward application of quantum electrodynamics. However, the cross section becomes extremely complicated due to recoil and exchange effects even in lowest-order perturbation theory (Section 5.3). The radiative collision of a free electron with an initially bound atomic electron is calculated using the impulse approximation (Section 5.4). The electron-photon coincidence technique enables one to differentiate completely between electron-electron and electron-nucleus bremsstrahlung. Experiments without regard to polarization variables (Section 6.1) yield the angular distribution of photons for fixed direction of outgoing electrons (Section 6.1.1) and the energy distribution (Section 6.1.2). A measurement of the photon linear polarization of the elementary process of bremsstrahlung (Section 6.2) is the only polarization measurement on electron-electron bremsstrahlung so far.

We do not discuss in detail the integrated cross sections (doubly differential angular distribution, singly differential energy spectrum, and total energy loss). Only a summary of this subject is given in Chapter 7. In addition, further bremsstrahlung processes are mentioned, such as positron-nucleus bremsstrahlung, electron-positron bremsstrahlung, two-photon bremsstrahlung, polarization bremsstrahlung, coherent bremsstrahlung from crystalline targets, bremsstrahlung from heavier particles, bremsstrahlung in nuclear decays, and bremsstrahlung in magnetic fields.

1.2 Short historical note

X-rays were discovered by W.C. Röntgen in 1895, but the separate study of the two components, the characteristic line spectrum and the continuous spectrum, started only 18 years later. For the continuous spectrum Sommerfeld proposed the name bremsstrahlung (braking radiation) with the approval of Röntgen himself. Although a coincidence experiment between emitted photons and outgoing electrons had been suggested as early as 1932 by Scherzer [1], the first measurements of the elementary bremsstrahlung process have been performed by Nakel [2; 3; 4], beginning in 1966, those of the elementary process of electron-electron bremsstrahlung by Nakel and Pankau [5; 6; 7], beginning in 1972.

Theoretical investigations of the bremsstrahlung process have followed the general developments in physics. Initial attempts were on the basis of classical electrodynamics. Semiclassical calculations of Kramers [8] in 1923 and of Wentzel [9] made use of the correspondence principle. The first quantum-mechanical cross-section formulae for the elementary process were derived in 1931 by Sommerfeld [10] in the nonrelativistic dipole approximation including retardation. With the advent of the Dirac theory Bethe and Heitler [11] achieved in 1934 the first relativistic calculation of bremsstrahlung in Born approximation. The resulting cross section is available in a relatively simple analytical form which has the advantage that the integration over the angles of the outgoing electrons and photons can be performed in closed form. Relativistic cross-section calculations with approximate Coulomb wave functions (Sommerfeld-Maue functions) were carried out in 1954 by Bethe and Maximon [12] in the high-energy, small-angle approximation, and by Elwert and Haug [13] without making these assumptions. The best available theory involves the use of 'exact' wave functions describing an electron in a screened nuclear Coulomb field. This method requires extensive numerical computations of the matrix element by means of the partial-wave expansion. Therefore this procedure was feasible only in 1971 when Tseng and Pratt [14] succeeded in computing the doubly differential cross section and photon spectra. The evaluation of the triply differential cross section is still more complicated, and it was not until the nineties of the last century that a few groups presented results on the elementary process [15; 16; 17; 18].

The history of the use of the term 'bremsstrahlung' in English and German literature has been traced by Pratt and Feng [19].

1.3 Notations and definitions

In this section we present the notations and define the various quantities which are used throughout this book.

For the process of electron-atom bremsstrahlung, usually just called bremsstrahlung, we will mostly use the term electron-nucleus bremsstrahlung in order to differentiate it clearly from electron-electron bremsstrahlung. In sections where the terms electron-nucleus and electron-electron bremsstrahlung occur frequently, we will use the abbreviations e-nucleus bremsstrahlung and e-e bremsstrahlung, respectively.

In all experiments on the elementary bremsstrahlung process coplanar geometry was used so far, i.e., the momenta of the incoming and outgoing electrons and of the photons all lie in the same plane, the so-called emission plane or reaction plane.

In order to present the results from various authors in a clearly arranged form, we have standardized the different notations as shown in Table 1.3 and in Fig. 1.1.

Table 1.3

The following definitions are given for the symbols and constants used in this book:

(1) Electron-nucleus bremsstrahlung

E_0, E_e	kinetic energy (keV) of the incident and outgoing electron
$\mathbf{p}_0, \mathbf{p}_e$	momentum of the incident and outgoing electron
$h\nu$	energy (keV) of the emitted photon
$\mathbf{k}$	momentum of the emitted photon
$\theta_k = \theta_1$	angle between the directions of the incident electron and the emitted photon
θ_2	angle between the directions of the outgoing electron and the emitted photon
θ_e	angle between the directions of the incident and the outgoing electron
ϵ_1, ϵ_2	total energy of the incident and outgoing electron in units of the electron rest energy mc^2

$\mathbf{p}_1, \mathbf{p}_2$	momentum (in units of mc) of the incident and outgoing electron
k	energy of the emitted photon in units of mc^2
$\mathbf{k}$	momentum (in units of mc) of the emitted photon
$\mathbf{q}$	recoil momentum (in units of mc) of the nucleus
$\mathbf{e}$	photon polarization vector
P_k	degree of photon polarization
P_l	degree of photon linear polarization
P_c	degree of photon circular polarization
P_e	degree of polarization of the incident electron
C_{lmn}	polarization coefficient
$\vec{\zeta}_1, \vec{\zeta}_2$	spin vector of the ingoing and outgoing electron
β_1, β_2	ratio of the incident and outgoing electron velocity v_1, v_2 to the velocity of light c
Z	atomic number of the target atom (nucleus)
$\alpha \approx \frac{1}{137}$	fine-structure constant
r_0	classical electron radius

Vectors with a hat, such as $\hat{\mathbf{p}}_1$, $\hat{\mathbf{p}}_2$, $\hat{\mathbf{k}}$, denote unit vectors.

(2) Electron-electron bremsstrahlung

(a) Arbitrary frame of reference

E_1, E_2	kinetic energy (keV) of the incident electrons
E_1', E_2'	kinetic energy (keV) of the outgoing electrons
ϵ_1, ϵ_2	total energy (in units of mc^2) of the incident electrons
ϵ_1', ϵ_2'	total energy (in units of mc^2) of the outgoing electrons
$\mathbf{p}_1, \mathbf{p}_2$	momentum (in units of mc) of the incident electrons
$\mathbf{p}_1', \mathbf{p}_2'$	momentum (in units of mc) of the outgoing electrons
$h\nu$	energy (keV) of the emitted photon
k	energy (in units of mc^2) of the emitted photon
$\underline{p_1}, \underline{p_2}$	energy-momentum four-vector of the incident electrons
$\underline{p_1'}, \underline{p_2'}$	energy-momentum four-vector of the outgoing electrons
$\underline{k}$	energy-momentum four-vector of the emitted photon
θ	angle between the momenta $\mathbf{p}_1$ and $\mathbf{k}$
θ_1	angle between the momenta $\mathbf{p}_1$ and $\mathbf{p}_1'$

φ	azimuth angle of the momentum $\mathbf{p}'_1$ with respect to the reaction plane $(\mathbf{p}_1, \mathbf{k})$
α	angle between the momenta $\mathbf{p}'_1$ and $\mathbf{k}$
$\vec{\zeta}_1, \vec{\zeta}_2$	spin vectors of the incident electrons
$\vec{\zeta}'_1, \vec{\zeta}'_2$	spin vectors of the outgoing electrons
$\mathbf{e}$	photon polarization vector

(b) Laboratory system

E_0	kinetic energy (keV) of the incident electron
E_e	kinetic energy (keV) of one of the outgoing electrons
$h\nu$	energy (keV) of the emitted photon
θ_e	angle between the directions of the incident electron and one of the outgoing electrons
θ_k	angle between the directions of the incident electron and the emitted photon

(3) Electron-positron bremsstrahlung

E_-, E_+	kinetic energy (keV) of the incident electron and positron, respectively
ϵ_-, ϵ_+	total energy (in units of mc^2) of the incident electron and positron, respectively
$\mathbf{p}, \mathbf{q}$	momentum (in units of mc) of the incident electron and positron, respectively
E'_-, E'_+	kinetic energy (keV) of the outgoing electron and positron, respectively
ϵ'_-, ϵ'_+	total energy (in units of mc^2) of the outgoing electron and positron, respectively
$\mathbf{p}', \mathbf{q}'$	momentum (in units of mc) of the outgoing electron and positron, respectively
k	energy (in units of mc^2) of the emitted photon
$\mathbf{k}$	momentum (in units of mc) of the emitted photon
$\underline{p}, \underline{q}$	energy-momentum four-vector of the incident electron

$\underline{p}', \underline{q}'$ and positron, respectively
energy-momentum four-vector of the outgoing electron
and positron, respectively
$\underline{k}$ energy-momentum four-vector of the emitted photon

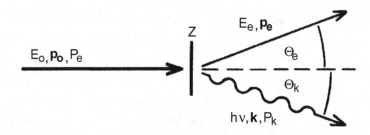

Fig. 1.1 Schematic diagram of coplanar geometry used in the experiments of electron-nucleus bremsstrahlung to introduce notations (in connection with Table 1.3).

Chapter 2

Classical and semiclassical considerations on the bremsstrahlung process

2.1 Electron-nucleus bremsstrahlung

A direct consequence of the classical theory of electrodynamics is that a charged particle, when accelerated, radiates an electromagnetic wave. In quantum mechanics there is a small but finite probability that a photon will be emitted each time a charged particle is accelerated. All experimental findings are in agreement with the results of quantum theory whereas even the most conspicuous feature of the bremsstrahlung spectrum, the sharp cut-off at short wavelengths, cannot be described classically. Nevertheless, classical and semiclassical calculations are presented in many textbooks (see, e.g., [20], [21]) and the limits of the applicability are discussed in detail. Also, numerical calculations on classical bremsstrahlung are published in original papers (see, e.g., [22]). Here we give only a qualitative view of the classical and semiclassical conception of bremsstrahlung emission from an accelerated charge.

We can understand graphically that an accelerated charge is radiating by considering the behaviour of the electric field as, for instance, described in detail in [23]. Figure 2.1 shows the field of an electron which had been moving (from left) with constant speed in a straight line until $t = 0$, at which time it reached $x = 0$. There it was abruptly stopped. Now the message that the electron was stopped cannot reach, by time t, any point farther away from the origin than ct. The field outside the sphere of radius ct must be that which would have prevailed if the electron had kept on moving at its original speed. That is why we see the outer part of the field lines in Fig. 2.1 pointing precisely down to the position where the electron would be if it had not stopped. The way to connect up the inner and outer field lines that is consistent with Gauss's law, is shown elsewhere (see, e.g., [23]). The result is a rather intense field in the transition region, with the field running mainly perpendicular to the radius vector from the origin. Similarly there is a transverse magnetic field. These two transverse fields propagating outward with the velocity of light, c, form the electromagnetic radiation emitted by the decelerated charge. At large distances from the region of deceleration the radiation has the well-known angular distribution of the electric dipole radiation. This picture illustrates the classical view of bremsstrahlung emission.

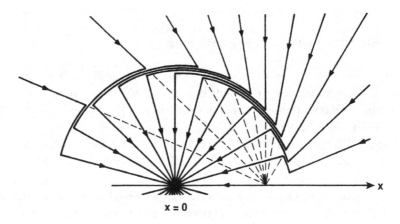

Fig. 2.1 The electric field lines surrounding a decelerated charge. The charge has been moving from left with constant velocity, reaches the point $x = 0$, where it is abruptly decelerated to rest, and remains at the point thereafter (adapted from [23]).

The quantitative relations of the classical radiation theory follow from Maxwell's equations (see, e.g., [21]). As a result the total radiation power of a charge e moving with velocity $v \ll c$ in an external field is [20]

$$I = \frac{1}{4\pi\epsilon_0} \frac{2e^2}{3c^3} a^2 \tag{2.1}$$

(Larmor formula) where a is the acceleration of the charge, and ϵ_0 denotes the permittivity of vacuum.

In the case of an electric dipole the radiation is determined by the second time derivative of the dipole moment, d,

$$I = \frac{1}{4\pi\epsilon_0} \frac{2}{3c^3} \ddot{d}^2 \,. \tag{2.2}$$

The angular distribution of a radiating dipole has the form

$$\frac{dI}{d\Omega} \propto \sin^2\theta \,, \tag{2.3}$$

where θ is the emission angle relative to the dipole axis. Radiation of this kind is called dipole radiation. Qualitative radiation patterns are shown in Figs. 2.2a and 2.2b. No energy is emitted forwards or backwards along the direction of acceleration and the energy emitted is maximum at right angles to this direction. In Fig. 2.2b the velocity of the electron is perpendicular to the acceleration vector as is the case when the electron is in a circular orbital motion. Again there is a toroidal radiation pattern with the direction of acceleration as symmetry axis.

The nonrelativistic dipole approximation and the resulting angular distributions are valid only for fairly low velocities of the incident electrons ($v \ll c$). At higher energies the distribution is no longer symmetric about $\theta = 90°$ and becomes increasingly forward peaked for increasing incident energy. Figures 2.2c and 2.2d show qualitative examples for the relativistic beaming with the acceleration parallel and perpendicular to the velocity, respectively. In the case of circular orbital motion the zeros of the radiation pattern are located in the orbital plane. Likewise, for arbitrary velocity and acceleration vectors of the electron, there are always two directions where the radiated intensity is zero.

It is interesting to consider the angular distribution of relativistic bremsstrahlung from a different point of view. Instead of using the laboratory frame where the nucleus is at rest we view the process as taking place in the rest frame of the incident particle. The radiation patterns as they appear in the two frames are shown schematically in Fig. 2.3. In its own frame of reference the incident particle emits in a characteristic dipole radiation pattern. The pertinent calculation by means of a Lorentz transformation is given by Jackson [20].

This type of bremsstrahlung may be of importance in astrophysics when energetic cosmic-ray protons collide with ambient electrons which are nearly at rest. The referring process is called proton-electron bremsstrahlung or inverse bremsstrahlung [24; 25].

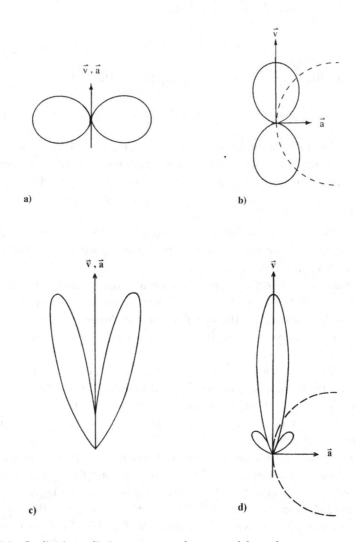

Fig. 2.2 Qualitative radiation patterns to be expected from electrons
(a) at low energy, acceleration **a** parallel to velocity **v**,
(b) at low energy, acceleration perpendicular to velocity,
(c) at relativistic energy, acceleration parallel to velocity,
(d) at relativistic energy, acceleration perpendicular to velocity.

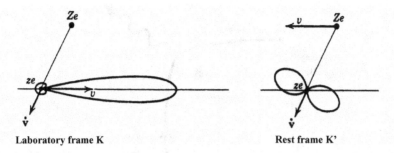

Fig. 2.3 Bremsstrahlung emitted during relativistic collisions viewed from the laboratory frame (nucleus at rest) and the frame where the incident particle is at rest (from Jackson [20]).

The classical theory of bremsstrahlung incorrectly predicts the emission of radiation in **every** collision in which an electron changes its velocity in magnitude or/and direction. In quantum mechanics, however, there is only a small probability that a photon of finite energy will be emitted. According to quantum electrodynamics each electron is surrounded by a cloud of virtual photons. Under certain conditions — as for instance in the collision with an atomic nucleus — the electron may shake off a photon. Because the radiative process involves the coupling of the electron with the electromagnetic field of the emitted photon the probability that the scattered electron will radiate is of the order of the fine-structure constant $\alpha \approx 1/137$. Therefore the cross section for bremsstrahlung is of the order of α times the cross section for elastic scattering. Most of the collisions of the incident electrons are elastic, only in rare events is a photon emitted. Figure 2.4 illustrates the elementary process of electron-nucleus bremsstrahlung on the photon view.

For incident electrons of not too high energy E_0 the recoil energy of the nucleus is negligible due to its large mass. Therefore the energy conservation has the form

$$E_0 = E_e + h\nu , \qquad (2.4)$$

and the photon can have any energy $h\nu$ up to the kinetic energy of the incident electron. The photon of highest energy will be emitted when an electron loses all its kinetic energy in one deceleration process. This shows that the very existence of a high-frequency limit is a quantum phenomenon.

14 *Classical and semiclassical considerations on the bremsstrahlung process*

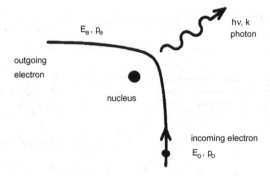

Fig. 2.4 The elementary process of electron-nucleus bremsstrahlung on the photon view.

Classical electromagnetic theory cannot account for this fact. In order to find the spectrum of the emitted radiation, the time-dependent field strengths must be developed in Fourier integrals [8] so that the spectrum extends up to infinity.

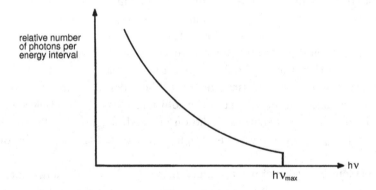

Fig. 2.5 Relative number of bremsstrahlung photons per energy interval as a function of the photon energy $h\nu$ showing the high-frequency limit at $h\nu_{\max}$.

In Fig. 2.5 the relative number of bremsstrahlung photons is shown as a function of the photon energy. The spectrum cuts off sharply at the maximum photon energy, $h\nu_{\max} = E_0$.

The initial momentum $\mathbf{p}_0$ of the incident electron is shared between the momenta of three particles, the outgoing electron, $\mathbf{p}_e$, the emitted photon, $\mathbf{k}$, and the nucleus, $\mathbf{q}$:

$$\mathbf{p}_0 = \mathbf{p}_e + \mathbf{k} + \mathbf{q}. \tag{2.5}$$

Due to its large mass the nucleus can take any recoil momentum, so that for a fixed momentum of the outgoing electron the photon can be emitted in any direction.

The well-known radiation patterns of bremsstrahlung are produced by superposition of all radiation processes with different scattering angles of the outgoing electrons. These patterns are rotationally symmetric about the primary electron beam analogous to those of Figs. 2.2a and c. When, on the other hand, the photons are detected in coincidence with the decelerated electrons scattered into a specified direction the photon angular distribution of the elementary bremsstrahlung process can be measured. In all these (coplanar) measurements (see Chapter 4) it is found that there is a strong correlation with the photons being predominantly emitted on the same side relative to the primary beam as the decelerated electrons (Fig. 2.6). This behaviour can be understood classically by considering the radiation patterns of the electrons along the hyperbolic orbit around the nucleus (Fig. 2.7). At the smallest distance from the nucleus the emission is strongest. Here the velocity and the acceleration vectors are perpendicular to each other. Because of relativistic beaming the radiation is predominantly emitted into a lobe along the direction of motion. Thus one obtains most of the radiation on the same side relative to the primary beam as the decelerated outgoing electrons (see Figs. 3.2 and 4.2).

At extreme relativistic energies both the photon and the outgoing electron tend to proceed in the same direction as the incident electron.

The transitions from classical electrodynamics to quantum mechanics and from nonrelativistic quantum mechanics to the fully relativistic theory of the bremsstrahlung process (see Chapter 3) is discussed in an instructive article of Pratt and Feng [19]. In addition, the authors enter into the question about which regions of the atomic system are probed in the bremsstrahlung process. This is important in assessing the accuracy of an approximation scheme; it is also important in understanding what one may hope to learn in a particular experimental arrangement.

16   *Classical and semiclassical considerations on the bremsstrahlung process*

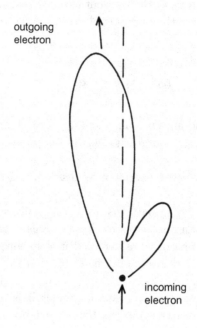

Fig. 2.6 Typical photon angular distribution of the elementary bremsstrahlung process for fixed direction of the outgoing decelerated electrons (schematically).

2.1.1 Screening effects

The screening of the nucleus by the charge distribution of atomic electrons affects the bremsstrahlung process which will be treated analytically in Section 3.3.1. The main consequence of screening is a reduction in cross section, corresponding to a reduction in the effective charge seen by the electron being scattered by the atom. The reduction is largest near the soft-photon end of the spectrum, characterized classically by large impact parameters for which the nuclear charge is nearly completely shielded; the reduction is least in the hard-photon region of the spectrum characterized by interior atomic distances, particularly for hard x-rays and soft γ-rays.

Numerical calculations for the classical bremsstrahlung spectrum, angular distribution, and polarization resulting from electrons scattering in screened atomic potentials have been performed by Kim and Pratt [22]. The calculations utilize the classical theory of electromagnetic radiation

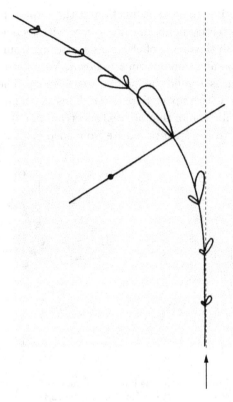

Fig. 2.7 Classical radiation pattern along the hyperbolic orbit of the electron (adapted from Elwert [26]).

from moving charges together with the classical mechanics of energy-loss-free orbits of charged particles in such potentials. These simple approaches are useful not only because they give appropriate predictions of the bremsstrahlung cross sections in certain situations, but also because they facilitate to understand the features of bremsstrahlung, tracing the origins of the properties of the process.

Here we will only show interesting differences in the calculated electron trajectories for non-screened and screened cases as described by Kim and Pratt [22]. The trajectories for electrons scattering from an attractive point Coulomb field are all hyperbolic (except the singular case when the impact parameter is exactly equal to zero). Figure 2.8a depicts the scattering orbits and the scattering angle as a function of the impact parameter. The scattering angle decreases monotonically to zero with increasing impact parameter. In Fig. 2.8b the orbits of electrons scattering from the charge distribution of a neutral aluminum atom are shown. Very-low-energy electrons can make several turns around the field before escaping. The scattering angle changes rapidly with impact parameter. This plays an important role in the angular distribution of the radiated spectrum in circumstances when most contributions to the spectrum arise from such orbits.

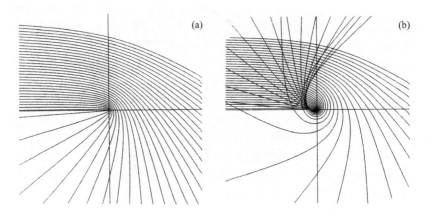

Fig. 2.8 Electron trajectories at various impact parameters for (a) the point Coulomb potential, (b) a screened potential (from Kim and Pratt [22]).

It is interesting to note that the high-frequency limit of classical bremsstrahlung ($h\nu/E_0 \to \infty$) for the point Coulomb potential will not be changed by the existence of screening.

2.1.2 Bremsstrahlung linear polarization

The degree of bremsstrahlung linear polarization is defined by the expression

$$P_l = \frac{I_\perp - I_\parallel}{I_\perp + I_\parallel}, \qquad (2.6)$$

where $I_\perp$ and $I_\parallel$ are the bremsstrahlung intensity components with electric vectors perpendicular and parallel, respectively, to the emission plane. The emission plane is given by the directions of the photon and the incident electron. Complete polarization with the electric vector parallel to the emission plane thus corresponds to $P_l = -1$.

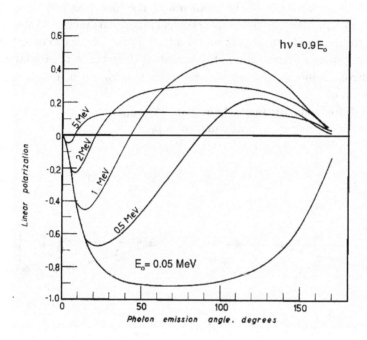

Fig. 2.9 Calculated angular dependence of the linear polarization of high-frequency bremsstrahlung (photon energies $h\nu = 0.9E_0$, where the initial electron kinetic energy, E_0, is equal to 0.05, 0.5, 1, 2, and 5 MeV). The polarization is derived from cross sections integrated over all angles of the outgoing electrons (from Motz and Placious [28]).

Bremsstrahlung produced by a beam of unpolarized electrons in the field of an atomic nucleus is, in general, partially linearly polarized. The polarization depends on the photon energy, the photon emission angle, the initial electron energy, and the atomic number of the target. Figure 2.9 illustrates the rather complex course of the polarization calculated in Born approximation by integrating over the directions of the decelerated outgoing electrons [27], corresponding to non-coincidence experiments. We will not discuss here the influence of the single parameters on the polarization (see, e.g., Motz and Placious [28]).

At nonrelativistic electron energies, the polarization is closely associated with the orbital motion of the electron. Low-energy bremsstrahlung consists overwhelmingly of electric dipole radiation (see Fig. 2.2a) with the electric vector in the direction of electron acceleration. The vector of acceleration is given by the difference of the momenta of the incoming and outgoing electrons, $\mathbf{p}_0 - \mathbf{p}_e$. We use this simple classical picture to describe in the first instance the general features of bremsstrahlung linear polarization provided that the outgoing electrons are not observed. A diagram of the initial and final electron momenta, $\mathbf{p}_0$ and $\mathbf{p}_e$, respectively, is shown in Fig. 2.10 for the emission of

(a) high frequency bremsstrahlung, $|\mathbf{p}_e| \ll |\mathbf{p}_0|$, and
(b) low frequency bremsstrahlung, $\mathbf{p}_e \approx \mathbf{p}_0$.

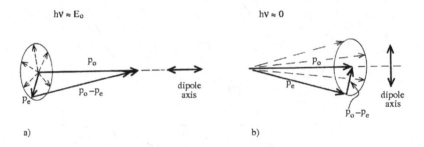

Fig. 2.10 Vector diagrams of the initial, $\mathbf{p}_0$, and final, $\mathbf{p}_e$, electron momenta for (a) high frequency photons where the photon energy, $h\nu$, is approximately equal to the initial electron kinetic energy, E_0 (or $|\mathbf{p}_e| \ll |\mathbf{p}_0|$), and (b) low frequency photons where $h\nu \approx 0$ (or $\mathbf{p}_e \approx \mathbf{p}_0$). For nonrelativistic electrons, the major dipole axis is given by the direction of the large component of the momentum difference, $\mathbf{p}_0 - \mathbf{p}_e$.

In case (a), the difference of electron momenta, $\mathbf{p}_0 - \mathbf{p}_e$, has a large component parallel to $\mathbf{p}_0$ at any outgoing electron angle, and the radiation source resembles an oscillating electric dipole parallel to $\mathbf{p}_0$.

In case (b), the bremsstrahlung process approaches the conditions in elastic scattering where most of the electrons have small scattering angles; therefore the difference of electron momenta, $\mathbf{p}_0 - \mathbf{p}_e$, has a large component perpendicular to $\mathbf{p}_0$, and the radiation source resembles an electric dipole perpendicular to $\mathbf{p}_0$.

Accordingly, if the radiation is observed in a direction perpendicular to $\mathbf{p}_0$ and to the plane of the diagram in Fig. 2.10, the polarization has a reversal feature: The high frequency photons tend to be polarized parallel to the emission plane, and the low frequency photons tend to be polarized perpendicular to the emission plane. If the radiation is observed in the forward or backward direction with respect to the incident beam, the degree of polarization is zero at all frequencies because of symmetry conditions.

This polarization behaviour is produced by averaging over many elementary bremsstrahlung processes with different directions of outgoing electrons. Picking out a single bremsstrahlung process by an electron-photon coincidence experiment one gets information on the elementary radiation process itself. According to the simple dipole picture, the radiation is expected to be completely linearly polarized with $P_l = -1$.

However, at relativistic electron energies, the dependence of the bremsstrahlung linear polarization on the orbital motion of the electron is complicated by the intimate relationship between the spin and momentum of the electron. Spin effects become important and cause interesting deviations from the otherwise complete polarization. Figure 2.11 shows the calculated linear polarization in the elementary process of bremsstrahlung as a function of the photon-emission angle for outgoing electrons of $\theta_e = +20°$ and the pertinent triply differential cross section for a primary energy of 300 keV (the theory is described in Section 3.7.2). The bremsstrahlung is almost completely polarized with the electric vector lying in the emission plane. Only near the minimum of the cross section a strong decrease of the polarization occurs.

For comparison with the elementary process, Fig. 2.11a depicts in addition the bremsstrahlung polarization integrated over all directions of the outgoing electrons (dotted curve). Here the polarization in the forward direction ($\theta_k = 0°$) is zero because of symmetry requirements, whereas the polarization in the elementary process is nearly complete in this direction. Moreover, the steep decrease of the polarization near the minimum of the triply differential cross section is absent due to the contributions of the electrons scattered into all directions. Altogether, the degree of polarization is significantly lower than in the elementary process.

The measurement of the polarization of the elementary bremsstrahlung process is a fine example of the power of the coincidence method. We will discuss this topic in Section 4.3.1.

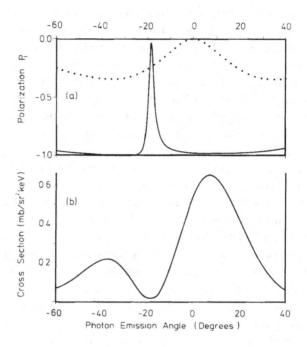

Fig. 2.11 (a) Calculated linear polarization in the elementary process of electron-nucleus bremsstrahlung as a function of the photon emission angle θ_k. The primary electrons, incident on a carbon target ($Z = 6$), have the energy $E_0 = 300$ keV, the outgoing electrons of 140 keV are scattered at $+20°$. For comparison with the elementary process, the dotted curve shows the bremsstrahlung polarization integrated over all directions of motion of the outgoing electrons. (b) Calculated triply differential cross section of electron-nucleus bremsstrahlung.

2.1.3 Bremsstrahlung from transversely polarized electrons

There is a left-right asymmetry in the bremsstrahlung emitted by transversely polarized electrons if the spin polarization is perpendicular to the emission plane. This phenomenon is similar to the elastic scattering of transversely polarized electrons (Mott scattering). In an analysis of Sobolak and Stehle [29] a classical argument is given for the origin of the bremsstrahlung asymmetry considering the force on a magnetic dipole moving in the Coulomb field of an atomic nucleus. In addition the authors show that the sign of the asymmetry depends on whether or not the electron spin flips in the bremsstrahlung emission process.

The detailed description of the results of the electron-photon coincidence experiments requires considerable theoretical efforts (see Secs. 3.6 and 4.3.2.

2.2 Electron-electron bremsstrahlung

In contrast to the electron-nucleus system the electron-electron system has no electric dipole moment. For the bremsstrahlung emission in the collision of two electrons it is evident that in the center-of-mass system the total electric dipole moment is initially zero and remains zero if the recoil due to the photon emission is neglected. Hence there is no electric dipole emission within the approximation made. The nonrelativistic electron-electron bremsstrahlung is expected to consist predominantly of electric quadrupole radiation with four lobes in the angular distribution. The radiation pattern of an electric quadrupole is shown in Fig. 2.12. In this connection we remind of a general finding in classical electrodynamics that a closed system of particles, for all of which the ratio of charge to mass is the same, cannot radiate by dipole radiation (see, e.g., [21]).

Fig. 2.12   Radiation pattern of an electric quadrupole.

At relativistic energies the cross section of electron-electron bremsstrahlung is found to be comparable with that for electron-proton bremsstrahlung. With decreasing energy in the nonrelativistic domain, however, the cross section decreases rapidly relative to the electron-proton bremsstrahlung cross section due to the zero dipole moment of the two-electron system. Thus radiative electron-electron collisions are usually ignored at low and moderate electron energies.

It is of interest to consider the maximum photon energy of electron-electron bremsstrahlung in the laboratory system which is reached for fixed emission angle θ_k with respect to the direction of the incident electron of energy E_0. In Section 5.2.2 it is shown* that

$$h\nu_{\max} = \frac{E_0 mc^2}{E_0 + 2mc^2 - \sqrt{E_0(E_0 + 2mc^2)}\cos\theta_k}. \qquad (2.7)$$

A qualitative comparison of the high-frequency limits of electron-electron and electron-nucleus bremsstrahlung is given in Fig. 2.13. The maximum photon energy for electron-electron bremsstrahlung is always smaller than for electron-nucleus bremsstrahlung because the energy of the recoil electron may be appreciable. In contrast to the electron-nucleus bremsstrahlung the maximum photon energy of electron-electron bremsstrahlung is dependent on the photon emission angle.

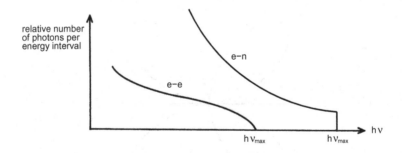

Fig. 2.13 Qualitative comparison of the high frequency limits of electron-electron and electron-nucleus bremsstrahlung (schematically).

Using the different upper limits of the bremsstrahlung spectra it was attempted to isolate the electron-electron contribution from the total spec-

*In Eq. (5.34) the energies are given in units of mc^2.

trum in non-coincidence experiments. However, even at moderate energies the cross section for electron-electron bremsstrahlung is much smaller than for the dipole radiation of the electron-nucleus system. Therefore, the experimental results are still inconclusive. The electron-photon coincidence technique, however, enables one to differentiate completely between electron-electron and electron-nucleus bremsstrahlung because of kinematical reasons (see Sec. 6.1).

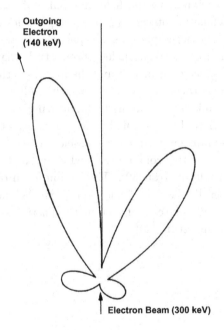

Fig. 2.14 Calculated angular distribution of the photons in the elementary process of electron-electron bremsstrahlung in the laboratory system for incident electrons of 300 keV and fixed energy ($E_e = 140$ keV) and angle ($\theta_e = 20°$) of the outgoing electron.

An example of a calculated angular distribution of the photons in the elementary process of electron-electron bremsstrahlung in the laboratory system is depicted in Fig. 2.14. The relativistic beaming is clearly to be seen. As in the case of electron-nucleus bremsstrahlung the main radiation lobe points to the same side relative to the primary beam as the decelerated outgoing electron.

2.3 Weizsäcker-Williams method of virtual quanta

The Weizsäcker-Williams method [30; 31] is a semiclassical method for calculating cross sections of electromagnetic interactions of relativistic particles. The method is based on the fact that the electric and magnetic fields of a rapidly moving charged particle are nearly transverse to the direction of motion and are similar to the field of a pulse of radiation, a swarm of virtual quanta. Thus the interactions between particles is simplified to that of the interactions between photons and particles (see, e.g., Ref. [20]).

The calculation of electron-nucleus bremsstrahlung uses a coordinate system where the electron is at rest and the nucleus is the incident particle. The struck system is the lighter particle since its radiation-scattering power is greater. The emission of bremsstrahlung can be viewed as Compton scattering of the virtual quanta of the moving nucleus by the electron.

For bremsstrahlung emitted in electron-electron collisions it is necessary to take the sum of the two contributions where each electron in turn is the struck system initially at rest [32]. The computation of such complicated radiative processes by means of the Weizsäcker-Williams method yields results quite easily at some cost of accuracy, in particular at the high-energy tail of the photon spectrum [24].

Chapter 3
Theory of the elementary process of electron-nucleus bremsstrahlung

3.1 Introduction

We consider an electron with initial kinetic energy E_0 and momentum $\mathbf{p}_1$ which is scattered by the Coulomb field of a nucleus or an atom at rest. There will be a certain probability that a photon with energy $h\nu$ and momentum $\mathbf{k}$ is emitted, whereby the electron makes a transition to a state with energy E_e and momentum $\mathbf{p}_2$. It is assumed that the initial and final states of the target atom coincide ('elastic' bremsstrahlung). The momentum transferred to the nucleus or atom is

$$\mathbf{q} = \mathbf{p}_1 - \mathbf{p}_2 - \mathbf{k} \,. \tag{3.1}$$

Since the mass M of the atomic nucleus is large compared to the electron mass m, the recoil energy of the nucleus, $E_r \approx \mathbf{q}^2/(2M)$, can be neglected for electron energies $E_0 \ll \frac{1}{2}Mc^2 \approx 469 A$ MeV, where A is the atomic mass of the nucleus. Thus energy conservation implies the simple relation

$$h\nu = E_0 - E_e \,. \tag{3.2}$$

The interaction causing the transition from the initial state to the final state consists of two parts:

(1) The interaction of the electron with the radiation field leading to the emission of the bremsstrahlung photon;
(2) the interaction of the electron with the electric field of the nucleus or atom.

28 Theory of the elementary process of electron-nucleus bremsstrahlung

In quantum electrodynamics the interaction of the electrons with the radiation field is treated as a small perturbation. For most applications it is sufficient to use the lowest non-vanishing order of an expansion in powers of the fine structure constant $\alpha \approx 1/137$ (or the square of the elementary charge, e^2). On the other hand, the interaction of the electron with the atomic field can, in principle, be handled rigorously by including it in the unperturbed Hamiltonian. For that purpose, exact solutions of the Dirac wave equation are required. It is, however, not possible to solve the Dirac equation in closed form for a free electron in a pure Coulomb field, let alone in a screened nuclear field. Therefore either approximate analytic solutions or numerical methods have to be used. In the nonrelativistic case the Schrödinger equation for an electron in the nuclear Coulomb field can be solved in closed form resulting in Sommerfeld's bremsstrahlung cross section.

3.2 Bremsstrahlung cross section

In the following we assume that the reader is familiar with the elementary theory of the Dirac wave equation. Therefore we repeat only the relevant items which are needed to derive the expression for the bremsstrahlung cross section, the starting point of our investigations.

The relativistic wave equation (Dirac equation) of an electron interacting with an electromagnetic field represented by its scalar potential φ and vector potential $\mathbf{A}$ is (see, e.g., [32; 33; 34])

$$i\hbar\frac{\partial \Psi}{\partial t} = i\hbar\dot{\Psi} = H\Psi = \left\{\vec{\alpha} \cdot (\mathbf{p} - e\mathbf{A})c + \beta mc^2 + e\varphi\right\}\Psi = 0 , \quad (3.3)$$

where $\mathbf{p} = -i\hbar\nabla$ is the momentum operator, e is the elementary charge, $\vec{\alpha}$ and β are 4×4 matrices satisfying the relations (δ_{ij} denotes the Kronecker symbol)

$$\alpha_i \alpha_j + \alpha_j \alpha_i = 2\delta_{ij} \quad (i,j = 1,2,3) , \quad (3.4)$$

$$\alpha_i \beta + \beta \alpha_i = 0 , \; \beta^2 = 1 . \quad (3.5)$$

Of course, the right-hand sides of Eqs. (3.4) and (3.5) contain a unit matrix.
The Dirac matrices $\vec{\alpha}$ and β can be represented by the 2×2 Pauli spin

matrices σ_i ($i = 1, 2, 3$) and the unit matrix 1,

$$\vec{\alpha} = \begin{pmatrix} 0 & \vec{\sigma} \\ \vec{\sigma} & 0 \end{pmatrix}, \quad \beta = \begin{pmatrix} 1 & 0 \\ 0 & -1 \end{pmatrix}. \quad (3.6)$$

The Pauli matrices satisfy the relations

$$\sigma_1 \sigma_2 = -\sigma_2 \sigma_1 = i\sigma_3, \text{ etc. }, \sigma_i^2 = 1 \quad (i = 1, 2, 3). \quad (3.7)$$

The wave function $\Psi(\mathbf{r}, t)$ consists of 4 components, Ψ_ρ, $\rho = 1, 2, 3, 4$. Ψ can be regarded as a column matrix, thus the matrix products $\alpha_i \Psi$ and $\beta \Psi$ result again in a column matrix.

The conditions (3.4) and (3.5) imply that the Hamiltonian H is hermitian, i.e., the eigenvalues of H are real. To obtain stationary solutions of the Dirac equation (3.3), we apply the usual harmonic ansatz

$$\Psi(\mathbf{r}, t) = \psi(\mathbf{r}) e^{-iEt/\hbar} \quad (3.8)$$

leading to the time-independent Dirac equation

$$\{-\vec{\alpha} \cdot (i\hbar \nabla + e\mathbf{A})c + \beta mc^2 + e\varphi\} \psi(\mathbf{r}) = E \psi(\mathbf{r}). \quad (3.9)$$

For any given value of the electron momentum there are four different eigensolutions corresponding to two different spin directions and to solutions with positive and negative energy. Denoting the type of eigensolutions by the index n, the functions $\psi_{\rho n}(\mathbf{r})$ satisfy the orthogonality relation

$$\sum_\rho \int \psi_{\rho n}^\dagger(\mathbf{r}) \psi_{\rho n'}(\mathbf{r}) d^3 r = \delta_{nn'} \quad (3.10)$$

and the completeness relation

$$\sum_n \psi_{\rho n}^\dagger(\mathbf{r}) \psi_{\rho' n}(\mathbf{r}') = \delta_{\rho \rho'} \delta(\mathbf{r} - \mathbf{r}'), \quad (3.11)$$

where $\psi^\dagger$ denotes the matrix adjoint to ψ.

Since Dirac's theory has a linearized form, the Hamiltonian describing the interaction of an electron with the radiation field is linearly composed of two parts, namely, (1) the Hamiltonian of the electron in the absence of the radiation field, and (2) the interaction with the radiation field:

$$H = H_{el} + H_{int}. \quad (3.12)$$

Considering only the part depending on the radiation field, we have

$$H_{int} = -ec(\vec{\alpha} \cdot \mathbf{A}) \,. \tag{3.13}$$

This corresponds to the term $-ec\mathbf{A}$ of the classical Hamiltonian describing the interaction of an electron with an electromagnetic field. After quantization of the radiation field the vector potential may be written as a series of plane waves enclosed in a cube of volume L^3. For the emission of one photon of energy $h\nu$ and momentum $\mathbf{k}$ and the electron transition $1 \to 2$ the matrix element of H_{int} in the Coulomb gauge ($\nabla \cdot \mathbf{A} = 0$) has the form

$$M_{1\to 2} = -\frac{e\hbar c}{\sqrt{2\epsilon_0 h\nu L^3}} \int \psi_2^\dagger(\mathbf{r})(\vec{\alpha} \cdot \mathbf{e}^*) e^{-i\mathbf{k}\cdot\mathbf{r}/\hbar} \psi_1(\mathbf{r}) \, d^3r \,, \tag{3.14}$$

where $\epsilon_0 \approx 8.8542 \cdot 10^{-12}$ F m^{-1} is the permittivity of vacuum and the unit vector $\mathbf{e}$ gives the direction of polarization which is always perpendicular to $\mathbf{k}$, i.e., $\mathbf{e} \cdot \mathbf{k} = 0$. The polarization vector $\mathbf{e}$ is taken to be complex in order to be able to include circularly polarized photons.

According to time-dependent perturbation theory the transition rate from the state $|i\rangle$ of a quantum mechanical system to a set of continuous final states $|f\rangle$ is given in first order by

$$W_{i\to f} = \frac{2\pi}{\hbar} |\langle i|V|f\rangle|^2 \rho_f \,, \tag{3.15}$$

where V is the perturbation defining the interaction of the eigenstates $|i\rangle$ and $|f\rangle$ of the unperturbed Hamiltonian and ρ_f is the number of final states per energy interval dE_f (Fermi's golden rule [35]). In the final state of the bremsstrahlung process there is a photon with momentum $\mathbf{k}$ and an electron with momentum $\mathbf{p}_2$. Since the mass of the atomic nucleus is large compared with the electron mass, $\mathbf{k}$ and $\mathbf{p}_2$ are independent of each other, and the density of final states ρ_f is equal to the product of the density functions ρ_k and ρ_{p_2} for the photon and electron.* Thus we have

$$\rho_f dE_2 = \frac{d^3p_2 \, d^3k}{(2\pi mc)^6} = \frac{p_2^2 \, dp_2 \, d\Omega_{p_2} \, k^2 \, dk \, d\Omega_k}{(2\pi mc)^6} \,. \tag{3.16}$$

Here $d\Omega_{p_2}$ and $d\Omega_k$ are the elements of solid angle in the direction of $\mathbf{p}_2$ and $\mathbf{k}$, respectively.

*The directions of the outgoing electron and the emitted photon are not restricted by the kinematics because the nucleus can take any recoil momentum.

From the relativistic energy-momentum relation

$$(E + mc^2)^2 = (pc)^2 + (mc^2)^2 \qquad (3.17)$$

we get $p_2 c^2 dp_2 = (E_2 + mc^2)\, dE_2$, thus

$$\rho_f = \frac{p_2(E_2 + mc^2)\, d\Omega_{p_2} k^2 dk\, d\Omega_k}{(2\pi m)^6 c^8} . \qquad (3.18)$$

The cross section of a process is defined as the number of interactions per unit time and per unit flux of the incident particles. Dividing $W_{i \to f}$ by the incident electron flux (i.e., the current density in a cube of side L),

$$J_1 = v_1/L^3 = \frac{1}{L^3} \frac{p_1 c^2}{E_1 + mc^2} \qquad (3.19)$$

and inserting Eqs. (3.14) and (3.18), we obtain the differential cross section for the elementary process of bremsstrahlung,

$$\frac{d^3\sigma}{dk\, d\Omega_k\, d\Omega_{p_2}} = (2\pi)^{-4} \frac{e^2 \hbar}{4\pi\epsilon_0} \frac{p_2(E_1 + mc^2)(E_2 + mc^2)k^2}{p_1 (h\nu) m^6 c^8} |M|^2 , \qquad (3.20)$$

where

$$M = \int \psi_2^\dagger(\mathbf{r}) (\vec{\alpha} \cdot \mathbf{e}^*) e^{-i\mathbf{k}\cdot\mathbf{r}/\hbar} \psi_1(\mathbf{r})\, d^3 r \qquad (3.21)$$

is the matrix element (3.14) without the factor before the integral. If we introduce the fine-structure constant $\alpha = e^2/(4\pi\epsilon_0 \hbar c) \approx 1/137$, the cross section takes the form

$$\frac{d^3\sigma}{dk\, d\Omega_k\, d\Omega_{p_2}} = \frac{\alpha}{(2\pi)^4} \left(\frac{\hbar}{mc}\right)^2 \frac{p_2(E_1 + mc^2)(E_2 + mc^2)k^2}{m^4 c^5 p_1 (h\nu)} |M|^2 . \qquad (3.22)$$

Now throughout the following treatment, unless otherwise specified, we will use dimensionless units where $\hbar = m = c = 1$. The units of energy, momentum, and length will then be mc^2, mc, and $\hbar/mc$, respectively. Denoting the total electron energy including rest energy by ϵ, i.e., $\epsilon = E/mc^2 + 1$, the relativistic relation between energy and momentum has the simple form

$$\epsilon^2 = p^2 + 1 , \qquad (3.23)$$

and for the photon the energy $k = h\nu/mc^2$ is equal to the absolute value of the momentum, $k = |\mathbf{k}|$.

In these units the bremsstrahlung cross section reads

$$\frac{d^3\sigma}{dk\, d\Omega_k\, d\Omega_{p_2}} = \frac{\alpha}{(2\pi)^4}\frac{\epsilon_1\epsilon_2 p_2 k}{p_1}\left(\frac{\hbar}{mc}\right)^2 |M|^2 \ . \tag{3.24}$$

The factor $(\hbar/mc)^2$ has been left to point out that the cross section has the dimension of an area.

3.3 Born approximation (Bethe-Heitler formula)

The simplest and most widely used method to calculate the bremsstrahlung matrix element (3.21) is the Born approximation where one effectively expands in the coupling to the external field, αZ. This corresponds to using free-particle wave functions (plane waves) for the incoming and outgoing electrons, i.e., the distortion of the electron wave functions by the nuclear or atomic field is neglected. Then, however, the matrix element (3.21) vanishes so that one has to employ one higher order in the expansion of the wave function.

For an electron moving in the electric field of a static nucleus with atomic number Z, the potential term is $e\varphi = -\alpha Z/r$ and the Dirac equation (3.9) reads

$$\{-i\vec{\alpha}\cdot\nabla + \beta - \epsilon - a/r\}\psi(\mathbf{r}) = 0 \ , \tag{3.25}$$

where $a = \alpha Z \approx Z/137$ is the Coulomb parameter. Equation (3.25) represents our standard form of the Dirac equation with Coulomb potential. In order to solve it, we expand the wave function $\psi(\mathbf{r})$ into a series in powers of a:

$$\psi(\mathbf{r}) = \Phi_0(\mathbf{r}) + a\Phi_1(\mathbf{r}) + a^2\Phi_2(\mathbf{r}) + \ldots \tag{3.26}$$

Φ_0 is the incoming (asymptotically plane) wave, while the other terms characterize the distortion by the potential with which the wave interacts. The basic assumption of the Born approximation is that the incident and outgoing waves are only weakly deformed from a plane wave. Without going into the details of Feynman formalism, we can say that in the case of bremsstrahlung each term in this expansion corresponds to a sum of Feynman graphs (Fig. 3.1) of the same order in a: The linear term in a corresponds to the exchange of one virtual photon, while the terms proportional to a^n correspond to graphs with exchange of n virtual photons between the electron and the target nucleus. An extension of the Feynman formalism proposed by Furry [36; 37] allows the inclusion of the effects of the external potential. The lepton wave functions are taken as solutions of the Dirac equation with the external potential rather than as free-particle solutions, and the interaction with the radiation field is taken into account using the Feynman rules. Then a graphical representation analogous to Feynman diagrams can be used. Here the lepton lines correspond to exact distorted wave functions. This is represented graphically by the use of double lines to visualize the particle interacting with the field. The Furry diagram corresponding to bremsstrahlung (Fig.3.1) can be depicted as a sum of Feynman diagrams in which the wavy lines ending on a cross correspond to the interaction with the external potential.

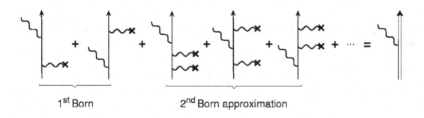

Fig. 3.1 Sum of Feynman-Dyson graphs for bremsstrahlung and the equivalent graph of the Furry picture.

In zeroth order Eq. (3.25) reads

$$\{-i\vec{\alpha} \cdot \nabla + \beta - \epsilon\}\Phi_0(\mathbf{r}) = 0 \,. \tag{3.27}$$

This is the force-free Dirac equation with the plane-wave solution

$$\Phi_0(\mathbf{r}) = e^{i\mathbf{p}\cdot\mathbf{r}}u(\mathbf{p}) \,, \tag{3.28}$$

where the free-electron spinor $u(\mathbf{p})$ is independent of $\mathbf{r}$ and t and obeys the relation

$$(\vec{\alpha} \cdot \mathbf{p} + \beta - \epsilon)\, u(\mathbf{p}) = 0 \,. \tag{3.29}$$

$u(\mathbf{p})$ can be represented as a 4-component quantity, and hence it will have, for any given value of the electron momentum $\mathbf{p}$, four different eigensolutions. These correspond to two different spin directions and solutions with positive and negative energy. Here we do not need the explicit form of $u(\mathbf{p})$.

Substituting the series (3.26) into the wave equation (3.25) and utilizing (3.27), the equation of first order in a has the form

$$\left(-i\vec{\alpha}\cdot\nabla + \beta - \epsilon\right)\Phi_1(\mathbf{r}) = \frac{1}{r}\Phi_0(\mathbf{r}) = \frac{1}{r}e^{i\mathbf{p}\cdot\mathbf{r}}u(\mathbf{p}) . \qquad (3.30)$$

Introducing the Green function $G(\mathbf{r}, \mathbf{r}')$ which satisfies the differential equation

$$\left(-i\vec{\alpha}\cdot\nabla + \beta - \epsilon\right)G(\mathbf{r}, \mathbf{r}') = \delta(\mathbf{r} - \mathbf{r}') , \qquad (3.31)$$

the solution of (3.30) is given by

$$\Phi_1(\mathbf{r}) = \int G(\mathbf{r}, \mathbf{r}')\frac{e^{i\mathbf{p}\cdot\mathbf{r}'}}{r'} d^3r' \, u(\mathbf{p}) . \qquad (3.32)$$

Now it is expedient to change to the exponential Fourier transforms $G(\mathbf{k}, \mathbf{r})$ and $\Phi_1(\mathbf{k})$ defined by

$$G(\mathbf{r}, \mathbf{r}') = \int G(\mathbf{k}, \mathbf{r}')\, e^{i\mathbf{k}\cdot\mathbf{r}}\, d^3k \qquad (3.33)$$

and

$$\Phi_1(\mathbf{r}) = \int \Phi_1(\mathbf{k})\, e^{i\mathbf{k}\cdot\mathbf{r}}\, d^3k . \qquad (3.34)$$

Substituting (3.33) into (3.32) gives

$$\Phi_1(\mathbf{r}) = \int d^3r' \frac{e^{i\mathbf{p}\cdot\mathbf{r}'}}{r'} \int d^3k\, G(\mathbf{k}, \mathbf{r}')\, e^{i\mathbf{k}\cdot\mathbf{r}}\, u(\mathbf{p}) \qquad (3.35)$$

and

$$\nabla\Phi_1(\mathbf{r}) = \int d^3r' \frac{e^{i\mathbf{p}\cdot\mathbf{r}'}}{r'} \int d^3k\, G(\mathbf{k}, \mathbf{r}')\, (i\mathbf{k})\, e^{i\mathbf{k}\cdot\mathbf{r}}\, u(\mathbf{p}) . \qquad (3.36)$$

Inserting these expressions in (3.30), we get

$$\int d^3r' \frac{e^{i\mathbf{p}\cdot\mathbf{r}'}}{r'} \int d^3k\, (\vec{\alpha}\cdot\mathbf{k} + \beta - \epsilon)\, G(\mathbf{k}, \mathbf{r}')\, e^{i\mathbf{k}\cdot\mathbf{r}}\, u(\mathbf{p}) = \frac{e^{i\mathbf{p}\cdot\mathbf{r}}}{r}\, u(\mathbf{p}) . \qquad (3.37)$$

The solution of this equation has the form

$$G(\mathbf{k}, \mathbf{r}) = \frac{1}{(2\pi)^3} \int G(\mathbf{r}, \mathbf{r}')\, e^{-i\mathbf{k}\cdot\mathbf{r}'}\, d^3r' = \lim_{\lambda\to 0} \frac{1}{(2\pi)^3} \frac{\vec{\alpha}\cdot\mathbf{k} + \beta + \epsilon}{k^2 - p^2 - i\lambda}\, e^{-i\mathbf{k}\cdot\mathbf{r}} , \qquad (3.38)$$

where the term $i\lambda$ guarantees the correct behaviour in complex integrations. This solution can easily be verified by substitution into (3.37) and employing the relations (3.4) and (3.5) of the Dirac matrices, Eq. (3.23), and the representation

$$\delta(\mathbf{r}) = \frac{1}{(2\pi)^3} \int e^{i\mathbf{k}\cdot\mathbf{r}} d^3k \qquad (3.39)$$

of the δ function:

$$\int d^3r' \frac{e^{i\mathbf{p}\cdot\mathbf{r}'}}{r'} \int d^3k \left(\vec{\alpha}\cdot\mathbf{k} + \beta - \epsilon\right) G(\mathbf{k}, \mathbf{r}') e^{i\mathbf{k}\cdot\mathbf{r}} u(\mathbf{p})$$

$$= \lim_{\lambda \to 0} \frac{1}{(2\pi)^3} \int d^3r' \frac{e^{i\mathbf{p}\cdot\mathbf{r}'}}{r'}$$

$$\cdot \int d^3k \frac{(\vec{\alpha}\cdot\mathbf{k} + \beta - \epsilon)(\vec{\alpha}\cdot\mathbf{k} + \beta + \epsilon)}{k^2 - p^2 - i\lambda} e^{i\mathbf{k}\cdot(\mathbf{r}-\mathbf{r}')} u(\mathbf{p})$$

$$= \frac{1}{(2\pi)^3} \int d^3r' \frac{e^{i\mathbf{p}\cdot\mathbf{r}'}}{r'} \int d^3k \frac{k^2 + 1 - \epsilon^2}{k^2 - p^2} e^{i\mathbf{k}\cdot(\mathbf{r}-\mathbf{r}')} u(\mathbf{p})$$

$$= \int \frac{e^{i\mathbf{p}\cdot\mathbf{r}'}}{r'} \delta(\mathbf{r}-\mathbf{r}') d^3r' \, u(\mathbf{p}) = \frac{e^{i\mathbf{p}\cdot\mathbf{r}}}{r} u(\mathbf{p}) \ . \qquad (3.40)$$

The reverse of the Fourier transform (3.34) is

$$\Phi_1(\mathbf{k}) = \frac{1}{(2\pi)^3} \int \Phi_1(\mathbf{r}) e^{-i\mathbf{k}\cdot\mathbf{r}} d^3r \ . \qquad (3.41)$$

Inserting (3.35) and (3.38) into (3.41) we obtain

$$\Phi_1(\mathbf{k}) = \lim_{\lambda \to 0} \frac{1}{(2\pi)^6} \int d^3r \, e^{-i\mathbf{k}\cdot\mathbf{r}} \int d^3r' \frac{e^{i\mathbf{p}\cdot\mathbf{r}'}}{r'}$$

$$\cdot \int d^3k' \frac{\vec{\alpha}\cdot\mathbf{k}' + \beta + \epsilon}{k'^2 - p^2 - i\lambda} e^{i\mathbf{k}'\cdot(\mathbf{r}-\mathbf{r}')} u(\mathbf{p})$$

$$= \lim_{\lambda \to 0} \frac{1}{(2\pi)^3} \int d^3r' \frac{e^{i\mathbf{p}\cdot\mathbf{r}'}}{r'} \int d^3k' \frac{\vec{\alpha}\cdot\mathbf{k}' + \beta + \epsilon}{k'^2 - p^2 - i\lambda} e^{-i\mathbf{k}'\cdot\mathbf{r}'} \delta(\mathbf{k}' - \mathbf{k}) u(\mathbf{p})$$

$$= \lim_{\lambda \to 0} \frac{1}{(2\pi)^3} \int \frac{e^{i(\mathbf{p}-\mathbf{k})\cdot\mathbf{r}'}}{r'} d^3r' \frac{\vec{\alpha}\cdot\mathbf{k} + \beta + \epsilon}{k^2 - p^2 - i\lambda} u(\mathbf{p}) \ . \qquad (3.42)$$

Using the integral

$$\int_\infty \frac{e^{i\mathbf{q}\cdot\mathbf{r}}}{r} d^3r = \lim_{\mu \to 0} 2\pi \int_0^\infty dr\, r\, e^{-\mu r} \int_{-1}^{+1} d(\cos\theta)\, e^{iqr\cos\theta}$$

$$= \lim_{\mu \to 0} \frac{2\pi}{iq} \int_0^\infty \left[e^{(iq-\mu)r} - e^{-(iq+\mu)r} \right] dr$$
$$= \frac{2\pi}{iq} \lim_{\mu \to 0} \left(\frac{1}{\mu - iq} - \frac{1}{\mu + iq} \right) = \frac{4\pi}{q^2}, \quad (3.43)$$

we get

$$\Phi_1(\mathbf{k}) = \lim_{\lambda \to 0} \frac{1}{2\pi^2} \frac{1}{(\mathbf{k}-\mathbf{p})^2} \frac{\vec{\alpha} \cdot \mathbf{k} + \beta + \epsilon}{k^2 - p^2 - i\lambda} u(\mathbf{p}). \quad (3.44)$$

From (3.30) and (3.34), $\Phi_1(\mathbf{k})$ obeys the wave equation

$$\int (\vec{\alpha} \cdot \mathbf{k} + \beta - \epsilon) \Phi_1(\mathbf{k}) e^{i\mathbf{k} \cdot \mathbf{r}} d^3k = \frac{e^{i\mathbf{p} \cdot \mathbf{r}}}{r} u(\mathbf{p}). \quad (3.45)$$

Substituting $\psi(\mathbf{r}) = \Phi_0(\mathbf{r}) + a\Phi_1(\mathbf{r})$ into the wave equation (3.25) we get by means of (3.27) and (3.30)

$$(-i\vec{\alpha} \cdot \nabla + \beta - \epsilon - a/r)[\Phi_0(\mathbf{r}) + a\Phi_1(\mathbf{r})] = -(a^2/r)\Phi_1(\mathbf{r}). \quad (3.46)$$

Hence the differential equation in second order of a has the form

$$(-i\vec{\alpha} \cdot \nabla + \beta - \epsilon) \Phi_2(\mathbf{r}) = (1/r)\Phi_1(\mathbf{r}). \quad (3.47)$$

In order to obtain the bremsstrahlung matrix element in first Born approximation we do not need $\Phi_2(\mathbf{r})$. We state without proof that the Fourier transform of $\Phi_2(\mathbf{r})$ is given by (see Problem 3.1)

$$\Phi_2(\mathbf{k}) = \lim_{\lambda \to 0} \frac{1}{(2\pi^2)^2} \frac{\vec{\alpha} \cdot \mathbf{k} + \beta + \epsilon}{k^2 - p^2 - i\lambda}$$
$$\cdot \int d^3k' \frac{\vec{\alpha} \cdot \mathbf{k}' + \beta + \epsilon}{(\mathbf{k}'-\mathbf{p})^2(\mathbf{k}'-\mathbf{k})^2(k'^2 - p^2 - i\lambda)} u(\mathbf{p}). \quad (3.48)$$

Incidentally, the same procedure may also be carried out for a screened Coulomb potential $e\varphi(r) = -(a/r)e^{-\mu r}$. Then the Fourier transform of the wave function $\Phi_1(\mathbf{r})$ has the form

$$\Phi_1(\mathbf{k}) = \lim_{\lambda \to 0} \frac{1}{2\pi^2} \frac{1}{(\mathbf{k}-\mathbf{p})^2 + \mu^2} \frac{\vec{\alpha} \cdot \mathbf{k} + \beta + \epsilon}{k^2 - p^2 - i\lambda} u(\mathbf{p}). \quad (3.49)$$

$\Phi_2(\mathbf{k})$ is obtained from (3.48) by corresponding changes in the denominator of the integral. The functions $\Phi_1(\mathbf{k})$ and $\Phi_2(\mathbf{k})$ for a screened Coulomb potential are required for the calculation of the matrix element in second Born approximation, because it involves divergent integrals for a pure Coulomb potential. But all the observable quantities turn out to be finite when the limit of zero screening is taken after the calculation has been performed [38].

Now we can evaluate the bremsstrahlung matrix element (3.21) where the indices 1 and 2 refer to the incoming and outgoing electrons, respectively, to the lowest order in the Coulomb parameter $a = \alpha Z$. Turning to the Fourier transforms of the wave functions, we get, using Eqs. (3.34) and (3.39),

$$M = \int d^3r\, e^{-i\mathbf{k}\cdot\mathbf{r}} \int d^3k_2\, \psi_2^\dagger(\mathbf{k}_2)\, e^{-i\mathbf{k}_2\cdot\mathbf{r}} (\vec{\alpha}\cdot\mathbf{e}^*) \int d^3k_1\, \psi_1(\mathbf{k}_1)\, e^{i\mathbf{k}_1\cdot\mathbf{r}}$$

$$= (2\pi)^3 \int d^3k_2\, \psi_2^\dagger(\mathbf{k}_2)\, (\vec{\alpha}\cdot\mathbf{e}^*) \int d^3k_1\, \psi_1(\mathbf{k}_1)\, \delta(\mathbf{k}_1 - \mathbf{k}_2 - \mathbf{k})$$

$$= (2\pi)^3 \int \psi_2^\dagger(\mathbf{k}_2)\, (\vec{\alpha}\cdot\mathbf{e}^*)\, \psi_1(\mathbf{k}_2 + \mathbf{k})\, d^3k_2\,. \tag{3.50}$$

The Fourier transform of $\Phi_0(\mathbf{r})$ is given by

$$\Phi_0(\mathbf{k}) = \frac{u(\mathbf{p})}{(2\pi)^3} \int e^{i(\mathbf{p}-\mathbf{k})\cdot\mathbf{r}}\, d^3r = \delta(\mathbf{p} - \mathbf{k})\, u(\mathbf{p})\,. \tag{3.51}$$

Thus, up to first order in a, the transformed wave functions have the form

$$\psi_1(\mathbf{k}_1) = \left\{ \delta(\mathbf{p}_1 - \mathbf{k}_1) + \lim_{\lambda\to 0} \frac{a}{2\pi^2} \frac{1}{(\mathbf{p}_1 - \mathbf{k}_1)^2} \frac{\vec{\alpha}\cdot\mathbf{k}_1 + \beta + \epsilon_1}{k_1^2 - p_1^2 - i\lambda} \right\} u(\mathbf{p}_1) \tag{3.52}$$

and

$$\psi_2^\dagger(\mathbf{k}_2) = u^\dagger(\mathbf{p}_2) \left\{ \delta(\mathbf{p}_2 - \mathbf{k}_2) + \lim_{\lambda\to 0} \frac{a}{2\pi^2} \frac{1}{(\mathbf{p}_2 - \mathbf{k}_2)^2} \frac{\vec{\alpha}\cdot\mathbf{k}_2 + \beta + \epsilon_2}{k_2^2 - p_2^2 + i\lambda} \right\}. \tag{3.53}$$

Inserting these functions into the matrix element (3.50) yields

$$M = (2\pi)^3 u^\dagger(\mathbf{p}_2) \int d^3k_2 \Big\{ \delta(\mathbf{p}_2 - \mathbf{k}_2)\, \delta(\mathbf{p}_1 - \mathbf{k} - \mathbf{k}_2)\, (\vec{\alpha}\cdot\mathbf{e}^*)$$

$$+ \frac{a}{2\pi^2} \lim_{\lambda\to 0} \bigg[\frac{\delta(\mathbf{p}_1 - \mathbf{k} - \mathbf{k}_2)}{(\mathbf{p}_2 - \mathbf{k}_2)^2} \frac{(\vec{\alpha}\cdot\mathbf{k}_2 + \beta + \epsilon_2)(\vec{\alpha}\cdot\mathbf{e}^*)}{k_2^2 - p_2^2 + i\lambda}$$

$$+ \frac{\delta(\mathbf{p}_2 - \mathbf{k}_2)}{(\mathbf{p}_1 - \mathbf{k} - \mathbf{k}_2)^2} \frac{(\vec{\alpha}\cdot\mathbf{e}^*)[\vec{\alpha}\cdot(\mathbf{k}_2 + \mathbf{k}) + \beta + \epsilon_1]}{(\mathbf{k} + \mathbf{k}_2)^2 - p_1^2 - i\lambda} \bigg] \Big\} u(\mathbf{p}_1) + O(a^2)$$

$$= (2\pi)^3 u^\dagger(\mathbf{p}_2) \Big\{ \delta(\mathbf{q})(\vec{\alpha}\cdot\mathbf{e}^*) + \frac{a}{2\pi^2} \bigg[\frac{[\vec{\alpha}\cdot(\mathbf{p}_1 - \mathbf{k}) + \beta + \epsilon_2](\vec{\alpha}\cdot\mathbf{e}^*)}{D_1 q^2}$$

$$-\frac{(\vec{\alpha}\cdot \mathbf{e}^*)[\vec{\alpha}\cdot(\mathbf{p}_2+\mathbf{k})+\beta+\epsilon_1]}{D_2 q^2}\bigg]\bigg\}u(\mathbf{p}_1)+O(a^2)\,,\qquad(3.54)$$

where $\mathbf{q}=\mathbf{p}_1-\mathbf{p}_2-\mathbf{k}$, $O(x)$ means order of x, and

$$D_1 = (\mathbf{p}_1-\mathbf{k})^2 - p_2^2 = 2(\epsilon_1 k - \mathbf{p}_1\cdot\mathbf{k})\,,\qquad(3.55)$$
$$D_2 = p_1^2 - (\mathbf{p}_2+\mathbf{k})^2 = 2(\epsilon_2 k - \mathbf{p}_2\cdot\mathbf{k})\,.\qquad(3.56)$$

We leave it as an exercise (see Problem 3.2) to show that the recoil momentum of the nucleus $\mathbf{q}$ never vanishes, hence $\delta(\mathbf{q})=0$. For that reason we had to calculate the wave function $\psi(\mathbf{r})$ up to the first order in a to obtain the lowest order of the matrix element. Physically this means that an electron cannot emit a bremsstrahlung photon in the absence of another particle's Coulomb field since energy and momentum conservation are not satisfied simultaneously. Therefore bremsstrahlung is a second-order process.

The expression (3.54) can still be reduced by utilizing the relation (3.1) and applying the wave equation (3.29) of the free-electron spinor $u(\mathbf{p})$ and its adjoint equation. Then the bremsstrahlung matrix element to lowest order in a (first Born approximation) takes the form

$$M = \frac{4\pi a}{q^2} u^\dagger(\mathbf{p}_2)\bigg\{2\bigg(\frac{\epsilon_2}{D_1}-\frac{\epsilon_1}{D_2}\bigg)(\vec{\alpha}\cdot\mathbf{e}^*)+\frac{(\vec{\alpha}\cdot\mathbf{q})(\vec{\alpha}\cdot\mathbf{e}^*)}{D_1}$$
$$+\frac{(\vec{\alpha}\cdot\mathbf{e}^*)(\vec{\alpha}\cdot\mathbf{q})}{D_2}\bigg\}u(\mathbf{p}_1)=\frac{4\pi a}{q^2}u^\dagger(\mathbf{p}_2)\,Q\,u(\mathbf{p}_1)\,,\qquad(3.57)$$

where Q is the operator in curly brackets. Clearly, Eq. (3.57) can also be derived from Feynman diagrams.

The triply differential cross section is proportional to the absolute square of the matrix element (3.57),

$$|M|^2 = \bigg(\frac{4\pi a}{q^2}\bigg)^2 u^\dagger(\mathbf{p}_1)\bigg[2\bigg(\frac{\epsilon_2}{D_1}-\frac{\epsilon_1}{D_2}\bigg)(\vec{\alpha}\cdot\mathbf{e})+\frac{(\vec{\alpha}\cdot\mathbf{e})(\vec{\alpha}\cdot\mathbf{q})}{D_1}$$
$$+\frac{(\vec{\alpha}\cdot\mathbf{q})(\vec{\alpha}\cdot\mathbf{e})}{D_2}\bigg]u(\mathbf{p}_2)u^\dagger(\mathbf{p}_2)\bigg[2\bigg(\frac{\epsilon_2}{D_1}-\frac{\epsilon_1}{D_2}\bigg)(\vec{\alpha}\cdot\mathbf{e}^*)$$
$$+\frac{(\vec{\alpha}\cdot\mathbf{q})(\vec{\alpha}\cdot\mathbf{e}^*)}{D_1}+\frac{(\vec{\alpha}\cdot\mathbf{e}^*)(\vec{\alpha}\cdot\mathbf{q})}{D_2}\bigg]u(\mathbf{p}_1)\qquad(3.58)$$

This expression refers to given spin directions of the incoming and outgoing electrons and to given polarization vector of the emitted radiation. The pertinent cross section represents the most detailed independently observable quantity in the bremsstrahlung process. It includes all polarization correlations, corresponding to coincidence measurements of directions of photon emission and recoil electron for a fixed partition of incident electron kinetic energy between them.

If it is assumed that the incident electrons are unpolarized and that the spin states of the outgoing electrons are not observed, $|M|^2$ has to be summed over the spin directions of the final state and averaged over the spin directions of the initial state (since there is equal probability of initial spin in either direction). These summations can be performed without explicit reference to the representations of $\vec{\alpha}$, β, and $u(\mathbf{p})$. We rather utilize the wave equation (3.29) of the free-electron spinor $u(\mathbf{p})$. Applying the projection operator $(H + |\epsilon|)/2|\epsilon|$ to the spinor $u(\mathbf{p})$, we get

$$\frac{H + |\epsilon|}{2|\epsilon|} u(\mathbf{p}) = \frac{\vec{\alpha} \cdot \mathbf{p} + \beta + |\epsilon|}{2|\epsilon|} u(\mathbf{p}) = \begin{cases} u(\mathbf{p}) & \text{for } \epsilon > 0, \\ 0 & \text{for } \epsilon < 0. \end{cases} \quad (3.59)$$

Therefore the summations $\mathbf{S}^p$ over the spin directions only, for positive sign of ϵ, can be replaced by a summation over all 4 states of the spinor $u(\mathbf{p})$ pertaining to the same momentum $\mathbf{p}$, which we will denote by $\sum^p$. Thus, if Q_1 and Q_2 denote operators composed of Dirac matrices $\vec{\alpha}$ and β, we have

$$\mathbf{S}^{p_2} \Big(u^\dagger(\mathbf{p}_1) Q_1 u(\mathbf{p}_2) \Big) \Big(u^\dagger(\mathbf{p}_2) Q_2 u(\mathbf{p}_1) \Big)$$
$$= \frac{1}{2\epsilon_2} \sum^{p_2} \Big(u^\dagger(\mathbf{p}_1) Q_1 (\vec{\alpha} \cdot \mathbf{p}_2 + \beta + \epsilon_2) u(\mathbf{p}_2) \Big) \Big(u^\dagger(\mathbf{p}_2) Q_2 u(\mathbf{p}_1) \Big) .$$
$$(3.60)$$

The completeness relation (3.11) reduces to

$$\sum\nolimits^p u_\rho^\dagger(\mathbf{p}) u_{\rho'}(\mathbf{p}) = \delta_{\rho\rho'} , \quad (3.61)$$

and we get readily

$$\sum\nolimits^{p_2} \Big(u^\dagger(\mathbf{p}_1) Q_1 u(\mathbf{p}_2) \Big) \Big(u^\dagger(\mathbf{p}_2) Q_2 u(\mathbf{p}_1) \Big) = \Big(u^\dagger(\mathbf{p}_1) Q_1 Q_2 u(\mathbf{p}_1) \Big)$$
$$(3.62)$$

and
$$\sum\nolimits^p \Big(u^\dagger(\mathbf{p}) Q u(\mathbf{p}) \Big) = \operatorname{Tr} Q = \sum_\rho Q_{\rho\rho} . \quad (3.63)$$

Applying these relations to $|M|^2$ [Eq. (3.58)] we obtain

$$\mathbf{S}^{p_2}|M|^2 = \left(\frac{4\pi a}{q^2}\right)^2 \frac{1}{2\epsilon_2}\sum^{p_2}\left(u^\dagger(\mathbf{p}_1)\,Q^\dagger\,(\vec{\alpha}\cdot\mathbf{p}_2+\beta+\epsilon_2)\,u(\mathbf{p}_2)\right)$$
$$\cdot \left(u^\dagger(\mathbf{p}_2)\,Q\,u(\mathbf{p}_1)\right)$$
$$= \frac{8\pi^2 a^2}{\epsilon_2 q^4}\left(u^\dagger(\mathbf{p}_1)\,Q^\dagger\,(\vec{\alpha}\cdot\mathbf{p}_2+\beta+\epsilon_2)\,Q\,u(\mathbf{p}_1)\right) \quad (3.64)$$

and

$$\tfrac{1}{2}\mathbf{S}^{p_1}\mathbf{S}^{p_2}|M|^2$$
$$= \frac{2\pi^2 a^2}{\epsilon_1\epsilon_2 q^4}\sum^{p_1}\left(u^\dagger(\mathbf{p}_1)\,Q^\dagger(\vec{\alpha}\cdot\mathbf{p}_2+\beta+\epsilon_2)\,Q\,(\vec{\alpha}\cdot\mathbf{p}_1+\beta+\epsilon_1)\,u(\mathbf{p}_1)\right)$$
$$= \frac{2\pi^2 a^2}{\epsilon_1\epsilon_2 q^4}\,\mathrm{Tr}\{Q^\dagger(\vec{\alpha}\cdot\mathbf{p}_2+\beta+\epsilon_2)\,Q\,(\vec{\alpha}\cdot\mathbf{p}_1+\beta+\epsilon_1)\}\,. \quad (3.65)$$

Q is the operator defined in (3.57); its hermitian conjugate, $Q^\dagger$, results from reversing the order of the factors containing matrices, e.g., $(\vec{\alpha}\cdot\mathbf{q})(\vec{\alpha}\cdot\mathbf{e}^*)$, since the α_i are self-adjoint. Hence we get

$$\tfrac{1}{2}\mathbf{S}^{p_1}\mathbf{S}^{p_2}|M|^2 = \frac{2\pi^2 a^2}{\epsilon_1\epsilon_2 q^4}\,\mathrm{Tr}\Bigg\{\left[2\left(\frac{\epsilon_2}{D_1}-\frac{\epsilon_1}{D_2}\right)(\vec{\alpha}\cdot\mathbf{e}) + \frac{(\vec{\alpha}\cdot\mathbf{e})(\vec{\alpha}\cdot\mathbf{q})}{D_1}\right.$$
$$+ \left.\frac{(\vec{\alpha}\cdot\mathbf{q})(\vec{\alpha}\cdot\mathbf{e})}{D_2}\right](\vec{\alpha}\cdot\mathbf{p}_2+\beta+\epsilon_2)\left[2\left(\frac{\epsilon_2}{D_1}-\frac{\epsilon_1}{D_2}\right)(\vec{\alpha}\cdot\mathbf{e}^*)\right.$$
$$+ \left.\frac{(\vec{\alpha}\cdot\mathbf{q})(\vec{\alpha}\cdot\mathbf{e}^*)}{D_1} + \frac{(\vec{\alpha}\cdot\mathbf{e}^*)(\vec{\alpha}\cdot\mathbf{q})}{D_2}\right](\vec{\alpha}\cdot\mathbf{p}_1+\beta+\epsilon_1)\Bigg\}$$
$$= \frac{2\pi^2 a^2}{\epsilon_1\epsilon_2 q^4}\Bigg\{4\left(\frac{\epsilon_2}{D_1}-\frac{\epsilon_1}{D_2}\right)^2\mathrm{Tr}\big[(\vec{\alpha}\cdot\mathbf{e})(\vec{\alpha}\cdot\mathbf{p}_2+\beta+\epsilon_2)$$
$$\cdot(\vec{\alpha}\cdot\mathbf{e}^*)(\vec{\alpha}\cdot\mathbf{p}_1+\beta+\epsilon_1)\big]$$
$$+ 2\left(\frac{\epsilon_2}{D_1}-\frac{\epsilon_1}{D_2}\right)\mathrm{Tr}\bigg[\frac{1}{D_1}(\vec{\alpha}\cdot\mathbf{e})(\vec{\alpha}\cdot\mathbf{p}_2+\beta+\epsilon_2)(\vec{\alpha}\cdot\mathbf{q})$$
$$\cdot(\vec{\alpha}\cdot\mathbf{e}^*)(\vec{\alpha}\cdot\mathbf{p}_1+\beta+\epsilon_1)$$
$$+ \frac{1}{D_2}(\vec{\alpha}\cdot\mathbf{e})(\vec{\alpha}\cdot\mathbf{p}_2+\beta+\epsilon_2)(\vec{\alpha}\cdot\mathbf{e}^*)(\vec{\alpha}\cdot\mathbf{q})(\vec{\alpha}\cdot\mathbf{p}_1+\beta+\epsilon_1)$$
$$+ \frac{1}{D_1}(\vec{\alpha}\cdot\mathbf{e})(\vec{\alpha}\cdot\mathbf{q})(\vec{\alpha}\cdot\mathbf{p}_2+\beta+\epsilon_2)(\vec{\alpha}\cdot\mathbf{e}^*)(\vec{\alpha}\cdot\mathbf{p}_1+\beta+\epsilon_1)$$
$$+ \frac{1}{D_2}(\vec{\alpha}\cdot\mathbf{q})(\vec{\alpha}\cdot\mathbf{e})(\vec{\alpha}\cdot\mathbf{p}_2+\beta+\epsilon_2)(\vec{\alpha}\cdot\mathbf{e}^*)(\vec{\alpha}\cdot\mathbf{p}_1+\beta+\epsilon_1)\bigg]$$

$$+ \frac{1}{D_1^2}\text{Tr}\bigl[(\vec{\alpha}\cdot\mathbf{e})(\vec{\alpha}\cdot\mathbf{q})(\vec{\alpha}\cdot\mathbf{p}_2+\beta+\epsilon_2)(\vec{\alpha}\cdot\mathbf{q})(\vec{\alpha}\cdot\mathbf{e}^*)(\vec{\alpha}\cdot\mathbf{p}_1+\beta+\epsilon_1)\bigr]$$
$$+ \frac{1}{D_2^2}\text{Tr}\bigl[(\vec{\alpha}\cdot\mathbf{q})(\vec{\alpha}\cdot\mathbf{e})(\vec{\alpha}\cdot\mathbf{p}_2+\beta+\epsilon_2)(\vec{\alpha}\cdot\mathbf{e}^*)(\vec{\alpha}\cdot\mathbf{q})(\vec{\alpha}\cdot\mathbf{p}_1+\beta+\epsilon_1)\bigr]$$
$$+ \frac{1}{D_1 D_2}\text{Tr}\bigl[(\vec{\alpha}\cdot\mathbf{e})(\vec{\alpha}\cdot\mathbf{q})(\vec{\alpha}\cdot\mathbf{p}_2+\beta+\epsilon_2)(\vec{\alpha}\cdot\mathbf{e}^*)(\vec{\alpha}\cdot\mathbf{q})(\vec{\alpha}\cdot\mathbf{p}_1+\beta+\epsilon_1)$$
$$+ (\vec{\alpha}\cdot\mathbf{q})(\vec{\alpha}\cdot\mathbf{e})(\vec{\alpha}\cdot\mathbf{p}_2+\beta+\epsilon_2)(\vec{\alpha}\cdot\mathbf{q})(\vec{\alpha}\cdot\mathbf{e}^*)(\vec{\alpha}\cdot\mathbf{p}_1+\beta+\epsilon_1)\bigr]\biggr\}.$$
(3.66)

The traces of the operators can be easily evaluated. From Eqs. (3.4) and (3.5) it follows immediately that the traces of all products with an odd number of factors of either α_i or β vanish. By employing repeatedly the relation

$$(\vec{\alpha}\cdot\mathbf{a})(\vec{\alpha}\cdot\mathbf{b}) + (\vec{\alpha}\cdot\mathbf{b})(\vec{\alpha}\cdot\mathbf{a}) = 2(\mathbf{a}\cdot\mathbf{b})\,\mathbf{1} \tag{3.67}$$

following from (3.6) and the commutator (3.7), we obtain

$$\text{Tr}(\vec{\alpha}\cdot\mathbf{a})(\vec{\alpha}\cdot\mathbf{b}) = \text{Tr}(\vec{\alpha}\cdot\mathbf{b})(\vec{\alpha}\cdot\mathbf{a}) = 4\,\mathbf{a}\cdot\mathbf{b} \tag{3.68}$$

and

$$\text{Tr}\{(\vec{\alpha}\cdot\mathbf{a})(\vec{\alpha}\cdot\mathbf{b})(\vec{\alpha}\cdot\mathbf{c})(\vec{\alpha}\cdot\mathbf{d})\} = 4\bigl[(\mathbf{a}\cdot\mathbf{b})(\mathbf{c}\cdot\mathbf{d}) + (\mathbf{a}\cdot\mathbf{d})(\mathbf{b}\cdot\mathbf{c}) - (\mathbf{a}\cdot\mathbf{c})(\mathbf{b}\cdot\mathbf{d})\bigr]. \tag{3.69}$$

If the relations (3.68) and (3.69) are applied to (3.66) this yields

$$\tfrac{1}{2}\mathbf{S}^{p_1}\mathbf{S}^{p_2}|M|^2$$
$$= \frac{8\pi^2 a^2}{\epsilon_1\epsilon_2 q^4}\biggl\{4\Bigl(\frac{\epsilon_2}{D_1}-\frac{\epsilon_1}{D_2}\Bigr)^2\bigl[\epsilon_1\epsilon_2 - 1 - \mathbf{p}_1\cdot\mathbf{p}_2 + 2\text{Re}(\mathbf{e}\cdot\mathbf{p}_1)(\mathbf{e}^*\cdot\mathbf{p}_2)\bigr]$$
$$+ 4\Bigl(\frac{\epsilon_2}{D_1}-\frac{\epsilon_1}{D_2}\Bigr)\biggl[\frac{\epsilon_2}{D_1}\{2\text{Re}(\mathbf{e}\cdot\mathbf{p}_1)(\mathbf{e}^*\cdot\mathbf{q}) - \mathbf{q}\cdot\mathbf{p}_1\}$$
$$+ \frac{\epsilon_1}{D_2}\{2\text{Re}(\mathbf{e}\cdot\mathbf{p}_2)(\mathbf{e}^*\cdot\mathbf{q}) - \mathbf{q}\cdot\mathbf{p}_2\} + \frac{\epsilon_1}{D_1}(\mathbf{q}\cdot\mathbf{p}_2) + \frac{\epsilon_2}{D_2}(\mathbf{q}\cdot\mathbf{p}_1)\biggr]$$
$$+ \frac{1}{D_1^2}\bigl[\{\epsilon_1\epsilon_2 + \mathbf{p}_1\cdot\mathbf{p}_2 + 1 - 2\text{Re}(\mathbf{e}\cdot\mathbf{p}_1)(\mathbf{e}^*\cdot\mathbf{p}_2)\}q^2$$
$$+ 2(\mathbf{q}\cdot\mathbf{p}_2)\{2\text{Re}(\mathbf{e}\cdot\mathbf{p}_1)(\mathbf{e}^*\cdot\mathbf{q}) - \mathbf{q}\cdot\mathbf{p}_1\}\bigr]$$

$$+ \frac{1}{D_2^2}\Big[\{\epsilon_1\epsilon_2 + \mathbf{p}_1 \cdot \mathbf{p}_2 + 1 - 2\mathrm{Re}(\mathbf{e}\cdot\mathbf{p}_1)(\mathbf{e}^*\cdot\mathbf{p}_2)\}q^2$$

$$+ 2(\mathbf{q}\cdot\mathbf{p}_1)\{2\mathrm{Re}(\mathbf{e}\cdot\mathbf{p}_2)(\mathbf{e}^*\cdot\mathbf{q}) - \mathbf{q}\cdot\mathbf{p}_2\}\Big]$$

$$+ \frac{2}{D_1 D_2}\Big[(\epsilon_1\epsilon_2 + 1)(2|\mathbf{e}\cdot\mathbf{q}|^2 - q^2) + 2(\mathbf{p}_1\cdot\mathbf{p}_2)|\mathbf{e}\cdot\mathbf{q}|^2$$

$$- 2(\mathbf{q}\cdot\mathbf{p}_1)\mathrm{Re}(\mathbf{e}\cdot\mathbf{p}_2)(\mathbf{e}^*\cdot\mathbf{q}) + q^2\{2\mathrm{Re}(\mathbf{e}\cdot\mathbf{p}_1)(\mathbf{e}^*\cdot\mathbf{p}_2) - \mathbf{p}_1\cdot\mathbf{p}_2\}$$

$$+ 2(\mathbf{q}\cdot\mathbf{p}_1)(\mathbf{q}\cdot\mathbf{p}_2) - 2(\mathbf{q}\cdot\mathbf{p}_2)\mathrm{Re}(\mathbf{e}\cdot\mathbf{p}_1)(\mathbf{e}^*\cdot\mathbf{q})\Big]\Big\}$$

$$= \frac{8\pi^2 a^2}{\epsilon_1\epsilon_2 q^4}\Big\{\frac{1}{D_1^2}\Big[4\epsilon_2^2\{2|\mathbf{e}\cdot\mathbf{p}_1|^2 + \underbrace{\epsilon_1\epsilon_2 - 1 - \mathbf{p}_1\cdot\mathbf{p}_2 - \mathbf{q}\cdot\mathbf{p}_1}_{-D_1/2}\}$$

$$+ \{\epsilon_1\epsilon_2 + 1 + \mathbf{p}_1\cdot\mathbf{p}_2 - 2\mathrm{Re}(\mathbf{e}\cdot\mathbf{p}_1)(\mathbf{e}^*\cdot\mathbf{p}_2)\}q^2$$

$$+ 2(\mathbf{q}\cdot\mathbf{p}_2)\{2\epsilon_1\epsilon_2 - \mathbf{q}\cdot\mathbf{p}_1 + 2\mathrm{Re}(\mathbf{e}\cdot\mathbf{p}_1)(\mathbf{e}^*\cdot\mathbf{q})\}\Big]$$

$$+ \frac{1}{D_2^2}\Big[4\epsilon_1^2\{2|\mathbf{e}\cdot\mathbf{p}_2|^2 + \underbrace{\epsilon_1\epsilon_2 - 1 - \mathbf{p}_1\cdot\mathbf{p}_2 + \mathbf{q}\cdot\mathbf{p}_2}_{D_2/2}\}$$

$$+ \{\epsilon_1\epsilon_2 + 1 + \mathbf{p}_1\cdot\mathbf{p}_2 - 2\mathrm{Re}(\mathbf{e}\cdot\mathbf{p}_1)(\mathbf{e}^*\cdot\mathbf{p}_2)\}q^2$$

$$- 2(\mathbf{q}\cdot\mathbf{p}_1)\{2\epsilon_1\epsilon_2 + \mathbf{q}\cdot\mathbf{p}_2 - 2\mathrm{Re}(\mathbf{e}\cdot\mathbf{p}_2)(\mathbf{e}^*\cdot\mathbf{q})\}\Big]$$

$$+ \frac{2}{D_1 D_2}\Big[2\epsilon_1\epsilon_2\{\underbrace{2\mathbf{p}_1\cdot\mathbf{p}_2 + 2 - 2\epsilon_1\epsilon_2 + \mathbf{q}\cdot\mathbf{p}_1 - \mathbf{q}\cdot\mathbf{p}_2}_{-\mathbf{k}\cdot\mathbf{q}}$$

$$- 4\mathrm{Re}(\mathbf{e}\cdot\mathbf{p}_1)(\mathbf{e}^*\cdot\mathbf{p}_2) - |\mathbf{e}\cdot\mathbf{q}|^2\}$$

$$- q^2\{\epsilon_1\epsilon_2 + 1 + \mathbf{p}_1\cdot\mathbf{p}_2 - 2\mathrm{Re}(\mathbf{e}\cdot\mathbf{p}_1)(\mathbf{e}^*\cdot\mathbf{p}_2)\}$$

$$+ 2(\mathbf{q}\cdot\mathbf{p}_1)\{\epsilon_2^2 - \mathrm{Re}(\mathbf{e}\cdot\mathbf{p}_2)(\mathbf{e}^*\cdot\mathbf{q})\}$$

$$- 2(\mathbf{q}\cdot\mathbf{p}_2)\{\epsilon_1^2 + \mathrm{Re}(\mathbf{e}\cdot\mathbf{p}_1)(\mathbf{e}^*\cdot\mathbf{q})\}$$

$$+ 2(\mathbf{q}\cdot\mathbf{p}_1)(\mathbf{q}\cdot\mathbf{p}_2) + 2(1+\mathbf{p}_1\cdot\mathbf{p}_2)|\mathbf{e}\cdot\mathbf{q}|^2\Big]\Big\}$$

$$= \frac{16\pi^2 a^2}{\epsilon_1\epsilon_2 q^4}\Big\{\frac{1}{D_1^2}\Big[\epsilon_2^2(4|\mathbf{e}\cdot\mathbf{p}_1|^2 - D_1) + (\mathbf{q}\cdot\mathbf{p}_2)\{2\epsilon_1\epsilon_2 - \mathbf{q}\cdot\mathbf{p}_1$$

$$+ 2\mathrm{Re}(\mathbf{e}\cdot\mathbf{p}_1)(\mathbf{e}^*\cdot\mathbf{q})\}\Big] + \frac{1}{D_2^2}\Big[\epsilon_1^2(4|\mathbf{e}\cdot\mathbf{p}_2|^2 + D_2)$$

$$- (\mathbf{q}\cdot\mathbf{p}_1)\{2\epsilon_1\epsilon_2 + \mathbf{q}\cdot\mathbf{p}_2 - 2\mathrm{Re}(\mathbf{e}\cdot\mathbf{p}_2)(\mathbf{e}^*\cdot\mathbf{q})\}\Big]$$

$$-\frac{2}{D_1 D_2}\Big[\epsilon_1\epsilon_2\{\mathbf{k}\cdot\mathbf{q}+|\mathbf{e}\cdot\mathbf{q}|^2+4\operatorname{Re}(\mathbf{e}\cdot\mathbf{p}_1)(\mathbf{e}^*\cdot\mathbf{p}_2)\}$$
$$-(\mathbf{q}\cdot\mathbf{p}_1)\{\epsilon_2^2-\operatorname{Re}(\mathbf{e}\cdot\mathbf{p}_2)(\mathbf{e}^*\cdot\mathbf{q})\}+(\mathbf{q}\cdot\mathbf{p}_2)\{\epsilon_1^2+\operatorname{Re}(\mathbf{e}\cdot\mathbf{p}_1)(\mathbf{e}^*\cdot\mathbf{q})\}$$
$$-(\mathbf{q}\cdot\mathbf{p}_1)(\mathbf{q}\cdot\mathbf{p}_2)-(1+\mathbf{p}_1\cdot\mathbf{p}_2)|\mathbf{e}\cdot\mathbf{q}|^2\Big]$$
$$+\tfrac{1}{2}q^2\Big(\frac{1}{D_1}-\frac{1}{D_2}\Big)^2\Big[\epsilon_1\epsilon_2+1+\mathbf{p}_1\cdot\mathbf{p}_2-2\operatorname{Re}(\mathbf{e}\cdot\mathbf{p}_1)(\mathbf{e}^*\cdot\mathbf{p}_2)\Big]\Big\}.$$
(3.70)

By means of the relations

$$\mathbf{e}\cdot\mathbf{q}=\mathbf{e}\cdot\mathbf{p}_1-\mathbf{e}\cdot\mathbf{p}_2,\tag{3.71}$$
$$2\mathbf{q}\cdot\mathbf{p}_1=q^2+D_2,\tag{3.72}$$
$$2\mathbf{q}\cdot\mathbf{p}_2=D_1-q^2,\tag{3.73}$$
$$D_2-D_1=2(\mathbf{q}\cdot\mathbf{k}),\tag{3.74}$$

and

$$q^2=D_1-D_2+2(\epsilon_1\epsilon_2-\mathbf{p}_1\cdot\mathbf{p}_2-1)\tag{3.75}$$

we finally arrive at

$$\tfrac{1}{2}\mathbf{S}^{p_1}\mathbf{S}^{p_2}|M|^2=\frac{16\pi^2 a^2}{\epsilon_1\epsilon_2 q^4}\Big\{\frac{4\epsilon_2^2-q^2}{D_1^2}|\mathbf{e}\cdot\mathbf{p}_1|^2+\frac{4\epsilon_1^2-q^2}{D_2^2}|\mathbf{e}\cdot\mathbf{p}_2|^2$$
$$-2\frac{4\epsilon_1\epsilon_2-q^2}{D_1 D_2}\operatorname{Re}(\mathbf{e}\cdot\mathbf{p}_1)(\mathbf{e}^*\cdot\mathbf{p}_2)+\frac{(\mathbf{q}\times\mathbf{k})^2}{D_1 D_2}\Big\}.\tag{3.76}$$

It is seen that the first three terms of (3.76) depend on the photon polarization but that the last term does not. Further we note that the cross section in first Born approximation for photons emitted by unpolarized electrons does not have any terms correlating the momenta of the particles with the circular polarization of the photons. Thus, in first Born approximation unpolarized electrons can only produce linearly polarized photons (cf. Sec. 3.7.3).

If we are not interested in the photon polarization, we sum over the two linear polarization directions (for linear polarization the vector $\mathbf{e}$ is real)

$$\mathbf{e}_\perp=\frac{\mathbf{p}_1\times\mathbf{k}}{|\mathbf{p}_1\times\mathbf{k}|},\quad \mathbf{e}_\parallel=\frac{\mathbf{k}\times(\mathbf{p}_1\times\mathbf{k})}{k|\mathbf{p}_1\times\mathbf{k}|}=\frac{k^2\mathbf{p}_1-(\mathbf{k}\cdot\mathbf{p}_1)\mathbf{k}}{k|\mathbf{p}_1\times\mathbf{k}|}.\tag{3.77}$$

Applying the vector calculus formula

$$[\mathbf{a} \cdot (\mathbf{b} \times \mathbf{c})][\mathbf{e} \cdot (\mathbf{f} \times \mathbf{g})] = (\mathbf{a} \cdot \mathbf{e})[(\mathbf{b} \cdot \mathbf{f})(\mathbf{c} \cdot \mathbf{g}) - (\mathbf{b} \cdot \mathbf{g})(\mathbf{c} \cdot \mathbf{f})]$$
$$+ (\mathbf{a} \cdot \mathbf{f})[(\mathbf{b} \cdot \mathbf{g})(\mathbf{c} \cdot \mathbf{e}) - (\mathbf{b} \cdot \mathbf{e})(\mathbf{c} \cdot \mathbf{g})] + (\mathbf{a} \cdot \mathbf{g})[(\mathbf{b} \cdot \mathbf{e})(\mathbf{c} \cdot \mathbf{f}) - (\mathbf{b} \cdot \mathbf{f})(\mathbf{c} \cdot \mathbf{e})] \tag{3.78}$$

we get for arbitrary vectors $\mathbf{x}$ and $\mathbf{y}$ the sum

$$\sum_{\mathbf{e}} (\mathbf{e} \cdot \mathbf{x})(\mathbf{e} \cdot \mathbf{y}) = \mathbf{x} \cdot \mathbf{y} - (\mathbf{x} \cdot \hat{\mathbf{k}})(\mathbf{y} \cdot \hat{\mathbf{k}}) = (\mathbf{x} \times \hat{\mathbf{k}}) \cdot (\mathbf{y} \times \hat{\mathbf{k}}), \tag{3.79}$$

where $\hat{\mathbf{k}} = \mathbf{k}/k$ denotes the unit vector in photon direction. Then the summed matrix element squared becomes

$$\tfrac{1}{2} \sum_{\mathbf{e}} \mathbf{S}^{p_1} \mathbf{S}^{p_2} |M|^2 = \frac{16 \pi^2 a^2}{\epsilon_1 \epsilon_2 q^4} \left\{ \frac{4\epsilon_2^2 - q^2}{D_1^2} (\mathbf{p}_1 \times \hat{\mathbf{k}})^2 + \frac{4\epsilon_1^2 - q^2}{D_2^2} (\mathbf{p}_2 \times \hat{\mathbf{k}})^2 \right.$$
$$\left. - 2 \frac{4\epsilon_1 \epsilon_2 - q^2}{D_1 D_2} (\mathbf{p}_1 \times \hat{\mathbf{k}}) \cdot (\mathbf{p}_2 \times \hat{\mathbf{k}}) + 2 \frac{(\mathbf{q} \times \mathbf{k})^2}{D_1 D_2} \right\}. \tag{3.80}$$

From (3.24) the differential cross section for the elementary process of bremsstrahlung in Born approximation (Bethe-Heitler formula) has the form [11]

$$\frac{d^3 \sigma_B}{dk \, d\Omega_k \, d\Omega_{p_2}} = \frac{\alpha Z^2 r_0^2}{\pi^2} \frac{p_2}{k p_1 q^4} \left\{ \frac{4\epsilon_2^2 - q^2}{D_1^2} (\mathbf{p}_1 \times \mathbf{k})^2 + \frac{4\epsilon_1^2 - q^2}{D_2^2} (\mathbf{p}_2 \times \mathbf{k})^2 \right.$$
$$\left. - 2 \frac{4\epsilon_1 \epsilon_2 - q^2}{D_1 D_2} (\mathbf{p}_1 \times \mathbf{k}) \cdot (\mathbf{p}_2 \times \mathbf{k}) + \frac{2k^2}{D_1 D_2} (\mathbf{q} \times \mathbf{k})^2 \right\}, \tag{3.81}$$

where $r_0 = (\hbar/mc)\alpha \approx 2.818 \cdot 10^{-15}$ m is the classical electron radius. The corresponding cross section including photon polarization is obtained by substituting the curly brackets of Eq. (3.76) for that of (3.81).

With the cross section written as

$$\frac{d^3 \sigma_B}{dk \, d\Omega_k \, d\Omega_{p_2}} = \frac{\alpha Z^2 r_0^2}{\pi^2} \frac{p_2}{k p_1 q^4} \left\{ \left(\frac{2\epsilon_2}{D_1} \mathbf{p}_1 \times \mathbf{k} - \frac{2\epsilon_1}{D_2} \mathbf{p}_2 \times \mathbf{k} \right)^2 \right.$$
$$\left. - q^2 \left(\frac{\mathbf{p}_1 \times \mathbf{k}}{D_1} - \frac{\mathbf{p}_2 \times \mathbf{k}}{D_2} \right)^2 + \frac{2k^2}{D_1 D_2} (\mathbf{q} \times \mathbf{k})^2 \right\} \tag{3.82}$$

one sees that the Bethe-Heitler cross section depends on the three vectors

$$\mathbf{A} = \frac{2\epsilon_2}{D_1} (\mathbf{p}_1 \times \mathbf{k}) - \frac{2\epsilon_1}{D_2} (\mathbf{p}_2 \times \mathbf{k}), \tag{3.83}$$

$$\mathbf{B} = q\left(\frac{\mathbf{p}_1 \times \mathbf{k}}{D_1} - \frac{\mathbf{p}_2 \times \mathbf{k}}{D_2}\right), \tag{3.84}$$

and

$$\mathbf{C} = \frac{k}{\sqrt{D_1 D_2}}(\mathbf{q} \times \mathbf{k}) = \frac{k}{\sqrt{D_1 D_2}}(\mathbf{p}_1 \times \mathbf{k} - \mathbf{p}_2 \times \mathbf{k}). \tag{3.85}$$

It is convenient to introduce a system of polar coordinates of which the z axis has the direction $\hat{\mathbf{k}}$ of the emitted photon and where the incoming electron is moving in the x-z plane. Then the momenta $\mathbf{p}_1$, $\mathbf{p}_2$, and $\mathbf{k}$ have the form

$$\begin{aligned}\mathbf{p}_1 &= p_1\{\sin\theta_1, 0, \cos\theta_1\}, \\ \mathbf{p}_2 &= p_2\{\sin\theta_2 \cos\varphi, \sin\theta_2 \sin\varphi, \cos\theta_2\}, \\ \mathbf{k} &= k\{0, 0, 1\},\end{aligned} \tag{3.86}$$

and the differential cross section becomes

$$\begin{aligned}\frac{d^3\sigma_B}{dk\, d\Omega_k\, d\Omega_{p_2}} = \frac{\alpha Z^2 r_0^2}{4\pi^2} \frac{p_2}{k p_1 q^4} &\Bigg\{ (4\epsilon_2^2 - q^2)\frac{p_1^2 \sin^2\theta_1}{(\epsilon_1 - p_1 \cos\theta_1)^2} \\ &+ (4\epsilon_1^2 - q^2)\frac{p_2^2 \sin^2\theta_2}{(\epsilon_2 - p_2 \cos\theta_2)^2} \\ &- (4\epsilon_1\epsilon_2 - q^2 + 2k^2)\frac{2p_1 p_2 \sin\theta_1 \sin\theta_2 \cos\varphi}{(\epsilon_1 - p_1 \cos\theta_1)(\epsilon_2 - p_2 \cos\theta_2)} \\ &+ 2k^2 \frac{p_1^2 \sin^2\theta_1 + p_2^2 \sin^2\theta_2}{(\epsilon_1 - p_1 \cos\theta_1)(\epsilon_2 - p_2 \cos\theta_2)} \Bigg\}\end{aligned} \tag{3.87}$$

with

$$\begin{aligned}q^2 = 2k\big[(\epsilon_1 - p_1 \cos\theta_1) - (\epsilon_2 - p_2 \cos\theta_2)\big] \\ + 2\big[\epsilon_1\epsilon_2 - p_1 p_2(\cos\theta_1 \cos\theta_2 + \sin\theta_1 \sin\theta_2 \cos\varphi) - 1\big].\end{aligned} \tag{3.88}$$

The triply differential cross section gives the probability that an electron of total energy ϵ_1 is scattered into the direction $\hat{\mathbf{p}}_2$ under emission of a photon of momentum $\mathbf{k}$ (elementary process of bremsstrahlung). Since the Bethe-Heitler formula is obtained by means of a first-order perturbation method for scattering in the Coulomb potential $\alpha Z/r$ the cross section is proportional to the square of the atomic number Z and tends to zero at the short-wavelength limit $p_2 = 0$ where the photon receives the total kinetic energy of the incident electron. The Born approximation only gives correct results if the de Broglie wavelength of the electron, $\hbar/mv$, is large compared to the 'size' d of the Coulomb field given by $Ze^2/(4\pi\epsilon_0 d) \approx mv^2$. Therefore the condition for validity is

$$\frac{\alpha Z}{\beta_1} \ll 1 \quad \text{and} \quad \frac{\alpha Z}{\beta_2} \ll 1, \tag{3.89}$$

where $\beta_1 = p_1/\epsilon_1$ and $\beta_2 = p_2/\epsilon_2$ are the electron velocities in units of the light velocity c. Since the fine-structure constant $\alpha \approx 1/137$ is a small quantity, the condition (3.89) is satisfied for elements of low atomic number Z if the electron energies are relativistic, i.e., $\beta_{1,2} \approx 1$. But even for energetic electrons with $\epsilon_1 \gg 1$ the second constraint (3.89) is violated near the short-wavelength limit where $\beta_2 \ll 1$. Indeed, the cross section does not vanish for $p_2 = 0$ if the distortion of the electron wave functions by the nuclear Coulomb field is taken into account (see Sec. 3.5.3). It is easy to understand why the Bethe-Heitler cross section tends to zero in the limit $p_2 \to 0$. According to Eqs. (3.26) and (3.28) the final electron in Born approximation is described by a plane wave and a first correction. Both terms remain finite as $p_2 \to 0$. The cross section is proportional to $p_2|M|^2$ [see Eq. (3.24)] where the matrix M element is finite at the short-wavelength limit, so that the cross section vanishes with p_2. By contrast, the exact point-Coulomb wave function does not remain finite in the limit $p_2 \to 0$ but diverges as $p_2^{-1/2}$ thus leading to a finite cross section. The Coulomb correction can be allowed for approximately by multiplying the cross section (3.81) by the Elwert factor [39]

$$F_E = \frac{a_2}{a_1} \frac{1 - e^{-2\pi a_1}}{1 - e^{-2\pi a_2}}, \tag{3.90}$$

the square of the ratio of final to initial continuum-wave-function normalization, where

$$a_1 = \frac{\alpha Z}{\beta_1} = \alpha Z \frac{\epsilon_1}{p_1}, \quad a_2 = \frac{\alpha Z}{\beta_2} = \alpha Z \frac{\epsilon_2}{p_2} \tag{3.91}$$

are the Coulomb parameters. For $p_2 = 0$ the factor a_2 of (3.90) compensates the factor p_2 of the cross section so that it tends to a finite value. On the other hand, at the long-wavelength limit $p_2 \approx p_1$, and the factor (3.90) is $F_E \approx 1$, in agreement with the fact that the Born approximation is accurate near the low-energy end of the bremsstrahlung spectrum. The Elwert factor is always larger than unity since $a_2 > a_1$ and the function $x/\{1 - \exp(-x)\}$

is monotonically increasing. Reasons for the success of the factor (3.90) at low atomic numbers are given by Pratt and Tseng [40].

3.3.1 Atomic screening

The cross section (3.81) has been derived on the assumption that the bremsstrahlung production takes place in a pure Coulomb potential. In most real cases, however, the field of the atomic nucleus is shielded by shell electrons. In order to evaluate the effect of screening we follow the argument of Heitler [32] and take note that in Born approximation the bremsstrahlung matrix element (3.57) is proportional to[†]

$$V(\mathbf{q}) = \int V(r) e^{i\mathbf{q}\cdot\mathbf{r}} d^3r = -\alpha Z \int \frac{1}{r} e^{i\mathbf{q}\cdot\mathbf{r}} d^3r = -\frac{4\pi}{q^2}\alpha Z, \qquad (3.92)$$

where $\mathbf{q}$ is the momentum transferred to the atom [see Eq. (3.1)]. The region which essentially determines the magnitude of this integral is given by $r \approx 1/q$. Actually, for large values of r the oscillations of $\exp(i\mathbf{q}\cdot\mathbf{r})$ reduce the contributions to the integral, whereas low values of r are negligible due to the small volume element d^3r. Obviously the screening effect will play no part if the maximum value of r, corresponding to the minimum value of q, is much smaller than the size d of the atom: $r_{\max} \ll d$. Conversely the screening will be complete for $r_{\max} \gg d$. The minimum value of q is

$$q_{\min} = p_1 - p_2 - k = \sqrt{\epsilon_1^2 - 1} - \sqrt{p_1^2 - 2\epsilon_1 k + k^2} - k. \qquad (3.93)$$

For nonrelativistic energies ($\epsilon_1 \approx 1$, $p_1^2 \ll 1$, $k \ll p_1$) we have $p_2 \approx p_1 - k/p_1$, hence $q_{\min} \approx k/p_1$.

At extremely relativistic energies, $\epsilon_1 \gg 1$, $\epsilon_2 \gg 1$, Eq. (3.93) yields

$$q_{\min} \approx \epsilon_1 - \frac{1}{2\epsilon_1} - \left(\epsilon_2 - \frac{1}{2\epsilon_2}\right) - (\epsilon_1 - \epsilon_2) = \frac{k}{2\epsilon_1\epsilon_2}. \qquad (3.94)$$

Therefore the maximum value of r is given by

$$r_{\max} \approx \begin{cases} p_1/k & \text{for } p_1 \ll 1,\ k \ll p_1, \\ 2\epsilon_1\epsilon_2/k & \text{for } \epsilon_1 \gg 1,\ \epsilon_2 \gg 1. \end{cases} \qquad (3.95)$$

[†]The integral (3.92) is not convergent. To get a meaningful result one has to modify the long-range Coulomb potential by a factor $\exp(-\mu r)$, evaluate the integral and then take the limit $\mu \to 0$ [see Eq. (3.43)].

It is seen that $1/q_{\min}$ can exceed the dimensions of the atom for sufficiently low photon energies. Likewise, the screening becomes appreciable at high incident energies; for a given ratio ϵ_2/k, $r_{\max}$ is larger at higher ϵ_1. This leads to a significant reduction of the bremsstrahlung cross section.

Screening of the nuclear field by the atomic electrons may be taken into account by means of a form factor $F(\mathbf{q})$. The atomic potential is given by

$$V(\mathbf{r}) = -\frac{\alpha Z}{r} + \alpha \int \frac{\rho(\mathbf{r}')}{|\mathbf{r}-\mathbf{r}'|} d^3r' , \qquad (3.96)$$

where $-\rho(\mathbf{r})$ denotes the electron charge density normalized to Z,

$$\int \rho(\mathbf{r}) d^3r = Z . \qquad (3.97)$$

Then the bremsstrahlung matrix element is proportional to

$$\begin{aligned}V(\mathbf{q}) &= \int V(\mathbf{r}) e^{i\mathbf{q}\cdot\mathbf{r}} d^3r = -\alpha \left\{ \frac{4\pi Z}{q^2} - \int d^3r' \rho(\mathbf{r}') \int d^3r \frac{e^{i\mathbf{q}\cdot\mathbf{r}}}{|\mathbf{r}-\mathbf{r}'|} \right\} \\ &= -\alpha \left\{ \frac{4\pi Z}{q^2} - \int d^3r' \rho(\mathbf{r}') e^{i\mathbf{q}\cdot\mathbf{r}'} \underbrace{\int d^3r \frac{e^{i\mathbf{q}\cdot(\mathbf{r}-\mathbf{r}')}}{|\mathbf{r}-\mathbf{r}'|}}_{4\pi/q^2} \right\} \\ &= -\frac{4\pi \alpha Z}{q^2} \left\{ 1 - \frac{1}{Z} \int \rho(\mathbf{r}) e^{i\mathbf{q}\cdot\mathbf{r}} d^3r \right\} . \end{aligned} \qquad (3.98)$$

For a pure Coulomb potential the curly bracket is equal to one. Hence the matrix element has to be multiplied by $1 - F(\mathbf{q})$ in order to allow for nuclear screening, where the atomic form factor is defined as

$$F(\mathbf{q}) = \frac{1}{Z} \int \rho(\mathbf{r}) e^{i\mathbf{q}\cdot\mathbf{r}} d^3r . \qquad (3.99)$$

If, as is a common practice, the charge density $\rho(\mathbf{r})$ is taken to be spherically symmetric, $\rho(\mathbf{r}) = \rho(r)$, the form factor is a function of the magnitude q, $F(\mathbf{q}) = F(q)$.

The simplest case is a Coulomb potential with exponential screening, $V(r) = -\alpha Z e^{-\mu r}/r$ (Yukawa potential). Then we have [cf. Eq. (3.43)]

$$V(q) = -\frac{4\pi \alpha Z}{q^2 + \mu^2} = -\frac{4\pi \alpha Z}{q^2}[1 - F(q)] \qquad (3.100)$$

resulting in

$$1 - F(q) = \frac{q^2}{q^2 + \mu^2} . \qquad (3.101)$$

This simple model gives a qualitative account of screening. It can be improved by choosing a linear superposition of Yukawa potentials,

$$1 - F(q) = \sum_i b_i \frac{q^2}{q^2 + \mu_i^2} \qquad (3.102)$$

with $\sum_i b_i = 1$ [41; 42]. For target atoms with large nuclear charge Z the Thomas-Fermi approximation [35] to the effective potential $V(r)$ is sufficiently accurate.

Equation (3.101) shows that $1 - F(q)$ tends to unity for $q^2 \gg \mu^2$. This condition is certainly satisfied if the minimum value of q [Eq. (3.93)] is large against μ^2. Assuming a Thomas-Fermi model for the atom the reciprocal atomic radius is given by $\mu = \alpha Z^{1/3}$. That implies screening is negligible [$F(q) \ll 1$] for $q_{\min}^2 \gg \alpha^2 Z^{2/3}$. On the other hand, screening is nearly complete if $q_{\min}^2 \ll \mu^2$ (the differential cross section (3.81) is highest near $q = q_{\min}$). Since $dq_{\min}/d\epsilon_2 = 1 - \epsilon_2/p_2 < 0$, $q_{\min}$ decreases continuously towards the long-wavelength end of the spectrum, $\epsilon_2 \approx \epsilon_1$, where screening effects are most pronounced. Conversely, screening is largely negligible around the short-wavelength limit, $\epsilon_2 = 1$, in particular at high initial energies. From its origin in the Born approximation the form-factor approach can be expected to assess the screening best for low-Z elements. Indeed, by comparing the results with those derived by means of the partial-wave method (Section 3.5) it is found that the change due to screening is well predicted for low Z and poorly predicted for high Z. In addition, the form-factor description is a high-energy approximation; low-energy screening is quantitatively incorrect.

The intensity of the emitted bremsstrahlung radiation is given by the total bremsstrahlung cross section multiplied by the photon energy, $k\,d\sigma/dk$; it diverges logarithmically at the long-wavelength end of the spectrum, $k \to 0$. This divergence is characteristic of the pure Coulomb field with its infinite range. If the screening of the nuclear field by the atomic electrons is taken into account, the divergence is removed.

3.3.2 Nuclear recoil effects

The energy conservation (3.2), referring to an infinitely heavy scattering centre, has been used throughout. In Born approximation, additional calculations of the bremsstrahlung cross section have been performed where the finite mass of the scattering center was taken into account [43]. The recoil terms are reduced relative to the leading terms of the Bethe-Heitler formula by a factor $(m/M)q \approx v/c$ where m is the electron mass, and M, q, and v are, respectively, the mass, recoil momentum, and recoil velocity of the scattering center. The correction terms of order $(m/M)q$ result from both kinematical and dynamical considerations. The kinematical corrections originate from the equations of momentum and energy conservation whereas the dynamical corrections comprise the contributions from the two Feynman graphs with emission of the bremsstrahlung photon by the recoiling nucleus (classically the acceleration of the nucleus is smaller than that of the electron by a factor m/M). For the energies considered here these corrections are negligible except for the experiments on wide-angle bremsstrahlung at very high incident energies (Section 4.2.4). The recoil energy of the scattering atom at the direction $\hat{\mathbf{q}} = \mathbf{q}/q$ is given for fixed photon momentum $\mathbf{k}$ by

$$\epsilon_r - 1 = \frac{E_r}{Mc^2} = \frac{(m/M)^2}{\left[1 + (m/M)(\epsilon_1 - k)\right]^2 - (m/M)^2\left[(\mathbf{p}_1 - \mathbf{k}) \cdot \hat{\mathbf{q}}\right]^2}$$
$$\cdot \left\{ \left[(\mathbf{p}_1 - \mathbf{k}) \cdot \hat{\mathbf{q}}\right]^2 - \left[1 + (m/M)(\epsilon_1 - k)\right](\epsilon_1 k - \mathbf{p}_1 \cdot \mathbf{k}) \right.$$
$$\pm (\mathbf{p}_1 - \mathbf{k}) \cdot \hat{\mathbf{q}} \left[\{(\mathbf{p}_1 - \mathbf{k}) \cdot \hat{\mathbf{q}}\}^2 - 2(\epsilon_1 k - \mathbf{p}_1 \cdot \mathbf{k})\{1 + (m/M)(\epsilon_1 - k)\}\right.$$
$$\left.\left. + (m/M)^2(\epsilon_1 k - \mathbf{p}_1 \cdot \mathbf{k})^2\right]^{1/2} \right\}. \tag{3.103}$$

At incident electron energies of some GeV the kinetic energy E_r of the recoiling atom may amount to a substantial fraction of E_0 and thus has to be taken into account. One notes that the recoil energy of the atom is dependent on its direction $\hat{\mathbf{q}}$. The same is true for the photon energy,

$$k = \frac{\epsilon_1 - \epsilon_2 - (m/M)(\epsilon_1\epsilon_2 - \mathbf{p}_1 \cdot \mathbf{p}_2 - 1)}{1 + (m/M)(\epsilon_1 - \hat{\mathbf{k}} \cdot \mathbf{p}_1 - \epsilon_2 + \hat{\mathbf{k}} \cdot \mathbf{p}_2)}. \tag{3.104}$$

For $m \ll M$ and small angles θ_e, θ_k this reduces to the customary relation $k = \epsilon_1 - \epsilon_2$.

3.3.3 Effects of nuclear structure

In the preceding calculations the target nucleus was assumed to be a point charge. At extremely high incident electron energies and large momentum transfers, however, the effects of nuclear size and magnetic moment on the bremsstrahlung production have to be considered. In scattering of electrons by a nucleus the emission of photons depends on the nuclear magnetic moment as well as on the nuclear charge. Magnetic scattering by nonzero-spin nuclei will be appreciable under conditions in which charge scattering is not completely dominant. The relative order of magnitude of the magnetic and Coulomb interactions can be easily evaluated [44]. The measure of the strength of the Coulomb interaction is Ze; the measure of the magnetic interaction is $q\mu$ where q is the momentum transfer and μ the magnetic moment of the nucleus. The relative size of the two effects will therefore be of the order

$$\frac{q\mu}{eZ} = \frac{q(\mu/\mu_N)}{2(m_p/m)Z}, \qquad (3.105)$$

where $\mu_N = e\hbar/2m_p$ denotes the nuclear magneton and m_p the proton mass (note that we use units where $\hbar = m = c = 1$ and q is dimensionless). It can be shown that the fractional increase in the differential cross section of bremsstrahlung is of the order of the square of the above ratio. Evidently, magnetic scattering is of greatest importance for energetic electrons, the effect decreasing with Z. Unless the momentum transfer q is comparable with the ratio of nuclear mass and electron mass, the magnetic properties of the nucleus are almost completely masked by the nuclear charge.

Sarkar [45] has calculated the bremsstrahlung cross section in Born approximation corresponding to a spin-independent (i.e., classical) nuclear magnetic moment. The cross section was first derived for target nuclei with definite direction of the magnetic moment, and then it was averaged over all possible directions. It is obvious that the effect of the magnetic-moment interaction is more pronounced for spin-polarized nuclei than for magnetic moments with random directions. As was to be expected, the effect of the magnetic moment is only appreciable for large values of the recoil momentum q.

Ginsberg and Pratt [46] have extended the calculations to distributions of nuclear charge and magnetic moment. The scalar potential in momentum space arising from the nuclear charge distribution (normalized to Ze), $\rho(\mathbf{r})$, is given by (cf. Problem 3.3)

$$\varphi(\mathbf{q}) = \frac{1}{4\pi\epsilon_0} \frac{4\pi}{q^2} \int \rho(\mathbf{r}) \, e^{i\mathbf{q}\cdot\mathbf{r}} \, d^3r \, . \tag{3.106}$$

Analogously, the vector potential in momentum space due to the magnetic contribution to the nuclear current density, $\mathbf{j}(\mathbf{r}) = \nabla \times \vec{\mu}(\mathbf{r})$, has the form

$$\mathbf{A}(\mathbf{q}) = \mu_0 \frac{4\pi}{q^2} \int \mathbf{j}(\mathbf{r}) \, e^{i\mathbf{q}\cdot\mathbf{r}} \, d^3r \tag{3.107}$$

(μ_0 is the permeability of vacuum). The nuclear magnetic-moment distribution is specified by the function

$$\vec{\mu}(\mathbf{r}) = (\mathbf{I}/I)\mu(\mathbf{r}) \, , \tag{3.108}$$

where $\mathbf{I}$ denotes the nuclear spin and

$$\mu_n = \int \mu(\mathbf{r}) \, d^3r \tag{3.109}$$

is the nuclear magnetic moment. The vector potential (3.107) may be transformed to

$$\begin{aligned}\mathbf{A}(\mathbf{q}) &= \frac{4\pi\mu_0}{q^2} \int \nabla \times \vec{\mu}(\mathbf{r}) \, e^{i\mathbf{q}\cdot\mathbf{r}} \, d^3r \\ &= \frac{4\pi\mu_0}{q^2} \left\{ \int \nabla \times [\vec{\mu}(\mathbf{r}) \, e^{i\mathbf{q}\cdot\mathbf{r}}] \, d^3r - i \int \mathbf{q} \times \vec{\mu}(\mathbf{r}) \, e^{i\mathbf{q}\cdot\mathbf{r}} \, d^3r \right\} . \end{aligned} \tag{3.110}$$

The first integral on the right-hand side can be converted to a surface integral which vanishes. Thus, using Eq. (3.108), $\mathbf{A}(\mathbf{q})$ takes the form

$$\mathbf{A}(\mathbf{q}) = i \frac{4\pi\mu_0}{q^2} \frac{\mathbf{I} \times \mathbf{q}}{I} \int \mu(\mathbf{r}) \, e^{i\mathbf{q}\cdot\mathbf{r}} \, d^3r \, . \tag{3.111}$$

The form factor which is characteristic of the nuclear charge distribution $\rho(\mathbf{r})$ is defined as

$$F_c(\mathbf{q}) = \frac{1}{Ze} \int \rho(\mathbf{r}) \, e^{i\mathbf{q}\cdot\mathbf{r}} \, d^3r \, , \tag{3.112}$$

such that it has the value unity for zero momentum transfer. Correspondingly, the magnetic form factor is defined as

$$F_m(\mathbf{q}) = \frac{1}{\mu_n} \int \mu(\mathbf{r}) \, e^{i\mathbf{q}\cdot\mathbf{r}} \, d^3r \, . \tag{3.113}$$

By these definitions the expressions (3.106) and (3.107) are just the potentials due to a point charge Ze and point magnetic moment μ_n, multiplied by the form factors for the nuclear-charge and magnetic-moment distributions, respectively. The specific effects of the magnetic moment interaction are due to its non-central nature, i.e., the fact that it contains a preferred direction. If the nuclear charge and magnetic-moment distributions are spherically symmetric, as is commonly assumed, the form factors will be functions of the magnitude q only.

The nuclear-charge and magnetic form factors defined by Eqs. (3.112) and (3.113) are identical to the form factors which occur in the differential cross section for elastic electron scattering [46]. In fact, it is by means of electron scattering experiments that these form factors are known [47].

Corresponding to the Bethe-Heitler cross section for a point-Coulomb field, the bremsstrahlung cross section for the scattering of an electron by finite, static distributions of charge and magnetic moment is given in Born approximation by [46]

$$\frac{d^3\sigma}{dk\,d\Omega_k\,d\Omega_{p_2}} = |F_c|^2 \frac{d^3\sigma_B}{dk\,d\Omega_k\,d\Omega_{p_2}} + \left(\frac{\mu_n}{Ze}\right)^2 \frac{I+1}{3I} q^2 |F_m|^2 \frac{d^3\sigma_m}{dk\,d\Omega_k\,d\Omega_{p_2}}.$$
(3.114)

The factor $(I+1)/3I$ of the second term originates from averaging the initial state and summing the final state over the nuclear spin projection.

The magnetic term $d^3\sigma_m/(dk\,d\Omega_k\,d\Omega_{p_2})$ is similar in appearance to the Bethe-Heitler cross section, but is essentially different in character. There is no interference term in (3.114) between the electric and magnetic contributions because of the nuclear spin sums. Use of the Born approximation is justified for light nuclei except when the momentum transfer is near a zero of the form factors. For small values of q, the bremsstrahlung cross section (3.114) reduces to the Bethe-Heitler formula.

In an experiment involving extreme-relativistic electrons and hard photons the momentum transfers associated with characteristic angles $\theta \approx \epsilon_1^{-1}$ (where the cross section is maximum) are so small that nuclear-size and magnetic-moment effects are negligible. However, the form factor $F_c(\mathbf{q})$ may be considerably less than unity for values of the momentum transfer q large compared to the reciprocal nuclear radius (classically, this corresponds to impact parameters small compared to the nuclear radius). Due to finite nuclear size effects, thus, the bremsstrahlung cross section may be reduced by several orders of magnitude below that expected from a point nucleus, for collisions involving a large momentum transfer q, i.e., large angles of the outgoing electron and photon with respect to the incident electron (here screening by atomic electrons can be neglected). Likewise, the magnetic structure of the nucleus may contribute significantly to the cross section of wide-angle bremsstrahlung.

In bremsstrahlung studies the role of the nucleus has been essentially subsidiary. While its presence is necessary to conserve momentum, its internal structure has not been directly involved in the description of the process. Nevertheless this internal nuclear structure can be revealed if the possibility of transitions to excited states on electron impact is taken into account. A more general cross section for bremsstrahlung from the scattering of an electron by a nuclear charge and current distribution has been given by Maximon and Isabelle [48]. This cross section is applicable also to the bremsstrahlung accompanying nuclear excitation by the electrons, in addition to the case where the nucleus is in its ground state both initially and finally.

Nuclear structure effects on bremsstrahlung were investigated in Born approximation by Hubbard and Rose [49], taking into account photon emission by the nucleus by virtue of the possibility of transitions to excited states on electron impact[‡] in addition to the photon emission by the electron. The cross section consists, accordingly, of a sum of three terms: The cross section (3.114) for the emission by the electron, a part for electron scattering with the nuclear emission of a bremsstrahlung photon and, since these effects are coherent, an interference term. In the Feynman diagrams for the emission of a bremsstrahlung photon by the nucleus the intermediate nuclear state is entirely undetermined, except for the requirement of momentum conservation at each vertex. Therefore it is necessary to sum over all the intrinsic states available to the nucleus in order to evaluate the matrix elements corresponding to these diagrams. It is assumed that the nucleus is in its ground state both initially and finally. The results were expressed in terms of various form factors similar to those in Eqs. (3.112) and (3.113) which are applicable to an arbitrary nucleus.

In the specific case of ^{16}O and incident electron kinetic energy $E_0 = 51.1\,\text{MeV} = 100\,mc^2$, evaluated in the long-wavelength limit, effects due to

---

[‡]It must be emphasized that this is not a recoil effect since radiation arising from acceleration of the nucleus is negligible.

the intrinsic structure of the nucleus are found to be strikingly apparent in the coincidence arrangement where scattered electron and emitted photon are observed simultaneously. This is manifested in the form of peaks superimposed on the background from electron bremsstrahlung. Analysis of these peaks, their amplitude, position, and width would provide information on the various nuclear states involved. The relative magnitude of the nuclear effects depends strongly on the scattering angle and on the photon energy, essentially a resonance effect, and varies over a wide range from one nuclear level to another [49].

3.4 Approximations with the Sommerfeld-Maue wave function

3.4.1 Sommerfeld-Maue wave function

In the process of bremsstrahlung the electron interacts both with the radiation field and with the atomic (nuclear) field. In quantum electrodynamics the interaction with the radiation field is considered as a small perturbation and for most applications it is sufficient to use the lowest order of an expansion in powers of the fine-structure constant α (or e^2). On the other hand, the interaction of the electron with the atomic field can be taken into account rigorously by including it in the unperturbed Hamiltonian, i.e., by employing exact solutions of the Dirac equation in the matrix element (3.21). This corresponds to Furry's extension of the usual Feynman-Dyson formulation of quantum electrodynamics (Fig. 3.1). In the nonrelativistic case the Schrödinger equation for an electron in the nuclear Coulomb field can be solved in closed form [see Eq. (3.118)] resulting in Sommerfeld's bremsstrahlung cross section [10]. This is, however, not possible in the relativistic theory. The solution of the Dirac equation for the unscreened nuclear Coulomb field derived by Darwin [50] is not available in closed form but only as an expansion in Legendre polynomials (see Sec. 3.5.3). Using this expression one has to compute huge numbers of matrix elements corresponding to transitions from an initial angular momentum of the electron to a final one. The case becomes even worse if numerical solutions of the Dirac equation for an electron in the screened nuclear field are employed to calculate the matrix element (3.21). For the elementary process of bremsstrahlung this problem was only solved recently (see Sec. 3.5.2). Thus it is simpler to use an approximate solution of the Dirac equation with Coulomb potential found by Sommerfeld and Maue [51], even though this method has the drawback that it yields accurate results only for low atomic numbers of the target nucleus and that screening can be allowed for only approximately. In the point Coulomb potential the results are expected to be valid for $\alpha Z \ll 1$ at all energies [40], or for all αZ at sufficiently high incident and outgoing electron energies at small scattering angles [12]. Before the first computations by means of the partial-wave method were available, the cross-section formula was also applied to target elements of higher Z for comparison with experiments, in the absence of better options.

In order to derive the Sommerfeld-Maue (SM) wave function we transform the Dirac equation with Coulomb potential,

$$(i\vec{\alpha} \cdot \nabla - \beta + \epsilon + a/r)\psi(\mathbf{r}) = 0 , \qquad (3.115)$$

to an equation of second order by applying the operator $(-i\vec{\alpha}\cdot\nabla+\beta+\epsilon+a/r)$ resulting in

$$(\nabla^2 + p^2 + 2\epsilon a/r)\psi = [i\vec{\alpha} \cdot \nabla(a/r) - a^2/r^2]\psi . \qquad (3.116)$$

The left-hand side of (3.116) has the structure of the nonrelativistic Schrödinger equation for a Coulomb field of which the solution is well known. The right-hand side of (3.116) is of order $a = \alpha Z$, i.e., it becomes small for low atomic numbers Z. Therefore we solve (3.116) approximately by writing $\psi(\mathbf{r})$ as a series in terms of the parameter a:[§]

$$\psi(\mathbf{r}) = \psi_0(\mathbf{r}) + a\psi_1(\mathbf{r}) + a^2\psi_2(\mathbf{r}) + \ldots \qquad (3.117)$$

To lowest order in a the solution of $(\nabla^2 + p^2 + 2\epsilon a/r)\psi_0 = 0$ is

$$\psi_0(\mathbf{r}) = N e^{i\mathbf{p}\cdot\mathbf{r}} F(ia\epsilon/p; 1; ipr - i\mathbf{p} \cdot \mathbf{r}) u(\mathbf{p}) , \qquad (3.118)$$

where N is a normalization factor and $F(ia\epsilon/p; 1; ipr - i\mathbf{p} \cdot \mathbf{r})$ denotes the confluent hypergeometric function satisfying the differential equation (see Problem 3.4)

$$(\nabla^2 + 2i\mathbf{p} \cdot \nabla + 2\epsilon a/r) F(ia\epsilon/p; 1; ipr - i\mathbf{p} \cdot \mathbf{r}) = 0 . \qquad (3.119)$$

[§]In contrast to the procedure in the case of the Born approximation (Sec. 3.3) this is not a strict series expansion in terms of the Coulomb parameter a since the potential term proportional to a remains on the left-hand side of Eq. (3.116). Hence the lowest-order wave function ψ_0 represents a distorted wave (dependent on a) rather than a plane wave.

The free electron spinor $u(\mathbf{p})$ (see Sec. 3.3) ensures that ψ_0 is asymptotically a plane wave $\exp(i\mathbf{p}\cdot\mathbf{r})\,u(\mathbf{p})$.

The function $\psi_1(\mathbf{r})$ has to satisfy the equation

$$\left(\nabla^2 + p^2 + 2\epsilon a/r\right)\psi_1(\mathbf{r}) = i\vec{\alpha}\cdot\nabla(1/r)\,\psi_0 \,. \tag{3.120}$$

Its solution is readily found by applying the operator ∇ to Equation (3.119). We get

$$\left(\nabla^2 + 2i\mathbf{p}\cdot\nabla + 2\epsilon a/r\right)\nabla F = -2\epsilon a F\,\nabla(1/r) \,. \tag{3.121}$$

Substituting (3.118) and the ansatz

$$\psi_1(\mathbf{r}) = C\,e^{i\mathbf{p}\cdot\mathbf{r}}\vec{\alpha}\cdot\nabla F\,u(\mathbf{p}) \tag{3.122}$$

into (3.120) results in

$$C\,e^{i\mathbf{p}\cdot\mathbf{r}}\{\nabla^2 + 2i\,\mathbf{p}\cdot\nabla + 2\epsilon a/r\}\vec{\alpha}\cdot\nabla F\,u(\mathbf{p}) = iN\vec{\alpha}\cdot\nabla(1/r)\,e^{i\mathbf{p}\cdot\mathbf{r}}F\,u(\mathbf{p}) \,. \tag{3.123}$$

Since $\vec{\alpha}\cdot\nabla$ commutes with the first two terms in the curly brackets this can be written as

$$C\,e^{i\mathbf{p}\cdot\mathbf{r}}\Big\{(\vec{\alpha}\cdot\nabla)(\nabla^2 + 2i\,\mathbf{p}\cdot\nabla + 2\epsilon a/r)F - 2\epsilon a\,\vec{\alpha}\cdot\nabla(1/r)\,F\Big\}u(\mathbf{p})$$
$$= iN e^{i\mathbf{p}\cdot\mathbf{r}}F\vec{\alpha}\cdot\nabla(1/r)\,u(\mathbf{p}) = i\vec{\alpha}\cdot\nabla(1/r)\,\psi_0(\mathbf{r}) \,. \tag{3.124}$$

Applying (3.119) the first term on the left-hand side vanishes, yielding

$$C = -(i/2\epsilon a)N \,. \tag{3.125}$$

Thus the function $\psi_1(\mathbf{r})$ has the form

$$\psi_1(\mathbf{r}) = -Ne^{i\mathbf{p}\cdot\mathbf{r}}(i/2\epsilon a)\,\vec{\alpha}\cdot\nabla F(ia\epsilon/p;1;ipr - i\mathbf{p}\cdot\mathbf{r})\,u(\mathbf{p}) \,. \tag{3.126}$$

The factor $1/a$ in the function (3.126) cancels out since

$$\nabla F(ia\epsilon/p;1;ipr - i\mathbf{p}\cdot\mathbf{r}) = -\epsilon a F(1 + ia\epsilon/p;2;ipr - i\mathbf{p}\cdot\mathbf{r})\,(\hat{\mathbf{r}} - \hat{\mathbf{p}}) \,, \tag{3.127}$$

where $\hat{\mathbf{r}}$ and $\hat{\mathbf{p}}$ denote unit vectors.

Unfortunately, it is not possible to continue this iteration procedure to second order in a because the pertinent differential equation

$$(\nabla^2 + p^2 + 2\epsilon a/r)\psi_2(\mathbf{r}) = i\,[\vec{\alpha}\cdot\nabla(1/r)]\psi_1(\mathbf{r}) - \psi_0(\mathbf{r})/r^2$$
$$= -\frac{1}{r^2}\{\psi_0(\mathbf{r}) + i\,\vec{\alpha}\cdot\hat{\mathbf{r}}\,\psi_1(\mathbf{r})\} \tag{3.128}$$

cannot be solved analytically. The Sommerfeld-Maue wave function is composed of the terms $\psi_0 + a\psi_1$,¶

$$\psi_{\text{SM}}(\mathbf{r}) = Ne^{i\mathbf{p}\cdot\mathbf{r}}\left(1 - \frac{i}{2\epsilon}\vec{\alpha}\cdot\nabla\right)F(ia\epsilon/p; 1; ipr - i\mathbf{p}\cdot\mathbf{r})\, u(\mathbf{p})\,. \qquad (3.129)$$

Inserting $\psi_{\text{SM}}(\mathbf{r})$ into the Dirac equation with Coulomb potential one obtains, using (3.29), (3.67), and (3.119),

$$\begin{aligned}
&(i\vec{\alpha}\cdot\nabla - \beta + \epsilon + a/r)\,\psi_{\text{SM}}(\mathbf{r}) \\
&= Ne^{i\mathbf{p}\cdot\mathbf{r}}\big\{(\epsilon + a/r - \beta - \vec{\alpha}\cdot\mathbf{p})[1 - (i/2\epsilon)\vec{\alpha}\cdot\nabla] \\
&\quad + i\,\vec{\alpha}\cdot\nabla + \nabla^2/2\epsilon\big\}F\,u(\mathbf{p}) \\
&= Ne^{i\mathbf{p}\cdot\mathbf{r}}\big\{(a/r)[1 - (i/2\epsilon)\vec{\alpha}\cdot\nabla] - (i/2\epsilon)\vec{\alpha}\cdot\nabla\,(\epsilon + \beta + \vec{\alpha}\cdot\mathbf{p}) \\
&\quad + (i/\epsilon)\,\mathbf{p}\cdot\nabla + i\,\vec{\alpha}\cdot\nabla + \nabla^2/2\epsilon\big\}F\,u(\mathbf{p}) \\
&= -Ne^{i\mathbf{p}\cdot\mathbf{r}}\frac{ia}{2\epsilon r}\vec{\alpha}\cdot\nabla F(ia\epsilon/p; 1; ipr - i\mathbf{p}\cdot\mathbf{r})\,u(\mathbf{p}) \\
&= iNe^{i\mathbf{p}\cdot\mathbf{r}}\frac{a^2}{2r}F(1 + ia\epsilon/p; 2; ipr - i\mathbf{p}\cdot\mathbf{r})\,\vec{\alpha}\cdot(\hat{\mathbf{r}} - \hat{\mathbf{p}})\,u(\mathbf{p})\,. \qquad (3.130)
\end{aligned}$$

Thus the SM wave function satisfies the wave equation (3.115) to terms of order $(a^2/r)\sin(\vartheta/2)$.‖ If the angle ϑ between the vectors $\hat{\mathbf{r}}$ and $\hat{\mathbf{p}}$ tends to zero, it solves the Dirac equation exactly. Everywhere in space the wave equation (3.115) is satisfied to the first power in $a = \alpha Z$, and in the region of small angles ϑ — within the caustic of the classical trajectories — it is fulfilled to a high degree of approximation. At high energies the dominant contribution to the bremsstrahlung cross section arises from events in which both the scattered electron and the emitted photon emerge within an angle of order $1/\epsilon$ of the incident electron direction. Therefore the recoil momentum $\mathbf{q}$ imparted to the atom is low. This implies large impact parameters r of the order of ϵ. Hence the term on the right-hand side of (3.130) is small and it is to be expected that at high energies the SM wave function is a good approximation even for large values of a (high atomic numbers Z). Comparing ψ_{SM} with the exact solution of (3.115), given by Darwin [50]

¶This approximate solution of the Dirac equation with Coulomb potential was found earlier by Furry [52]. However, Furry wrote his wave function in a less convenient form using the four-component representation instead of the Dirac matrices and the free-electron spinor $u(\mathbf{p})$.

‖Even though the errors of the SM wave function are of order a^2, this does not hold for the bremsstrahlung cross section calculated by means of the SM function which is accurate only to lowest order in $a = \alpha Z$ (see Sec. 3.4.3).

in the form of an expansion in spherical harmonics, Bethe and Maximon [12] could show more precisely that all partial waves with $\kappa^2 \gg \alpha^2 Z^2$ are represented correctly for any energy, where $\kappa = \mp(j+\frac{1}{2})$ as $j = l \pm \frac{1}{2}$ is the relativistic angular momentum and parity quantum number (see Sec. 3.5.1). At high energies, partial waves with large $|\kappa|$ are dominant. Thus we can understand why the SM wave function works well for large Z and high energies but not for high Z and low energies.

The confluent hypergeometric function $F(a;c;z)$ behaves asymptotically for large values of $|z|$ as [53]

$$F(a;c;z) \sim \frac{\Gamma(c)}{\Gamma(c-a)} e^{\pm i\pi a} z^{-a} + \frac{\Gamma(c)}{\Gamma(a)} e^{z} z^{a-c}, \qquad (3.131)$$

where Γ denotes the gamma function and the plus sign in the exponential is for $\text{Im}\, z > 0$ and the minus sign for $\text{Im}\, z < 0$. Therefore the asymptotic behaviour of $\exp(i\mathbf{p}\cdot\mathbf{r})\, F(ia\epsilon/p; 1; ipr - i\mathbf{p}\cdot\mathbf{r})$ is

$$e^{i\mathbf{p}\cdot\mathbf{r}} F(ia\epsilon/p; 1; ipr - i\mathbf{p}\cdot\mathbf{r}) \sim \frac{e^{-\pi a\epsilon/2p}}{\Gamma(1 - ia\epsilon/p)} e^{i\mathbf{p}\cdot\mathbf{r} - ia(\epsilon/p)\ln(pr - \mathbf{p}\cdot\mathbf{r})}$$
$$- i \frac{e^{-\pi a\epsilon/2p}}{\Gamma(ia\epsilon/p)} \frac{e^{ipr + ia(\epsilon/p)\ln(pr - \mathbf{p}\cdot\mathbf{r})}}{pr - \mathbf{p}\cdot\mathbf{r}}. \qquad (3.132)$$

Since the Coulomb field decreases slowly the incoming plane wave is distorted even at large distances r from the scattering center. This becomes apparent in the logarithmic terms of the phase of both the plane wave and the spherical wave. The difference compared to the usual asymptotic form of scattering wave functions,

$$\psi \sim e^{i\mathbf{p}\cdot\mathbf{r}} + f(\hat{\mathbf{r}})\, e^{ipr}/r, \qquad (3.133)$$

is, however, insignificant because the resulting corrections in the flux density tend to zero for $r \to \infty$. The reason for the occurrence of the logarithmic terms can be explained classically. We consider the classical hyperbolic trajectories of nonrelativistic electrons with parallel directions of incidence, as shown in Fig. 2.8a. At large distances from the nucleus the equation for the surface perpendicular to the hyperbolas, which represents the wave front of the incident wave, is $\mathbf{p}\cdot\mathbf{r} - (a/p)\ln(pr - \mathbf{p}\cdot\mathbf{r}) = const$ rather than $\mathbf{p}\cdot\mathbf{r} = const$. This surface is just the area of constant phase of the plane wave in (3.132).

In calculating transition probabilities between stationary states, a scattered particle in the final state has to be represented by a wave function ψ_f^-

whose asymptotic form is plane wave plus *ingoing* spherical wave, rather than by the usual scattering wave function ψ_f^+ behaving asymptotically as plane wave plus *outgoing* spherical wave [Eq. (3.133)]. This may be seen as follows. For large negative values of $\mathbf{p}\cdot\mathbf{r}$ the incoming electron is represented by a plane wave. As a result of the scattering process it is superposed by a spherical wave originating from the atom. On the other hand, we want to have a scattered electron represented asymptotically by a plane wave without the superposition of an outgoing spherical wave. This can be achieved by requiring that the final electron has to satisfy the asymptotic boundary condition of a plane wave plus ingoing spherical wave converging towards the atom. Then we have only the outgoing plane wave in the direction $+\hat{\mathbf{p}}$ corresponding to the scattered electron (the incoming spherical wave does not matter). The need for using the ingoing spherical wave solution for the scattered electron has been discussed by Sommerfeld [54], Breit and Bethe [55], and Mott and Massey [56]. This particular choice of final-state wave function is only required if the differential cross section for a given direction of the scattered particle is to be calculated. If it is asked for the cross section integrated over the angles and summed over the spins of the scattered particle, one can forget about the difference between ψ_f^+ and ψ_f^- and equally well use the outgoing type of wave function for the final state [57]. In Born approximation the distinction between outgoing and ingoing spherical waves disappears in the matrix element (3.57).

The proper wave function ψ_2 of the final electron occurring in the matrix element (3.21) for bremsstrahlung may be derived from (3.129) by taking the complex conjugate to F and reversing the momentum vector $\mathbf{p}$ to $-\mathbf{p}$. The first step changes the outgoing spherical wave $\propto e^{ipr}/r$ to an incoming spherical wave $\propto e^{-ipr}/r$, and the plane wave takes the form $\propto e^{-i\mathbf{p}\cdot\mathbf{r}}$. The second step is necessary to retain the plane wave in the direction $+\hat{\mathbf{p}}$. Since the adjunct function $\psi_2^\dagger$ appears in the matrix element, the confluent hypergeometric function in $\psi_2^\dagger$ differs from that in (3.129) by just the sign of $\mathbf{p}$, and we have

$$\psi_2^\dagger(\mathbf{r}) = u^\dagger(\mathbf{p}_2) N_2^* e^{-i\mathbf{p}_2\cdot\mathbf{r}} \left(1 + \frac{i}{2\epsilon}\vec{\alpha}\cdot\nabla\right) F(ia\epsilon_2/p_2; 1; ip_2 r + i\mathbf{p}_2\cdot\mathbf{r}) \,. \quad (3.134)$$

The wave functions ψ_1 and ψ_2 have to be normalized to unit amplitudes. Since the exponentials and spinors $u(\mathbf{p})$ are normalized to unity the asymptotic behaviour of the confluent hypergeometric function [Eq. (3.132)] yields

$$N_1 = \Gamma(1 - ia_1) e^{\pi a_1/2} \,, \quad N_2 = \Gamma(1 + ia_2) e^{\pi a_2/2} \,, \quad (3.135)$$

where
$$a_1 = a\epsilon_1/p_1, \quad a_2 = a\epsilon_2/p_2. \tag{3.136}$$

Using the relations $\Gamma(x)\Gamma(1-x) = \pi/\sin(\pi x)$ and $\sin(ix) = i\sinh x$, we get

$$|\Gamma(1 \pm ia)|^2 = \Gamma(1+ia)\Gamma(1-ia) = ia\Gamma(ia)\Gamma(1-ia)$$
$$= \frac{\pi a}{\sinh(\pi a)} = \frac{2\pi a}{e^{\pi a} - e^{-\pi a}}, \tag{3.137}$$

so that

$$|N_1|^2 = |\Gamma(1-ia_1)|^2 e^{\pi a_1} = \frac{2\pi a_1}{1 - e^{-2\pi a_1}},$$
$$|N_2|^2 = |\Gamma(1+ia_2)|^2 e^{\pi a_2} = \frac{2\pi a_2}{1 - e^{-2\pi a_2}}. \tag{3.138}$$

3.4.2 Spin formalism

The SM wave function (3.129) contains the plane wave spinor $u(\mathbf{p})$. For subsequent applications it is useful to introduce the split spinor representation [58] which gives a simple method for separating the features depending on the electron polarization. In this representation the plane wave function $\psi(\mathbf{r}) = e^{i\mathbf{p}\cdot\mathbf{r}} u(\mathbf{p})$ is expressed in terms of two-component wave functions corresponding to Pauli spin states $v(\vec{\zeta})$ where $\vec{\zeta}$ is the polarization vector in the electron's rest system. To this end the Dirac operator $\vec{\alpha}$ is replaced by the Pauli spin matrices $\vec{\sigma} = (\sigma_1, \sigma_2, \sigma_3)$ [Eq. (3.6)], and the spinors $u(\mathbf{p})$ of the force-free Dirac equation (3.29) are replaced by the two-component Pauli spinors v and w:

$$u(\mathbf{p}) = N' \begin{pmatrix} v \\ w \end{pmatrix}. \tag{3.139}$$

N' is a normalization factor.

Inserting (3.6) and (3.139) into the free-particle wave equation

$$(\vec{\alpha} \cdot \mathbf{p} + \beta - \epsilon) u(\mathbf{p}) = 0, \tag{3.140}$$

we get the two coupled equations for v and w

$$(\vec{\sigma} \cdot \mathbf{p})w - (\epsilon - 1)v = 0, \tag{3.141}$$
$$(\vec{\sigma} \cdot \mathbf{p})v - (\epsilon + 1)w = 0. \tag{3.142}$$

Elimination of w by means of the second equation results in the representation of the spinor

$$u(\mathbf{p}) = N' \begin{pmatrix} v \\ \frac{\vec{\sigma} \cdot \mathbf{p}}{\epsilon+1} v \end{pmatrix} = N' \begin{pmatrix} 1 \\ \frac{\vec{\sigma} \cdot \mathbf{p}}{\epsilon+1} \end{pmatrix} v, \qquad (3.143)$$

where **1** denotes the 2×2 unit matrix. In this way any matrix element between four-component states u can be expressed by matrix elements between two-component states v.

Assuming that v is normalized to unity, $v^\dagger v = 1$, N' has the value

$$N' = \sqrt{\frac{\epsilon+1}{2\epsilon}}. \qquad (3.144)$$

Denoting the spin direction in the rest system of the electron by the unit vector $\vec{\zeta} = \mathbf{s}/s$ ($s = \frac{1}{2}$), the Pauli spin state $v(\vec{\zeta})$ obeys the equation

$$\vec{\sigma} v(\vec{\zeta}) = \vec{\zeta} v(\vec{\zeta}) \qquad (3.145)$$

or

$$(\vec{\zeta} \cdot \vec{\sigma}) v(\vec{\zeta}) = v(\vec{\zeta}), \qquad (3.146)$$

i.e., $v(\vec{\zeta})$ is the eigenstate of the component of $\vec{\sigma}$ along $\vec{\zeta}$. The explicit solution of (3.146) is well known (see Problem 3.7), however, we do not need it. Multiplying the equation adjoint to (3.146) by v from the left we have

$$v v^\dagger (\vec{\zeta} \cdot \vec{\sigma} - \mathbf{1}) = 0. \qquad (3.147)$$

For normalized v the solution of (3.147) is the projection operator

$$v v^\dagger = \tfrac{1}{2}(\mathbf{1} + \vec{\zeta} \cdot \vec{\sigma}) \qquad (3.148)$$

which can be easily verified if one multiplies (3.148) by $v^\dagger$ from the left and by v from the right.

The diagonal matrix element $(v, \vec{\sigma} v)$ is obtained by multiplying (3.145) by $v^\dagger$,

$$(v, \vec{\sigma} v) = v^\dagger \vec{\sigma} v = \vec{\zeta}, \qquad (3.149)$$

i.e., $\vec{\zeta}$ is the expectation value of $\vec{\sigma}$ for the eigenstate v.

3.4.3 Cross section

Inserting the wave functions (3.129) and (3.134) into the matrix element (3.21) results in

$$M = N_1 N_2^* u^\dagger(\mathbf{p}_2) \int e^{-i\mathbf{p}_2 \cdot \mathbf{r}} \left(1 + \frac{i}{2\epsilon_2} \vec{\alpha} \cdot \nabla\right) F(ia_2; 1; ip_2 r + i\mathbf{p}_2 \cdot \mathbf{r})$$

$$\cdot (\vec{\alpha} \cdot \mathbf{e}^*) e^{i(\mathbf{p}_1 - \mathbf{k}) \cdot \mathbf{r}} \left(1 - \frac{i}{2\epsilon_1} \vec{\alpha} \cdot \nabla\right) F(ia_1; 1; ip_1 r - i\mathbf{p}_1 \cdot \mathbf{r}) d^3r \, u(\mathbf{p}_1)$$

$$= N_1 N_2^* \{ u^\dagger(\mathbf{p}_2) (\vec{\alpha} \cdot \mathbf{e}^*) u(\mathbf{p}_1) I_1 + u^\dagger(\mathbf{p}_2) (\vec{\alpha} \cdot \mathbf{e}^*) (\vec{\alpha} \cdot \mathbf{I}_2) u(\mathbf{p}_1)$$

$$+ u^\dagger(\mathbf{p}_2) (\vec{\alpha} \cdot \mathbf{I}_3) (\vec{\alpha} \cdot \mathbf{e}^*) u(\mathbf{p}_1) + u^\dagger(\mathbf{p}_2) I_4 u(\mathbf{p}_1) \} \qquad (3.150)$$

with the notations

$$I_1 = \int e^{i\mathbf{q} \cdot \mathbf{r}} F(ia_2; 1; ip_2 r + i\mathbf{p}_2 \cdot \mathbf{r}) F(ia_1; 1; ip_1 r - i\mathbf{p}_1 \cdot \mathbf{r}) d^3r, \qquad (3.151)$$

$$\mathbf{I}_2 = -\frac{i}{2\epsilon_1} \int e^{i\mathbf{q} \cdot \mathbf{r}} F(ia_2; 1; ip_2 r + i\mathbf{p}_2 \cdot \mathbf{r})$$

$$\cdot \{\nabla F(ia_1; 1; ip_1 r - i\mathbf{p}_1 \cdot \mathbf{r})\} d^3r, \qquad (3.152)$$

$$\mathbf{I}_3 = \frac{i}{2\epsilon_2} \int e^{i\mathbf{q} \cdot \mathbf{r}} \{\nabla F(ia_2; 1; ip_2 r + i\mathbf{p}_2 \cdot \mathbf{r})\}$$

$$\cdot F(ia_1; 1; ip_1 r - i\mathbf{p}_1 \cdot \mathbf{r}) d^3r, \qquad (3.153)$$

$$I_4 = \frac{1}{4\epsilon_1 \epsilon_2} \int e^{i\mathbf{q} \cdot \mathbf{r}} \{\vec{\alpha} \cdot \nabla F(ia_2; 1; ip_2 r + i\mathbf{p}_2 \cdot \mathbf{r})\} (\vec{\alpha} \cdot \mathbf{e}^*)$$

$$\cdot \{\vec{\alpha} \cdot \nabla F(ia_1; 1; ip_1 r - i\mathbf{p}_1 \cdot \mathbf{r})\} d^3r. \qquad (3.154)$$

According to (3.130) the SM function agrees with the exact solution of the Dirac equation with Coulomb field up to first order in a. The integrals I_1, $\mathbf{I}_2$, and $\mathbf{I}_3$ are all of order a (see below) whereas I_4 is of order a^2. Besides the part of the matrix element containing I_4, other contributions of order a^2 have to be considered for large Z which result from higher-order terms of the correct wave function. Since these integrals cannot be calculated generally, it is consistent to neglect the term of the matrix element containing I_4. In fact, Bethe and Maximon [12] have shown that in the high-energy, small-angle approximation the contribution of I_4 is of order $1/\epsilon$ relative to I_1, $\mathbf{I}_2$, and $\mathbf{I}_3$, and thus is smaller than contributions from terms of the exact wave function which have been omitted in the SM approximation.

If the term containing I_4 is dropped, all the remaining integrals can be expressed in terms of the single integral

$$I_0 = \int \frac{1}{r} e^{i\mathbf{q}\cdot\mathbf{r}-\lambda r} F(ia_1; 1; ip_1 r - i\mathbf{p}_1\cdot\mathbf{r}) F(ia_2; 1; ip_2 r + i\mathbf{p}_2\cdot\mathbf{r}) d^3r \quad (3.155)$$

which can be calculated exactly. By differentiation with respect to the real positive parameter λ we get

$$I_1 = -\lim_{\lambda\to 0} \frac{\partial I_0}{\partial \lambda}, \quad (3.156)$$

and by means of the relation

$$\nabla(pr \pm \mathbf{p}\cdot\mathbf{r}) = \pm\frac{p}{r}\nabla_p(pr \pm \mathbf{p}\cdot\mathbf{r}), \quad (3.157)$$

where ∇_p denotes the gradient operator in $\mathbf{p}$ space,

$$\mathbf{I}_2 = \frac{ip_1}{2\epsilon_1} \lim_{\lambda\to 0} \nabla_{p_1} I_0 \quad (3.158)$$

and

$$\mathbf{I}_3 = \frac{ip_2}{2\epsilon_2} \lim_{\lambda\to 0} \nabla_{p_2} I_0. \quad (3.159)$$

In the last two relations it is understood that ∇_{p_1} and ∇_{p_2} do not act on $\mathbf{q}$, a_1, and a_2. The integral (3.155) can be solved by contour integration methods using the integral representations of the confluent hypergeometric functions. The result is [59]

$$I_0 = \frac{2\pi}{\alpha'} e^{-\pi a_1} \left(\frac{\alpha'}{\gamma}\right)^{ia_1} \left(\frac{\alpha'}{\alpha'+\beta}\right)^{ia_2} {}_2F_1\left(ia_1, ia_2; 1; \frac{\beta\gamma - \alpha'\delta}{\gamma(\alpha'+\beta)}\right) \quad (3.160)$$

with

$$\alpha' = \tfrac{1}{2}(q^2 + \lambda^2), \quad \beta = \mathbf{p}_2\cdot\mathbf{q} - i\lambda p_2,$$
$$\gamma = \mathbf{p}_1\cdot\mathbf{q} - \alpha' + i\lambda p_1, \quad \delta = p_1 p_2 + \mathbf{p}_1\cdot\mathbf{p}_2 - \beta. \quad (3.161)$$

${}_2F_1$ denotes the hypergeometric function. For $\lambda \to 0$ the four quantities α' to δ are real. Therefore the two powers with imaginary exponents in (3.160) will be of absolute magnitude unity.

We now differentiate I_0 as required in (3.156), (3.158), and (3.159) yielding

$$I_1 = 2K_1\left\{\frac{V}{q^2}\left(\frac{\epsilon_2}{D_1} - \frac{\epsilon_1}{D_2}\right) + i\frac{W}{D_1 D_2}\left[a_2\epsilon_1\left(\frac{\mu}{D_2} - 1\right) - a_1\epsilon_2\left(\frac{\mu}{D_1} + 1\right)\right]\right\}$$
$$= 2K_1\{\epsilon_2(A_1 - B_1) - \epsilon_1(A_2 + B_2)\}, \quad (3.162)$$

$$\mathbf{I}_2 = K_1 \left\{ \frac{V\mathbf{q}}{D_2 q^2} + i\, a_2 \frac{W}{D_1 D_2} \left[\frac{\mathbf{P}}{p_1} - \mathbf{q}\left(\frac{\mu}{D_2} - 1\right) \right] \right\}$$
$$= K_1 \{ A_2 \mathbf{q} + B_2 (\mathbf{P}/p_1 + \mathbf{q}) \} , \qquad (3.163)$$
$$\mathbf{I}_3 = K_1 \left\{ \frac{V\mathbf{q}}{D_1 q^2} + i\, a_1 \frac{W}{D_1 D_2} \left[\frac{\mathbf{P}}{p_2} - \mathbf{q}\left(\frac{\mu}{D_1} + 1\right) \right] \right\}$$
$$= K_1 \{ A_1 \mathbf{q} + B_1 (\mathbf{P}/p_2 - \mathbf{q}) \} \qquad (3.164)$$

with the following definitions:

$$D_1 = \lim_{\lambda \to 0} 2(\alpha' + \beta) = 2\mathbf{q} \cdot \mathbf{p}_2 + q^2 = 2(\epsilon_1 k - \mathbf{p}_1 \cdot \mathbf{k}) , \qquad (3.165)$$
$$D_2 = \lim_{\lambda \to 0} 2\gamma = 2\mathbf{q} \cdot \mathbf{p}_1 - q^2 = 2(\epsilon_2 k - \mathbf{p}_2 \cdot \mathbf{k}) , \qquad (3.166)$$
$$\mu = \lim_{\lambda \to 0} 2(\gamma + \delta) = (p_1 + p_2)^2 - k^2 = 2(\epsilon_1 \epsilon_2 + p_1 p_2 - 1), \qquad (3.167)$$
$$\mathbf{P} = p_1 \mathbf{p}_2 + p_2 \mathbf{p}_1 , \qquad (3.168)$$
$$K_1 = 4\pi a\, e^{-\pi a_1} \left(\frac{q^2}{D_2}\right)^{ia_1} \left(\frac{q^2}{D_1}\right)^{ia_2} , \quad |K_1| = 4\pi a\, e^{-\pi a_1} , \qquad (3.169)$$
$$A_1 = \frac{V - i a_1 (1-x) W}{D_1 q^2} , \quad A_2 = \frac{V - i a_2 (1-x) W}{D_2 q^2} , \qquad (3.170)$$
$$B_1 = \frac{i a_1 W}{D_1 D_2} , \quad B_2 = \frac{i a_2 W}{D_1 D_2} . \qquad (3.171)$$

V and W are the hypergeometric functions

$$V = {}_2F_1(ia_1, ia_2; 1; x) \qquad (3.172)$$

and

$$W = {}_2F_1(ia_1 + 1, ia_2 + 1; 2; x) = -\frac{1}{a_1 a_2} \frac{\partial V}{\partial x} , \qquad (3.173)$$

where

$$x = \lim_{\lambda \to 0} \left(1 - \frac{\alpha'}{\gamma} \frac{\gamma + \delta}{\alpha' + \beta} \right) = 1 - \frac{\mu q^2}{D_1 D_2}$$
$$= 2\frac{2(\mathbf{q} \cdot \mathbf{p}_1)(\mathbf{q} \cdot \mathbf{p}_2) - (p_1 p_2 + \mathbf{p}_1 \cdot \mathbf{p}_2) q^2}{D_1 D_2} . \qquad (3.174)$$

The argument of the hypergeometric functions V and W is never positive, $x \leq 0$. Its extreme values occur for coplanar geometry $(\mathbf{p}_1 \times \mathbf{p}_2) \cdot \mathbf{k} = 0$ and $D_2 |\hat{\mathbf{p}}_1 \times \hat{\mathbf{k}}| = D_1 |\hat{\mathbf{p}}_2 \times \hat{\mathbf{k}}|$ resulting in the maximum $x = 0$ and the minimum $x = 1 - (\mu/2k)^2$ independent of the angle θ_1 between $\mathbf{p}_1$ and $\mathbf{k}$ (see Problem 3.8).

For $|x| \leq 1$ the hypergeometric function V can be represented by the series expansion [53]

$$V = 1 + ia_1 ia_2 x + \frac{ia_1(ia_1+1)ia_2(ia_2+1)}{2!} \frac{x^2}{2!} + \cdots$$
$$= 1 - a_1 a_2 \{x + (ia_1+1)(ia_2+1)x^2/(2!)^2 + \ldots\}. \quad (3.175)$$

The property that x can never be positive is advantageous in the computation of the hypergeometric functions V and W because one can apply the transformation formulae [53] where the argument x is changed to $1/(1-x) \leq 1$ so that the hypergeometric series expansion can be used even for $|x| \gg 1$.

For target nuclei of low atomic numbers Z and sufficiently high energies ϵ_1 and ϵ_2, i.e., for $a_1 \ll 1$ and $a_2 \ll 1$, the function V tends to unity and the terms in (3.162) to (3.164) proportional to W tend to zero. Besides the factor

$$F = \frac{|N_1 N_2 K_1|^2}{(4\pi a)^2} = \frac{2\pi a_1}{e^{2\pi a_1} - 1} \frac{2\pi a_2}{1 - e^{-2\pi a_2}} \quad (3.176)$$

tends to unity. Then the integrals (3.162) to (3.164) reduce to

$$I_1 \approx \frac{2K_1}{q^2} \left(\frac{\epsilon_2}{D_1} - \frac{\epsilon_1}{D_2} \right), \quad (3.177)$$

$$\mathbf{I}_2 \approx \frac{K_1 \mathbf{q}}{D_2 q^2}, \quad (3.178)$$

$$\mathbf{I}_3 \approx \frac{K_1 \mathbf{q}}{D_1 q^2}, \quad (3.179)$$

and the matrix element takes the form (3.57) of the Born approximation.

All the integrals considered [Eqs. (3.162) to (3.164)] have the common factor K_1 and are thus of order $a = \alpha Z$. Hence the corrections to the matrix element M and to the cross section, due to the neglected terms, are of relative order a.

By applying Gauss's law to Eq. (3.153), one finds that the three integrals I_1, $\mathbf{I}_2$, and $\mathbf{I}_3$ are not independent of each other:

$$\mathbf{I}_3 = \frac{i}{2\epsilon_2} \int \nabla \{ e^{i\mathbf{q}\mathbf{r}} F(ia_1; 1; ip_1 r - i\mathbf{p}_1 \cdot \mathbf{r}) F(ia_2; 1; ip_2 r + i\mathbf{p}_2 \cdot \mathbf{r}) \} d^3 r$$
$$- \frac{i}{2\epsilon_2} \int F(ia_2; 1; ip_2 r + i\mathbf{p}_2 \cdot \mathbf{r}) \nabla \{ e^{i\mathbf{q}\mathbf{r}} F(ia_1; 1; ip_1 r - i\mathbf{p}_1 \cdot \mathbf{r}) \} d^3 r$$

$$= \frac{i}{2\epsilon_2} \int e^{i\mathbf{q}\cdot\mathbf{r}} F(ia_1; 1; ip_1 r - i\mathbf{p}_1\cdot\mathbf{r}) F(ia_2; 1; ip_2 r + i\mathbf{p}_2\cdot\mathbf{r}) d\mathbf{A}$$

$$- \frac{i}{2\epsilon_2} \int F(ia_2; 1; ip_2 r + i\mathbf{p}_2\cdot\mathbf{r}) e^{i\mathbf{q}\cdot\mathbf{r}} \{\nabla F(ia_1; 1; ip_1 r - i\mathbf{p}_1\cdot\mathbf{r})\} d^3 r$$

$$+ \frac{\mathbf{q}}{2\epsilon_2} \int F(ia_2; 1; ip_2 r + i\mathbf{p}_2\cdot\mathbf{r}) e^{i\mathbf{q}\cdot\mathbf{r}} F(ia_1; 1; ip_1 r - i\mathbf{p}_1\cdot\mathbf{r}) d^3 r \ .$$

(3.180)

The surface integral is equal to zero, so we get

$$\mathbf{I}_3 = \frac{\epsilon_1}{\epsilon_2} \mathbf{I}_2 + \frac{\mathbf{q}}{2\epsilon_2} I_1 \ . \tag{3.181}$$

The relation (3.181) can also be directly verified by inserting the expressions (3.162) to (3.164).

For the calculation of the cross section irrespective of electron spins the absolute square of the matrix element has to be averaged over the spin directions of the initial electron and to be summed over the spin directions of the final electron. Proceeding as in the case of the Born approximation (Sec. 3.3), we arrive at the expression

$$\begin{aligned}\tfrac{1}{2}\mathbf{S}^{p_1}\mathbf{S}^{p_2}|M|^2 = \frac{|N_1 N_2|^2}{2\epsilon_1\epsilon_2} &\Big\{ |I_1|^2 \big[\epsilon_1\epsilon_2 - \mathbf{p}_1\cdot\mathbf{p}_2 - 1 + 2\operatorname{Re}(\mathbf{e}\cdot\mathbf{p}_1)(\mathbf{e}^*\cdot\mathbf{p}_2)\big] \\ &+ \big(|\mathbf{I}_2|^2 + |\mathbf{I}_3|^2\big)\big[\epsilon_1\epsilon_2 + \mathbf{p}_1\cdot\mathbf{p}_2 + 1 - 2\operatorname{Re}(\mathbf{e}\cdot\mathbf{p}_1)(\mathbf{e}^*\cdot\mathbf{p}_2)\big] \\ &+ (\epsilon_1\epsilon_2 + \mathbf{p}_1\cdot\mathbf{p}_2 + 1)\big[|\mathbf{e}^*\cdot\mathbf{I}_2|^2 - |\mathbf{e}\cdot\mathbf{I}_2|^2 + |\mathbf{e}^*\cdot\mathbf{I}_3|^2 - |\mathbf{e}\cdot\mathbf{I}_3|^2 \\ &+ 2\operatorname{Re}\{(\mathbf{e}\cdot\mathbf{I}_2^*)(\mathbf{e}^*\cdot\mathbf{I}_3) + (\mathbf{e}^*\cdot\mathbf{I}_2^*)(\mathbf{e}\cdot\mathbf{I}_3) - \mathbf{I}_2^*\cdot\mathbf{I}_3\}\big] \\ &+ 2\epsilon_1\operatorname{Re}\big[I_1^*\{(\mathbf{e}\cdot\mathbf{p}_2)\mathbf{e}^*\cdot(\mathbf{I}_2+\mathbf{I}_3) + (\mathbf{e}^*\cdot\mathbf{p}_2)\mathbf{e}\cdot(\mathbf{I}_2-\mathbf{I}_3) \\ &- (\mathbf{I}_2-\mathbf{I}_3)\cdot\mathbf{p}_2\}\big] + 2\epsilon_2\operatorname{Re}\big[I_1^*\{(\mathbf{e}\cdot\mathbf{p}_1)\mathbf{e}^*\cdot(\mathbf{I}_2+\mathbf{I}_3) \\ &+ (\mathbf{e}^*\cdot\mathbf{p}_1)\mathbf{e}\cdot(\mathbf{I}_3-\mathbf{I}_2) - (\mathbf{I}_3-\mathbf{I}_2)\cdot\mathbf{p}_1\}\big] \\ &+ 2\operatorname{Re}\big[(\mathbf{e}^*\cdot\mathbf{p}_1)\{\mathbf{e}\cdot(\mathbf{I}_2-\mathbf{I}_3)\mathbf{p}_2\cdot(\mathbf{I}_2^*-\mathbf{I}_3^*) \\ &+ \mathbf{e}\cdot(\mathbf{I}_2^*+\mathbf{I}_3^*)\mathbf{p}_2\cdot(\mathbf{I}_3-\mathbf{I}_2)\} \\ &+ (\mathbf{e}^*\cdot\mathbf{p}_2)\{\mathbf{e}\cdot(\mathbf{I}_2-\mathbf{I}_3)\mathbf{p}_1\cdot(\mathbf{I}_2^*-\mathbf{I}_3^*) + \mathbf{e}\cdot(\mathbf{I}_2^*+\mathbf{I}_3^*)\mathbf{p}_1\cdot(\mathbf{I}_2-\mathbf{I}_3)\} \\ &+ \{(\mathbf{I}_2-\mathbf{I}_3)\cdot\mathbf{p}_1\}\{(\mathbf{I}_3^*-\mathbf{I}_2^*)\cdot\mathbf{p}_2\} \\ &+ (\mathbf{I}_2^*\cdot\mathbf{I}_3)\{(\mathbf{e}\cdot\mathbf{p}_1)(\mathbf{e}^*\cdot\mathbf{p}_2) + (\mathbf{e}^*\cdot\mathbf{p}_1)(\mathbf{e}\cdot\mathbf{p}_2)\}\big]\Big\} \ ,\end{aligned}$$

(3.182)

where Re denotes the real part. Finally, the summation over the polarizations of the emitted photon results in

$$\tfrac{1}{2} \mathbf{S}^{p_1} \mathbf{S}^{p_2} \sum_e |M|^2 = \frac{|N_1 N_2|^2}{\epsilon_1 \epsilon_2} \Big\{ |I_1|^2 [\epsilon_1 \epsilon_2 - 1 - (\mathbf{p}_1 \cdot \hat{\mathbf{k}})(\mathbf{p}_2 \cdot \hat{\mathbf{k}})]$$
$$+ (|\mathbf{I}_2|^2 + |\mathbf{I}_3|^2)[\epsilon_1 \epsilon_2 + 1 + (\mathbf{p}_1 \cdot \hat{\mathbf{k}})(\mathbf{p}_2 \cdot \hat{\mathbf{k}})]$$
$$- 2\,\mathrm{Re}\big[(\mathbf{I}_2 \cdot \hat{\mathbf{k}})(\mathbf{I}_3^* \cdot \hat{\mathbf{k}})\big](\epsilon_1 \epsilon_2 + 1 + \mathbf{p}_1 \cdot \mathbf{p}_2)$$
$$+ 2\,\mathrm{Re}\big[(\mathbf{I}_3^* - \mathbf{I}_2^*) \cdot \{\mathbf{p}_1(\mathbf{I}_2 \cdot \hat{\mathbf{k}})(\mathbf{p}_2 \cdot \hat{\mathbf{k}}) - \mathbf{p}_2(\mathbf{I}_3 \cdot \hat{\mathbf{k}})(\mathbf{p}_1 \cdot \hat{\mathbf{k}})\}\big]$$
$$+ (\mathbf{I}_2 \cdot \mathbf{p}_1)(\mathbf{I}_3^* \cdot \mathbf{p}_2) - (\mathbf{I}_2 \cdot \mathbf{p}_2)(\mathbf{I}_3^* \cdot \mathbf{p}_1)$$
$$+ \epsilon_1 I_1^* \{\mathbf{I}_3 \cdot \mathbf{p}_2 - (\mathbf{I}_2 \cdot \hat{\mathbf{k}})(\mathbf{p}_2 \cdot \hat{\mathbf{k}})\} + \epsilon_2 I_1^* \{\mathbf{I}_2 \cdot \mathbf{p}_1 - (\mathbf{I}_3 \cdot \hat{\mathbf{k}})(\mathbf{p}_1 \cdot \hat{\mathbf{k}})\}$$
$$+ (\mathbf{I}_2 \cdot \mathbf{I}_3^*)\{\mathbf{p}_1 \cdot \mathbf{p}_2 - (\mathbf{p}_1 \cdot \hat{\mathbf{k}})(\mathbf{p}_2 \cdot \hat{\mathbf{k}})\}\Big\} . \qquad (3.183)$$

Now the integrals I_1, $\mathbf{I}_2$, and $\mathbf{I}_3$ given in (3.162) to (3.164) are inserted in (3.183). The matrix element squared can be transformed by numerous applications of the relations (3.1), (3.165), (3.166), and (3.168). If it is substituted in Eq. (3.24) we get the cross section [13]

$$\frac{d^3\sigma}{dk\,d\Omega_k\,d\Omega_{p_2}} = \frac{2\pi a_1}{e^{2\pi a_1} - 1} \frac{2\pi a_2}{1 - e^{-2\pi a_2}} \frac{\alpha Z^2 r_0^2}{\pi^2} \frac{k p_2}{p_1} \{ X_1 |A_1|^2 + X_2 |A_2|^2$$
$$- 2 X_3 \mathrm{Re}(A_1^* A_2) - 2 Y_1 \mathrm{Re}(A_1^* B) - 2 Y_2 \mathrm{Re}(A_2^* B) + Y_3 |B|^2 \} . \qquad (3.184)$$

In this expression the following notations are used:

$$B = iaW/(D_1 D_2) , \qquad (3.185)$$
$$X_1 = (4\epsilon_2^2 - q^2)(\mathbf{p}_1 \times \hat{\mathbf{k}})^2 + \{ 2\,[(\mathbf{p}_2 \times \mathbf{k})^2 + k^2]/D_2$$
$$+ (\mathbf{p}_1 \times \hat{\mathbf{k}})^2 - (\mathbf{p}_1 \times \hat{\mathbf{k}}) \cdot (\mathbf{p}_2 \times \hat{\mathbf{k}}) \} D_1 , \qquad (3.186)$$
$$X_2 = (4\epsilon_1^2 - q^2)(\mathbf{p}_2 \times \hat{\mathbf{k}})^2 + \{ 2\,[(\mathbf{p}_1 \times \mathbf{k})^2 + k^2]/D_1$$
$$- (\mathbf{p}_2 \times \hat{\mathbf{k}})^2 + (\mathbf{p}_1 \times \hat{\mathbf{k}}) \cdot (\mathbf{p}_2 \times \hat{\mathbf{k}}) \} D_2 , \qquad (3.187)$$
$$X_3 = (4\epsilon_1 \epsilon_2 - q^2)(\mathbf{p}_1 \times \hat{\mathbf{k}}) \cdot (\mathbf{p}_2 \times \hat{\mathbf{k}})$$
$$+ \tfrac{1}{2} D_1 [(\mathbf{p}_1 \times \hat{\mathbf{k}}) \cdot (\mathbf{p}_2 \times \hat{\mathbf{k}}) - (\mathbf{p}_2 \times \hat{\mathbf{k}})^2]$$
$$+ \tfrac{1}{2} D_2 [(\mathbf{p}_1 \times \hat{\mathbf{k}})^2 - (\mathbf{p}_1 \times \hat{\mathbf{k}}) \cdot (\mathbf{p}_2 \times \hat{\mathbf{k}})]$$
$$+ 2\,[\mathbf{p}_1 \times \mathbf{k}) \cdot (\mathbf{p}_2 \times \mathbf{k}) + k^2] , \qquad (3.188)$$
$$Y_1 = k\rho[(\mathbf{p}_1 \times \hat{\mathbf{k}})^2 - (\mathbf{p}_1 \times \hat{\mathbf{k}}) \cdot (\mathbf{p}_2 \times \hat{\mathbf{k}})]$$
$$+ \tau\big[(\mathbf{k} \cdot \mathbf{p}_1)(\mathbf{p}_1 \cdot \mathbf{p}_2 - \mathbf{k} \cdot \mathbf{p}_2 + p_2^2) + 2k^2\big]$$
$$- (\tau p_1 p_2 - 2k/p_1)(\mathbf{k} \cdot \mathbf{p}_1 + \mathbf{k} \cdot \mathbf{p}_2 - k^2) , \qquad (3.189)$$

$$Y_2 = k\rho\big[(\mathbf{p}_2 \times \hat{\mathbf{k}})^2 - (\mathbf{p}_1 \times \hat{\mathbf{k}}) \cdot (\mathbf{p}_2 \times \hat{\mathbf{k}})\big]$$
$$+ \tau\big[(\mathbf{k} \cdot \mathbf{p}_2)(\mathbf{p}_1 \cdot \mathbf{p}_2 + \mathbf{k} \cdot \mathbf{p}_1 + p_1^2) - 2k^2\big]$$
$$- (\tau p_1 p_2 + 2k/p_2)(\mathbf{k} \cdot \mathbf{p}_1 + \mathbf{k} \cdot \mathbf{p}_2 + k^2), \qquad (3.190)$$

$$Y_3 = \mu\bigg\{k^2 - \frac{(\mathbf{k} \cdot \mathbf{p}_1)(\mathbf{k} \cdot \mathbf{p}_2)}{p_1 p_2}$$
$$+ \frac{p_1^2 - p_2^2}{p_1^2 p_2^2}\big[p_1^2 - p_2^2 + \mathbf{k} \cdot \mathbf{p}_1 + \mathbf{k} \cdot \mathbf{p}_2\big]\bigg\} - 2k^2\rho^2, \qquad (3.191)$$

$$\tau = \epsilon_1/p_1 + \epsilon_2/p_2, \quad \rho = 1/p_1 + 1/p_2. \qquad (3.192)$$

The formula (3.184) imitates Sommerfeld's [54] nonrelativistic cross section which also contains the quantities A_1 and A_2. So the transition to the nonrelativistic limit can be achieved in a particularly simple way (Section 3.4.4). The neglect of terms of order $a^2/|\kappa|$ in the wave functions leads to errors of the order $a/|\kappa|$ in the cross section because the lowest-order term in the matrix element is not $O(1)$ as in the wave function but $O(a)$. In a region of low and intermediate energies and low atomic numbers ($Z \le 13$) the cross section (3.184) is a significant improvement over the Born approximation [60]. In particular, the cross section remains finite at the short-wavelength limit, where the momentum p_2 of the final electron goes to zero, and the results differentiate between electron and positron bremsstrahlung as compared to the identical results in Born approximation (see Sec. 7.2).

As shown above the cross section (3.184) takes the form of the Bethe-Heitler formula for low atomic numbers Z and high energies ϵ_1 and ϵ_2. Then $V \approx 1$, $aW \approx 0$ and $F \approx 1$. A better approximation to (3.184), also valid for $a_2 \ll 1$ (which implies $a_1 \ll 1$) and avoiding the time-consuming computation of the hypergeometric functions, can be achieved by using the leading terms of the expansions in powers of $a = \alpha Z$ [see Eq. (3.175)]. Then

$$V \approx 1 - a_1 a_2\Big(x + \frac{x^2}{2^2} + \frac{x^3}{3^2} + \ldots\Big) = 1 - a_1 a_2 \mathrm{Li}_2(x) \qquad (3.193)$$

and

$$W \approx F(1,1;2;x) = -\frac{\ln(1-x)}{x}, \qquad (3.194)$$

where

$$\mathrm{Li}_2(x) = -\int_0^x \frac{\ln(1-t)}{t}\, dt \qquad (3.195)$$

is Euler's dilogarithm.

For comparison with experiments it is expedient to use a coordinate system where the z-axis is in the direction of the incident electron beam,

$$\hat{\mathbf{p}}_1 = \{0, 0, 1\},$$
$$\hat{\mathbf{p}}_2 = \{\sin\theta_e \cos\varphi_e, \sin\theta_e \sin\varphi_e, \cos\theta_e\},$$
$$\hat{\mathbf{k}} = \{\sin\theta_k \cos\varphi_k, \sin\theta_k \sin\varphi_k, \cos\theta_k\}. \quad (3.196)$$

Then the scalar and vector products can be represented by

$$\mathbf{p}_1 \cdot \mathbf{p}_2 = p_1 p_2 \cos\theta_e,$$
$$\mathbf{k} \cdot \mathbf{p}_1 = k p_1 \cos\theta_k,$$
$$\mathbf{k} \cdot \mathbf{p}_2 = k p_2 (\cos\theta_e \cos\theta_k + \sin\theta_e \sin\theta_k \cos\varphi), \quad (3.197)$$
$$(\mathbf{p}_1 \times \mathbf{k})^2 = k^2 p_1^2 \sin^2\theta_k,$$
$$(\mathbf{p}_2 \times \mathbf{k})^2 = k^2 p_2^2 \left[1 - (\cos\theta_e \cos\theta_k + \sin\theta_e \sin\theta_k \cos\varphi)^2\right].$$

The cross section depends only on the difference $\varphi = \varphi_e - \varphi_k$ of the azimuth angles. The polar angles θ_e and θ_k, as defined in (3.196), take values between 0° and 180°. In coplanar geometry, however, it is customary to use a negative value of θ_e or θ_k if the pertinent azimuth angle is equal to 180°.

The elements of solid angle are given by

$$d\Omega_k = \sin\theta_k\, d\theta_k\, d\varphi_k, \quad d\Omega_{p_2} = \sin\theta_e\, d\theta_e\, d\varphi_e. \quad (3.198)$$

Figure 3.2 shows examples of photon and electron angular distributions for aluminum targets and electron energies $E_0 = 300$ keV, $E_e = 100$ keV as calculated in coplanar geometry by means of Eq. (3.184). The photon distributions are given for various values of the electron angle θ_e, and the electron distributions pertain to various photon angles θ_k. One notes that even at the moderate impact energy of 300 keV the lobes are strongly beamed into forward direction. As discussed in Section 2.1 there is a strong correlation with the photons being predominantly emitted on the same side relative to the primary beam as the scattered electrons.

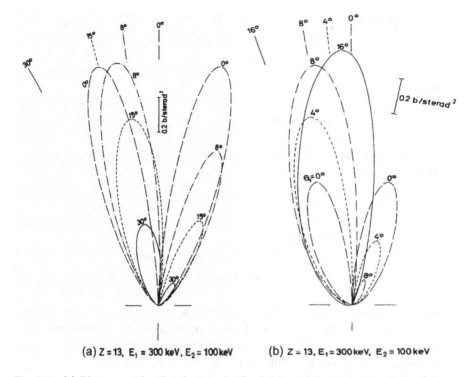

Fig. 3.2 (a) Photon angular distributions for fixed values of the electron angle θ_e. (b) Angular distributions of the outgoing electrons for fixed photon angles θ_k. Both sets of curves are calculated for $Z = 13$, $E_0 = 300$ keV, and $E_e = 100$ keV (from Elwert and Haug [13]).

3.4.4 Nonrelativistic approximation

The differential cross section for nonrelativistic energies can be obtained from (3.184) by taking the following limits:

$$\epsilon_{1,2} \to 1,\; p_{1,2} \to \beta_{1,2} \ll 1,\; k \to \tfrac{1}{2}(\beta_1^2 - \beta_2^2),\; \mu \to (\beta_1 + \beta_2)^2 \,. \quad (3.199)$$

Here β_1 and β_2 are the electron velocities in the initial and final state, respectively, in units of the velocity of light. Neglecting terms of order β^2 compared with unity, we obtain from Eqs. (3.184) to (3.192)

$$\frac{d^3\sigma_{nr}}{dk\, d\Omega_k\, d\Omega_{p_2}} = \frac{2\pi a/\beta_1}{e^{2\pi a/\beta_1} - 1} \frac{2\pi a/\beta_2}{1 - e^{-2\pi a/\beta_2}} \frac{\alpha Z^2 r_0^2}{\pi^2} \frac{\beta_2}{\beta_1 k q^4}$$

$$\cdot \left| \frac{\hat{\mathbf{p}}_1 \times \hat{\mathbf{k}}}{S_1} [\beta_1 V - ia(1-x)W] - \frac{\hat{\mathbf{p}}_2 \times \hat{\mathbf{k}}}{S_2} [\beta_2 V - ia(1-x)W] \right|^2 ,$$

(3.200)

where

$$S_1 = 1 - \hat{\mathbf{k}} \cdot \mathbf{p}_1 = 1 - \beta_1 \cos\theta_1 , \quad S_2 = 1 - \hat{\mathbf{k}} \cdot \mathbf{p}_2 = 1 - \beta_2 \cos\theta_2 \quad (3.201)$$

($\theta_{1,2}$ are the angles of the coordinate system (3.86) referring to the photon direction), and

$$V = {}_2F_1\left(\frac{ia}{\beta_1}, \frac{ia}{\beta_2}; 1; x\right), \quad W = {}_2F_1\left(\frac{ia}{\beta_1}+1, \frac{ia}{\beta_2}+1; 2; x\right),$$
$$x = 1 - \frac{(\mathbf{p}_1 - \mathbf{p}_2)^2}{(\beta_1 - \beta_2)^2 S_1 S_2} .$$

(3.202)

Equation (3.200) agrees with the cross section for bremsstrahlung of non-relativistic electrons as calculated with eigenfunctions of the Schrödinger equation taking retardation into account [10]. The quantities S_1 and S_2 containing the electron velocities to first order cause the angular distributions to be non-symmetric with respect to $\theta_{1,2} = 90°$ even for low electron energies. Neglecting, in addition, terms of order β compared with unity, one gets $S_1 \approx S_2 \approx 1$ and (3.200) reduces to

$$\frac{d^3\sigma_{nr}}{dk\, d\Omega_k\, d\Omega_{p_2}} = \frac{2\pi a/\beta_1}{e^{2\pi a/\beta_1} - 1} \frac{2\pi a/\beta_2}{1 - e^{-2\pi a/\beta_2}} \frac{\alpha Z^2 r_0^2}{\pi^2} \frac{\beta_2}{\beta_1 k q^4}$$
$$\cdot \left| (\hat{\mathbf{p}}_1 \times \hat{\mathbf{k}}) [\beta_1 V - ia(1-x)W] - (\hat{\mathbf{p}}_2 \times \hat{\mathbf{k}}) [\beta_2 V - ia(1-x)W] \right|^2 .$$

(3.203)

This equation corresponds to Sommerfeld's [10] method of matrix elements $\mathbf{M} = \int \psi_2^* \, \mathbf{r} \, \psi_1 \, d^3r$ in the earlier nonrelativistic calculation without retardation (dipole approximation).

3.4.5 The short-wavelength limit

It is appropriate to discuss in more detail the special situation of the short-wavelength limit ('tip region') of bremsstrahlung. Here the total kinetic energy E_0 of the incident electron is transferred to the photon so that the momentum $\mathbf{p}_2$ of the outgoing electron is zero. Then it follows that

$$k \to \epsilon_1 - 1, \; q^2 \to D_1, \; D_2 \to 2k, \; \mu \to 2k. \tag{3.204}$$

Further $a_2 = a\epsilon_2/p_2$ tends to infinity, and $x = 1 - \mu q^2/D_1D_2$ will be zero. However, a_2x remains finite, and the hypergeometric functions V and W transform into the confluent hypergeometric functions V_0 and W_0,

$$V = {}_2F_1(ia_1, ia_2; 1; x) \to V_0 = F(ia_1; 1; i\xi), \tag{3.205}$$
$$W = {}_2F_1(1 + ia_1, 1 + ia_2; 2; x) \to W_0 = F(1 + ia_1; 2; i\xi), \tag{3.206}$$

where, according to (3.174),

$$\xi = \lim_{p_2 \to 0} a_2 x = a\left(2\frac{\mathbf{q} \cdot \mathbf{p}_2}{q^2 p_2} - \frac{\mathbf{k} \cdot \mathbf{p}_2}{kp_2} - \frac{p_1}{k}\right). \tag{3.207}$$

One notes that the cross section depends on the direction of the outgoing electron, although it has zero velocity. Since some of the quantities in Eqs. (3.189) to (3.192) tend to infinity, we expediently do not use formula (3.184) to perform the limit $p_2 \to 0$ but rather go back to Eqs. (3.162) to (3.176). From (3.162) to (3.164) we get

$$I_{10} = \lim_{p_2 \to 0} I_1 = \frac{K_1}{q^2}\left\{\left(\frac{2}{q^2} - \frac{\epsilon_1}{k}\right)(V_0 - ia_1W_0) + \epsilon_1\frac{\mathbf{k} \cdot \mathbf{p}_2}{k^2 p_2}iaW_0\right\}, \tag{3.208}$$

$$\mathbf{I}_{20} = \lim_{p_2 \to 0} \mathbf{I}_2 = \frac{K_1}{2kq^2}\left\{\mathbf{q}(V_0 - ia_1W_0) + [\mathbf{k}/p_1 + \hat{\mathbf{p}}_2 - (\hat{\mathbf{k}} \cdot \hat{\mathbf{p}}_2)\mathbf{q}]iaW_0\right\}, \tag{3.209}$$

$$\mathbf{I}_{30} = \lim_{p_2 \to 0} \mathbf{I}_3 = \frac{K_1}{q^2}\left\{\frac{\mathbf{q}}{q^2}(V_0 - ia_1W_0) + \frac{\epsilon_1}{2k}(\mathbf{k}/p_1 + \hat{\mathbf{p}}_2)iaW_0\right\}. \tag{3.210}$$

Inserting these expressions into (3.183) and utilizing the relation

$$(\mathbf{p}_1 \times \mathbf{k})^2 = \epsilon_1 kq^2 - k^2 - \tfrac{1}{4}q^4, \tag{3.211}$$

we obtain for the differential cross section at the short-wavelength limit [13]

$$\frac{d^3\sigma_0}{dk\, d\Omega_k\, d\Omega_{p_2}} = \frac{2\pi a_1}{e^{2\pi a_1} - 1}\frac{\alpha^2 Z^3 r_0^2}{\pi k p_1 q^4}\left\{2\left(\frac{4}{q^2} + k - 1\right)\frac{(\mathbf{p}_1 \times \mathbf{k})^2}{q^2}|V_0 - ia_1W_0|^2\right.$$
$$+ \left[2\{k^2/p_1 + (1-k)(\hat{\mathbf{k}} \cdot \hat{\mathbf{p}}_2)\}\frac{(\mathbf{p}_1 \times \mathbf{k})^2}{q^2}\right.$$
$$+ \left.\left(2\frac{\epsilon_1^2 + \epsilon_1 + 2}{q^2} - 1\right)(\mathbf{p}_1 \times \mathbf{k}) \cdot (\hat{\mathbf{p}}_2 \times \hat{\mathbf{k}})\right]a[\mathrm{Im}(V_0W_0^*) - a_1|W_0|^2]$$

$$+ \left[(2\epsilon_1^2 - k)(\hat{\mathbf{p}}_2 \times \hat{\mathbf{k}})^2 + (\hat{\mathbf{k}} \cdot \hat{\mathbf{p}}_2)(\mathbf{p}_1 \times \mathbf{k}) \cdot (\hat{\mathbf{p}}_2 \times \hat{\mathbf{k}}) \right.$$
$$\left. + k^3 \{1 - (\hat{\mathbf{k}} \cdot \hat{\mathbf{p}}_1)(\hat{\mathbf{k}} \cdot \hat{\mathbf{p}}_2)\} \right] a^2 |W_0|^2 \bigg\} . \qquad (3.212)$$

The cross section (3.212) does not suffer from the failure of the Born approximation at the short-wavelength limit because the SM wave functions represent an approximation in αZ, not in $\alpha Z/\beta$. Therefore Eq. (3.212) is expected to be accurate for low atomic numbers Z.

Since the outgoing electron has zero kinetic energy, it cannot be detected in coincidence with the emitted photon. Therefore the elementary process at the short-wavelength limit is given by the doubly differential cross section, i.e., the expression (3.212) integrated over the angles of the outgoing electron.

Assuming $a_1 \ll 1$ and neglecting terms of order a_1^2 in the bracket, the cross section (3.212) reduces to

$$\frac{d^3\sigma_0}{dk \, d\Omega_k \, d\Omega_{p_2}} \approx \frac{2}{\pi} \alpha^2 Z^3 r_0^2 \frac{2\pi a_1}{e^{2\pi a_1} - 1} \frac{(\mathbf{p}_1 \times \mathbf{k})^2}{k p_1 q^6} \left(\frac{4}{q^2} + k - 1 \right). \qquad (3.213)$$

The same result is obtained if one multiplies the Born cross section (3.81) by the Elwert factor (3.90) and goes to the limit $p_2 \to 0$. Because the expression (3.213) does not depend on $\hat{\mathbf{p}}_2$, the corresponding doubly differential cross section is simply obtained if one multiplies by 4π.

At the short-wavelength limit the bremsstrahlung cross section is approximately proportional to the third power of the atomic number Z. The additional factor $a = \alpha Z$ arises from the second part of the normalization (3.176). Hence the relative strength of the high-energy end of the bremsstrahlung spectrum increases for targets of high atomic number Z. The shape of the photon angular distribution, given by $(\mathbf{p}_1 \times \mathbf{k})^2/q^6$, is that of a dipole radiation with relativistic beaming.

3.4.6 Approximation for the long-wavelength limit and for high electron energies

The argument x of the hypergeometric functions V and W is never positive. It is equal to zero at the short-wavelength limit as well as for

$$(\mathbf{p}_1 \times \mathbf{p}_2) \cdot \mathbf{k} = 0, \quad D_2 |\hat{\mathbf{p}}_1 \times \hat{\mathbf{k}}| = D_1 |\hat{\mathbf{p}}_2 \times \hat{\mathbf{k}}| . \qquad (3.214)$$

$|x|$ is large compared to unity on the following conditions:

(1) At the long-wavelength limit ($p_2 \approx p_1$, $k \ll p_2$) except $q \approx q_{\min} = p_1 - p_2 - k$.
(2) At high electron energies ($\epsilon_1 \gg 1$, $\epsilon_2 \gg 1$) except $q \approx q_{\min}$.

In both cases we have $a_1 \approx a_2$. Case (2) was investigated by Bethe and Maximon [12]. Now we suppose $|x| \gg 1$ in general. It is then convenient to make use of the following transformations for the hypergeometric functions [53]:

$$V = {}_2F_1(ia_1, ia_2; 1; x)$$
$$= \frac{iG}{a_1}(1-x)^{-ia_1} {}_2F_1(ia_1, 1-ia_2; 1-ia_2+ia_1; y)$$
$$+ \frac{iG^*}{a_2}(1-x)^{-ia_2} {}_2F_1(ia_2, 1-ia_1; 1+ia_2-ia_1; y), \quad (3.215)$$

$$W = {}_2F_1(1+ia_1, 1+ia_2; 2; x)$$
$$= \frac{G}{a_1 a_2}(1-x)^{-(1+ia_1)} {}_2F_1(1+ia_1, 1-ia_2; 1-ia_2+ia_1; y)$$
$$+ \frac{G^*}{a_1 a_2}(1-x)^{-(1+ia_2)} {}_2F_1(1+ia_2, 1-ia_1; 1+ia_2-ia_1; y),$$
$$(3.216)$$

where $y = 1/(1-x)$ and

$$G = \frac{\Gamma(ia_2 - ia_1)}{\Gamma(ia_2)\Gamma(-ia_1)}. \quad (3.217)$$

Expanding the hypergeometric functions ${}_2F_1$ to the order $y = (1-x)^{-1}$, performing the limit $a_2 \to a_1$, and using the well-known properties of the Γ function, one obtains

$$\lim_{a_2 \to a_1} \left| V - ia_1(1-x)W \right|^2 \approx \left(\frac{\sinh(\pi a_1)}{\pi a_1} \right)^2$$
$$\cdot \left\{ 1 - \frac{2a_1^2}{1-x}[\ln(1-x) + 1 - 2\gamma - 2\operatorname{Re}\psi(ia_1)] \right\}, \quad (3.218)$$

$$\lim_{a_2 \to a_1} |W|^2 \approx \left(\frac{\sinh(\pi a_1)}{\pi a_1} \right)^2 \left\{ \frac{\ln(1-x) - 2\gamma - 2\operatorname{Re}\psi(ia_1)}{1-x} \right\}^2$$
$$\cdot \left(1 + 2\frac{1+a_1^2}{1-x} \right), \quad (3.219)$$

76   *Theory of the elementary process of electron-nucleus bremsstrahlung*

and

$$\lim_{a_2 \to a_1} \{(1-x)a_1|W|^2 - \mathrm{Im}(VW^*)\}$$

$$\approx a_1 \left(\frac{\sinh(\pi a_1)}{\pi a_1}\right)^2 \frac{\ln(1-x) - 2\gamma - 2\,\mathrm{Re}\,\psi(ia_1)}{(1-x)^2}$$

$$\cdot [\ln(1-x) - 2\gamma - 2\,\mathrm{Re}\,\psi(ia_1) - 4a_1^2] \,, \qquad (3.220)$$

where

$$\psi(z) = \frac{\Gamma'(z)}{\Gamma(z)} \qquad (3.221)$$

is the logarithmic derivative of the gamma function, $\gamma \approx 0.5772$ is Euler's constant, and

$$\gamma + \mathrm{Re}\,\psi(ia_1) = a_1^2 \sum_{n=1}^{\infty} \frac{1}{n(n^2 + a_1^2)} \,. \qquad (3.222)$$

It is seen from Eqs. (3.219) and (3.220) that the terms of the bremsstrahlung cross section (3.184) proportional to $\mathrm{Re}(A_1^* B)$, $\mathrm{Re}(A_2^* B)$, and $|B|^2$ are of the order $[\ln(1-x)/(1-x)]^2$, so that we obtain up to order $\ln(1-x)/(1-x)$

$$\frac{d^3\sigma}{dk\,d\Omega_k\,d\Omega_{p_2}} \approx \frac{d^3\sigma_B}{dk\,d\Omega_k\,d\Omega_{p_2}} [\pi a_1/\sinh(\pi a_1)]^2 |V - ia_1(1-x)W|^2$$

$$\approx \frac{d^3\sigma_B}{dk\,d\Omega_k\,d\Omega_{p_2}} \{1 - [2a_1^2/(1-x)][\ln(1-x)$$

$$+ 1 - 2\gamma - 2\,\mathrm{Re}\,\psi(ia_1)]\} \,. \qquad (3.223)$$

That is, for $1 - x \gg 1$ and $(a_2 - a_1)/a_2 \ll 1$, the bremsstrahlung differential cross section in Born approximation is correct.

In case (2), i.e., if one sets $a_1 \approx a_2 \approx a$, the cross section in Born approximation has to be multiplied by the correction factor given in Eq. (3.223). This formula can be used to evaluate the screening correction to the bremsstrahlung cross section near the long-wavelength limit.

3.4.7   *Cross section in second Born approximation*

A relativistic scattering wave function which is valid to second order in $a = \alpha Z$ for all electron energies was developed by Johnson and Mullin [61].

The wave function has as its dominant term the SM function,

$$\psi(\mathbf{r}) = \psi_{\text{SM}}(\mathbf{r}) + \psi_c(\mathbf{r}) . \tag{3.224}$$

The correction term is given by its Fourier transform, which for the incident electron has the form [61]

$$\psi_{1c}(\mathbf{k}) = (2\pi)^{-3} \int \psi_{1c}(\mathbf{r}) e^{-i\mathbf{k}\cdot\mathbf{r}} d^3r$$

$$= \frac{N_1}{8\pi^2} \lim_{\lambda \to 0} \frac{a^2}{(k^2 - p_1^2 - i\lambda)k(\mathbf{p}_1 \times \mathbf{k})^2} \left\{ \frac{1}{p_1}[(\vec{\alpha}\cdot\mathbf{k})(\vec{\alpha}\cdot\mathbf{p}_1) + p_1^2] \right.$$

$$\cdot \left[\pi k p_1 s_1 + i\, k(\mathbf{p}_1 \cdot \mathbf{s}_1) \ln \frac{s_1^2}{k^2 - p_1^2 - 2ip_1\lambda} \right.$$

$$\left. + i\, p_1 (\mathbf{s}_1 \cdot \mathbf{k}) \ln \frac{p_1 + k}{p_1 - k + i\lambda} \right]$$

$$+ \frac{k^2 - p_1^2}{s_1} \left[\pi k(\mathbf{p}_1 \cdot \mathbf{s}_1) + i\, k p_1 s_1 \ln \frac{s_1^2}{k^2 - p_1^2 - 2ip_1\lambda} \right.$$

$$\left. \left. + i\, s_1 (\mathbf{p}_1 \cdot \mathbf{k}) \ln \frac{p_1 + k}{p_1 - k + i\lambda} \right] \right\} u(\mathbf{p}_1) , \tag{3.225}$$

where N_1 is the normalization factor (3.135) and

$$\mathbf{s}_1 = \mathbf{p}_1 - \mathbf{k} . \tag{3.226}$$

The wave function (3.225) can be used to calculate the correction terms of relative order αZ to the Bethe-Heitler formula (3.81) thereby avoiding the divergence problems inherent in the use of the second Born wave functions [38]. The bremsstrahlung matrix element correct to terms of order a^2 can be written as

$$M = M_{\text{SM}} + M_c , \tag{3.227}$$

where M_{SM} consists of the first three terms of (3.150) and M_c is composed of the last term of (3.150) and of

$$M_{21} = \int \psi_{2c}^\dagger(\mathbf{r}) (\vec{\alpha}\cdot\mathbf{e}^*) e^{-i\mathbf{k}\cdot\mathbf{r}} \psi_1(\mathbf{r}) d^3r , \tag{3.228}$$

$$M_{12} = \int \psi_2^\dagger(\mathbf{r}) (\vec{\alpha}\cdot\mathbf{e}^*) e^{-i\mathbf{k}\cdot\mathbf{r}} \psi_{1c}(\mathbf{r}) d^3r . \tag{3.229}$$

M_{SM} is of order a and M_c is of order a^2. Since the absolute square of the matrix element occurs in the cross section, its determination to within the given approximation requires only the evaluation of the expression

$$|M|^2 \approx |M_{\text{SM}}|^2 + 2\,\text{Re}(M_{\text{SM}}M_c^*)\,. \tag{3.230}$$

The first term $|M_{\text{SM}}|^2$, averaged over the spin directions of the initial electron and summed over the spin directions of the final electron and over the polarizations of the emitted photon, is given by (3.183). After a tedious but straightforward calculation, the result for the correction of order a to the Bethe-Heitler cross section has the form* [62]

$$\frac{d^3\sigma_c}{dk\,d\Omega_k\,d\Omega_{p_2}} = \frac{2\pi a_1}{e^{2\pi a_1}-1}\frac{2\pi a_2}{1-e^{-2\pi a_2}}\frac{\alpha^2 Z^3 r_0^2}{\pi}\frac{kp_2}{p_1 q^2}$$

$$\cdot\left\{\left(\frac{\epsilon_1}{D_2}-\frac{\epsilon_2}{D_1}\right)\left[\frac{1}{D_2}\left(\frac{1}{p_1}-\frac{2}{q}\right)-\frac{1}{D_1}\left(\frac{1}{s_1}-\frac{2}{q}\right)\right]\right.$$

$$+\frac{1}{2}\left(\frac{\epsilon_2}{D_1}\mathbf{s}_1+\frac{\epsilon_1}{D_2}\mathbf{s}_2\right)\frac{\hat{\mathbf{p}}_1+\hat{\mathbf{s}}_1}{p_1 s_1+\mathbf{p}_1\cdot\mathbf{s}_1}$$

$$+\frac{1}{D_1 D_2}\left[\epsilon_2(\mathbf{p}_1\cdot\mathbf{q})-\epsilon_1(\mathbf{p}_2\cdot\mathbf{q})\right]\left[\frac{\mathbf{p}_2\cdot(\hat{\mathbf{p}}_1+\hat{\mathbf{q}})}{p_1 q+\mathbf{p}_1\cdot\mathbf{q}}+\frac{\mathbf{s}_2\cdot(\hat{\mathbf{s}}_1+\hat{\mathbf{q}})}{s_1 q+\mathbf{s}_1\cdot\mathbf{q}}\right]$$

$$\left.-\frac{1}{D_1 D_2}\left[\frac{\epsilon_2}{p_1}(\mathbf{p}_2\cdot\mathbf{q})+\frac{\epsilon_1}{s_1}(\mathbf{s}_2\cdot\mathbf{q})\right]\right\}\,, \tag{3.231}$$

where

$$\mathbf{s}_1 = \mathbf{p}_1 - \mathbf{k} = \mathbf{p}_2 + \mathbf{q}\,,\ \mathbf{s}_2 = \mathbf{p}_2 + \mathbf{k} = \mathbf{p}_1 - \mathbf{q}\,. \tag{3.232}$$

Since the factor F [Eq. (3.176)] derives from the Coulomb normalization factors N_1 and N_2 [see Eq. (3.135)] which are common to the exact matrix element M, it has not been expanded in powers of a.

In comparisons of the second Born approximation with other theories only very limited agreement was found even where they would be expected to agree. This indicates that the Born series converges quite slowly.

At the short-wavelength limit the Born-approximation cross section including the correction of relative order a becomes, using the result (3.213),

$$\frac{d^3\sigma_0}{dk\,d\Omega_k\,d\Omega_{p_2}} = \frac{2\pi a_1}{e^{2\pi a_1}-1}\frac{\alpha^2 Z^3 r_0^2}{\pi}\frac{1}{kp_1 q^2}\left\{2\frac{(\mathbf{p}_1\times\mathbf{k})^2}{q^4}\left(\frac{4}{q^2}+k-1\right)\right.$$

---

*To lowest order in a the cross section (3.184) reduces to the Bethe-Heitler formula (3.81). Applying the approximations (3.193) and (3.194) the corrections to $|A_1|^2$, $|A_2|^2$,..., $|B|^2$ are of order a^2. Therefore Eq. (3.231) represents the correction of order a to the Bethe-Heitler cross section if the factor F [Eq. (3.176)] is not expanded in powers of a.

$$+ \pi \alpha Z \left[\frac{(\mathbf{p}_1 \times \mathbf{k})^2}{q^3} \left(\frac{2-k}{q^2} + 1 - \frac{1}{k} \right) \right.$$
$$\left. + (k-1) \left(\frac{k}{p_1 q^2} - \frac{1}{2q} - \frac{\epsilon_1}{2p_1} + \frac{q}{4k} \right) \right] \Bigg\} \,. \tag{3.233}$$

This cross section vanishes in forward and in backward direction, $\hat{\mathbf{k}} \cdot \hat{\mathbf{p}}_1 = \pm 1$.

3.5 Calculation using relativistic partial-wave expansions

3.5.1 Partial wave expansion

The best available approach to calculate the bremsstrahlung cross section utilizes the relativistic partial-wave approximation to describe the motion of the electron in the static field of the target atom and the multipole series expansion for the photon. It is assumed that the initial and final states of the atom coincide ('elastic' bremsstrahlung). If the exact numerical potential of the target atom, $V(\mathbf{r})$, is inserted in the unperturbed Hamiltonian the interaction of the electron is with the screened potential instead of the Coulomb potential. Then the Dirac equation is given by [see Eq. (3.25)]

$$\{-i\vec{\alpha} \cdot \nabla + \beta - \epsilon + V(\mathbf{r})\} \Phi(\mathbf{r}) = 0 \,. \tag{3.234}$$

In the following it is convenient to apply the split spinor representation. As in the case of the plane wave spinor (Sec. 3.4.2) the Dirac operators $\vec{\alpha}$ and β are replaced, respectively, by the Pauli spin matrices and the 2×2 unit matrices [Eq. (3.6)], and the wave function $\Phi(\mathbf{r})$ is replaced by the spinors v and w,

$$\Phi(\mathbf{r}) = \begin{pmatrix} v(\mathbf{r}) \\ w(\mathbf{r}) \end{pmatrix} \,. \tag{3.235}$$

In this way the wave equation (3.234) is split into the two coupled equations

$$i\vec{\sigma} \cdot \nabla w + [\epsilon - 1 - V(\mathbf{r})]v = 0 \,, \tag{3.236}$$
$$i\vec{\sigma} \cdot \nabla v + [\epsilon + 1 - V(\mathbf{r})]w = 0 \,. \tag{3.237}$$

Eliminating w by means of the second equation, the wave function takes the form

$$\Phi(\mathbf{r}) = \begin{pmatrix} 1 \\ -i(\epsilon + 1 - V)^{-1}\vec{\sigma}\cdot\nabla \end{pmatrix} v(\mathbf{r}) , \qquad (3.238)$$

where $v(\mathbf{r})$ is determined by the second-order differential equation

$$\nabla^2 v(\mathbf{r}) + (\epsilon + 1 - V)[\vec{\sigma}\cdot\nabla(\epsilon + 1 - V)^{-1}]\vec{\sigma}\cdot\nabla v(\mathbf{r})$$
$$+ [(\epsilon - V)^2 - 1]v(\mathbf{r}) = 0 . \qquad (3.239)$$

Equation (3.238) gives the split representation of $\Phi(\mathbf{r})$ for any static field V which does not necessarily need to be spherically symmetric. By this means any matrix element between four-component states may be reduced to matrix elements between two-component spinors v. Considering the case that V is a function of r only, the wave equation (3.239) is separable into radial and angular parts and we can set

$$v(\mathbf{r}) = g(r)\chi(\hat{\mathbf{r}}) . \qquad (3.240)$$

This is demonstrated by introducing the orbital angular momentum operator

$$\mathbf{L} = -i\mathbf{r}\times\nabla , \qquad (3.241)$$

which satisfies the equation

$$\vec{\sigma}\cdot\nabla = \sigma_r\left(\frac{\partial}{\partial r} - \frac{1}{r}\vec{\sigma}\cdot\mathbf{L}\right) , \qquad (3.242)$$

where $\sigma_r = \vec{\sigma}\cdot\hat{\mathbf{r}}$ denotes the component in $\hat{\mathbf{r}}$ direction of the Pauli matrix $\vec{\sigma}$. Equation (3.242) is easily proved by evaluating the quantity $(\vec{\sigma}\cdot\hat{\mathbf{r}})(\vec{\sigma}\cdot\mathbf{L})$. The products containing $\vec{\sigma}$-matrices can be reduced by means of the relation

$$(\vec{\sigma}\cdot\mathbf{a})(\vec{\sigma}\cdot\mathbf{b}) = \mathbf{a}\cdot\mathbf{b} + i\vec{\sigma}\cdot(\mathbf{a}\times\mathbf{b}) \qquad (3.243)$$

which follows from the commutation rules (3.7).

Applying Eq. (3.243) and the vector calculus relation

$$\mathbf{a}\times(\mathbf{b}\times\mathbf{c}) = (\mathbf{a}\cdot\mathbf{c})\mathbf{b} - (\mathbf{a}\cdot\mathbf{b})\mathbf{c} , \qquad (3.244)$$

we get

$$(\vec{\sigma}\cdot\hat{\mathbf{r}})(\vec{\sigma}\cdot\mathbf{L}) = -i(\vec{\sigma}\cdot\hat{\mathbf{r}})\vec{\sigma}\cdot(\mathbf{r}\times\nabla) = \vec{\sigma}\cdot[\hat{\mathbf{r}}\times(\mathbf{r}\times\nabla)] = r\left(\sigma_r\frac{\partial}{\partial r} - \vec{\sigma}\cdot\nabla\right). \qquad (3.245)$$

Now we express the angular part of the Laplace operator by $\mathbf{L}^2$ [35],

$$\nabla^2 = \frac{\partial^2}{\partial r^2} + \frac{2}{r}\frac{\partial}{\partial r} - \frac{1}{r^2}\mathbf{L}^2, \qquad (3.246)$$

and insert (3.242) into (3.239). Then we obtain, bearing in mind that $V = V(r)$ and $\sigma_r^2 = 1$,

$$\left(\frac{\partial^2}{\partial r^2} + \frac{2}{r}\frac{\partial}{\partial r} - \frac{1}{r^2}\mathbf{L}^2\right)v(\mathbf{r}) + \left\{(\epsilon - V)^2 - 1 + (\epsilon + 1 - V)\right.$$
$$\left. \cdot \left[\frac{d}{dr}(\epsilon+1-V)^{-1}\right]\left(\frac{\partial}{\partial r} - \frac{1}{r}\vec{\sigma}\cdot\mathbf{L}\right)\right\}v(\mathbf{r}) = 0. \qquad (3.247)$$

The total angular momentum $\mathbf{J}$ is composed of the orbital and the spin angular momentum. Squaring the total angular momentum operator

$$\mathbf{J} = \mathbf{L} + \tfrac{1}{2}\vec{\sigma} \qquad (3.248)$$

and using $\vec{\sigma}^2 = \sigma_1^2 + \sigma_2^2 + \sigma_3^2 = 3$, we get the relation

$$\vec{\sigma}\cdot\mathbf{L} = \mathbf{J}^2 - \mathbf{L}^2 - \tfrac{3}{4}. \qquad (3.249)$$

The wave equation (3.247) is separable if we choose $\chi(\hat{\mathbf{r}})$ in Eq. (3.240) to be the wave function corresponding to the composition (3.248). To obtain a simultaneous eigenfunction of $\mathbf{J}^2$ and J_z we have to take a linear combination of products of the form

$$\chi_{jj_z l}(\hat{\mathbf{r}}) = \sum_{m,s} C(j, j_z; l, m, \tfrac{1}{2}, s) Y_{lm}(\hat{\mathbf{r}}) v(\vec{\zeta}), \qquad (3.250)$$

where C denotes the vector addition (Clebsch-Gordan) coefficients [63], Y_{lm} are the normalized spherical harmonics which are eigenfunctions of $\mathbf{L}^2$ and L_z with eigenvalues $l(l+1)$ and m, respectively, and $v(\vec{\zeta})$ is the free-particle two-component spinor with polarization $\vec{\zeta}$ satisfying Eq. (3.146). The spin quantum number is $s = \pm\tfrac{1}{2}$, and j_z is the projection of the total angular momentum. The conservation of angular momentum is contained in the Clebsch-Gordan coefficients which vanish unless $j_z = m + s$ and $j = |l \pm \tfrac{1}{2}|$. Hence the double sum in (3.250) effectively reduces to a single sum of either m or s. With the explicit values of the vector addition coefficients [63; 64] the wave function χ has the form

$$\chi_{jj_z l}(\hat{\mathbf{r}}) = \begin{pmatrix} \sqrt{(j+j_z)/2j}\, Y_{l,j_z-1/2} \\ \sqrt{(j-j_z)/2j}\, Y_{l,j_z+1/2} \end{pmatrix} \qquad (3.251)$$

for $j = l + \frac{1}{2}$, and

$$\chi_{jj_z l}(\hat{\mathbf{r}}) = \begin{pmatrix} -\sqrt{(j+1-j_z)/(2j+2)}\, Y_{l,j_z-1/2} \\ \sqrt{(j+1+j_z)/(2j+2)}\, Y_{l,j_z+1/2} \end{pmatrix} \quad (3.252)$$

for $j = l - \frac{1}{2}$.

$\chi_{jj_z l}$ is an eigenstate of $\mathbf{J}^2$ and $\mathbf{L}^2$,

$$\mathbf{J}^2 \chi = j(j+1)\chi, \quad \mathbf{L}^2 \chi = l(l+1)\chi, \quad (3.253)$$

where $j = l \pm \frac{1}{2}$ for $l \neq 0$ and $j = \frac{1}{2}$ for $l = 0$.

When we insert (3.240) into (3.247) and apply the relations (3.249) and (3.253), this leads to the radial wave equation

$$\frac{d^2 g_\kappa}{dr^2} + \frac{2}{r}\frac{dg_\kappa}{dr} + \left\{ [\epsilon - V(r)]^2 - 1 - \frac{l(l+1)}{r^2} \right\} g_\kappa$$
$$+ \frac{dV}{dr}[\epsilon - V(r) + 1]^{-1}\left(\frac{d}{dr} + \frac{\kappa+1}{r}\right) g_\kappa = 0, \quad (3.254)$$

where the relativistic angular momentum and parity quantum number

$$\kappa = l(l+1) - j(j+1) - \tfrac{1}{4} \quad (3.255)$$

allows to find out how the orbital momentum l and the spin of the electron are coupled to the total angular momentum j. It is convenient to note that

$$\kappa = \begin{cases} -(l+1) & \text{for } j = l + \tfrac{1}{2}, \\ l & \text{for } j = l - \tfrac{1}{2}. \end{cases} \quad (3.256)$$

Thus a double sum over l and j may be replaced by a sum over all positive and negative values of κ (κ takes all integer values except zero). The use of the quantum number κ leads to an economical notation since its value gives both j and l:

$$j = |\kappa| - \tfrac{1}{2}, \quad l = \begin{cases} \kappa & \text{for } \kappa > 0, \\ -(\kappa+1) & \text{for } \kappa < 0. \end{cases} \quad (3.257)$$

Besides, it is easily verified from (3.256) that

$$\kappa(\kappa+1) = l(l+1). \quad (3.258)$$

Hence it is obvious that the radial wave function $g_\kappa(r)$ satisfying (3.254) depends only on the quantum number κ.

In Dirac theory, as in nonrelativistic theory, only one radial wave function has to be determined. Using (3.240), the wave function (3.238) can be recast as

$$\Phi_{\kappa j_z}(\mathbf{r}) = \Phi_{jj_z l}(\mathbf{r}) = \begin{pmatrix} 1 \\ -i(\epsilon+1-V)^{-1}\vec{\sigma}\cdot\nabla \end{pmatrix} g_\kappa(r)\chi_{\kappa j_z}(\hat{\mathbf{r}}), \quad (3.259)$$

where $\chi_{\kappa j_z}(\hat{\mathbf{r}}) = \chi_{jj_z l}(\hat{\mathbf{r}})$ is given by (3.250), and $g_\kappa(r)$ is determined by the radial wave equation (3.254). It is useful to note that $\vec{\sigma}\cdot\nabla$ may be written as an r dependent operator by means of the relations (3.242), (3.249), and (3.255):

$$\vec{\sigma}\cdot\nabla g_\kappa \chi_{\kappa j_z} = \sigma_r\left(\frac{\partial}{\partial r} + \frac{\kappa+1}{r}\right)g_\kappa \chi_{\kappa j_z}. \quad (3.260)$$

Then $\Phi_{\kappa j_z}(\mathbf{r})$ becomes

$$\Phi_{\kappa j_z}(\mathbf{r}) = \begin{pmatrix} 1 \\ -\frac{i}{\epsilon+1-V}\sigma_r\left(\frac{\partial}{\partial r} + \frac{\kappa+1}{r}\right) \end{pmatrix} g_\kappa \chi_{\kappa j_z}. \quad (3.261)$$

Further it is expedient to transform the second-order radial wave equation (3.254) to a pair of coupled differential equations of first order. To this end we define a function $f_\kappa(r)$ by

$$\left(\frac{d}{dr} + \frac{\kappa+1}{r}\right)g_\kappa(r) - (\epsilon+1-V)f_\kappa(r) = 0, \quad (3.262)$$

which gives

$$\Phi_{\kappa j_z}(\mathbf{r}) = \begin{pmatrix} g_\kappa(r) \\ -if_\kappa(r)\sigma_r \end{pmatrix} \chi_{\kappa j_z}(\hat{\mathbf{r}}). \quad (3.263)$$

The second of the first-order radial wave equations is obtained by inserting the derivatives of g_κ into (3.254) yielding

$$\left(\frac{d}{dr} - \frac{\kappa-1}{r}\right)f_\kappa(r) + (\epsilon-1-V)g_\kappa(r) = 0. \quad (3.264)$$

The functions $g_\kappa(r)$ and $f_\kappa(r)$ represent, respectively, the major and minor components of the Dirac radial functions. It is common to use functions which are r times these radial functions. By defining $\bar{g}_\kappa(r) = rg_\kappa(r)$ and $\bar{f}_\kappa(r) = rf_\kappa(r)$ the new radial wave functions satisfy the system of coupled differential equations

$$\left(\frac{d}{dr} + \frac{\kappa}{r}\right)\bar{g}_\kappa(r) - [\epsilon+1-V(r)]\bar{f}_\kappa(r) = 0, \quad (3.265)$$

$$\left(\frac{d}{dr} - \frac{\kappa}{r}\right)\bar{f}_\kappa(r) + [\epsilon - 1 - V(r)]\bar{g}_\kappa(r) = 0 , \qquad (3.266)$$

or, corresponding to (3.254),

$$\frac{d^2\bar{g}_\kappa}{dr^2} + \left\{[\epsilon - V(r)]^2 - 1 - \frac{l(l+1)}{r^2}\right\}\bar{g}_\kappa + \frac{dV}{dr}[\epsilon - V(r) + 1]^{-1}\left(\frac{d}{dr} + \frac{\kappa}{r}\right)\bar{g}_\kappa = 0 . \qquad (3.267)$$

The method of solution of the radial wave equation (3.267) with boundary conditions is the same as in the nonrelativistic case [56]. In the Coulomb case, $V = -a/r$, the radial equation (3.267) has a well-known analytic solution [64] in terms of the confluent hypergeometric function (see Sec. 3.5.3).

The calculation of the bremsstrahlung cross section requires wave functions behaving asymptotically as plane waves plus outgoing or ingoing spherical waves denoted by $\psi^+(\mathbf{p},\vec{\zeta},\mathbf{r})$ and $\psi^-(\mathbf{p},\vec{\zeta},\mathbf{r})$, respectively. For large values of r (3.267) reduces to $d^2\bar{g}_\kappa/dr^2 + p^2\bar{g}_\kappa = 0$ assuming that $V(r)$ tends to zero faster than $1/r$. Therefore the asymptotic form of any solution $g_\kappa(r) = \bar{g}_\kappa/r$ is

$$g_\kappa(r) \sim \frac{A}{pr} \sin(pr - \tfrac{1}{2}\pi l + \eta_\kappa) , \qquad (3.268)$$

where A is a constant. The phase shift η_κ depends on p and $V(r)$; it can in general be determined only numerically.[†] For the Coulomb field the phase shift can be calculated analytically (see Problem 3.12).

We construct the appropriate wave function $\psi^\pm(\mathbf{p},\vec{\zeta},\mathbf{r})$ by considering first the free-particle wave function $\psi_0(\mathbf{p},\vec{\zeta},\mathbf{r}) = e^{i\mathbf{p}\cdot\mathbf{r}}u(\mathbf{p})$ which is a solution of Eq. (3.29). In the split spinor representation [cf. Eqs. (3.143) and (3.144)] it takes the form

$$\psi_0(\mathbf{p},\vec{\zeta},\mathbf{r}) = \sqrt{\frac{\epsilon+1}{2\epsilon}} \begin{pmatrix} 1 \\ \frac{\vec{\sigma}\cdot\mathbf{p}}{\epsilon+1} \end{pmatrix} e^{i\mathbf{p}\cdot\mathbf{r}} v(\vec{\zeta})$$

$$= \sqrt{\frac{\epsilon+1}{2\epsilon}} \begin{pmatrix} 1 \\ -i\frac{\vec{\sigma}\cdot\vec{\nabla}}{\epsilon+1} \end{pmatrix} e^{i\mathbf{p}\cdot\mathbf{r}} v(\vec{\zeta}) . \qquad (3.269)$$

The familiar Rayleigh expansion of the plane wave in terms of Legendre

---

[†]It is only in the phase shifts η_κ that the detailed structure of the central field enters. The term $-\tfrac{1}{2}\pi l$ in (3.268) is added such that $\eta_\kappa = 0$ if $V(r) = 0$ [cf. Eq. (3.277)].

polynomials P_l is given by [56]

$$e^{i\mathbf{p}\cdot\mathbf{r}} = \sum_{l=0}^{\infty} i^l(2l+1)P_l(\hat{\mathbf{p}}\cdot\hat{\mathbf{r}})j_l(pr) , \qquad (3.270)$$

where $j_l(pr)$ is the spherical Bessel function of order l. Applying the addition theorem for spherical harmonics Y_{lm} [20],

$$P_l(\hat{\mathbf{k}}\cdot\hat{\mathbf{r}}) = \frac{4\pi}{2l+1}\sum_{m=-l}^{+l} Y_{lm}^*(\hat{\mathbf{k}})Y_{lm}(\hat{\mathbf{r}}) , \qquad (3.271)$$

Eq. (3.270) can be written as

$$e^{i\mathbf{p}\cdot\mathbf{r}} = 4\pi \sum_{l=0}^{\infty}\sum_{m=-l}^{+l} i^l j_l(pr)Y_{lm}^*(\hat{\mathbf{p}})Y_{lm}(\hat{\mathbf{r}}) . \qquad (3.272)$$

Hence we get

$$\psi_0(\mathbf{p},\vec{\zeta},\mathbf{r}) = 4\pi\sqrt{\frac{\epsilon+1}{2\epsilon}}\begin{pmatrix}1\\-i\frac{\vec{\sigma}\cdot\nabla}{\epsilon+1}\end{pmatrix} v(\vec{\zeta})\sum_{l,m} i^l j_l(pr)Y_{lm}^*(\hat{\mathbf{p}})Y_{lm}(\hat{\mathbf{r}}) . \qquad (3.273)$$

Now the spherical harmonic $Y_{lm}(\hat{\mathbf{r}})$ is expressed in terms of the angular part (3.250) of the wave function $\Phi(\mathbf{r})$. This is achieved if we multiply $\chi_{jj_z l}(\hat{\mathbf{r}})$ by $C(j,j_z;l,m',\frac{1}{2},s')$ and sum over j. Due to the relation

$$\sum_{j=l-1/2}^{l+1/2} C(j,j_z;l,m,\tfrac{1}{2},s)C(j,j_z;l,m',\tfrac{1}{2},s') = \delta_{mm'}\delta_{ss'} , \qquad (3.274)$$

which follows from the unitary character of the Clebsch-Gordan coefficients [64], we get

$$Y_{lm}(\hat{\mathbf{r}})v(\vec{\zeta}) = \sum_{j=l-1/2}^{l+1/2} C(j,j_z;l,m,\tfrac{1}{2},s)\chi_{jj_z l}(\hat{\mathbf{r}}) . \qquad (3.275)$$

Inserting (3.275) into (3.273) we arrive at

$$\psi_0(\mathbf{p},\vec{\zeta},\mathbf{r}) = 4\pi\sqrt{\frac{\epsilon+1}{2\epsilon}}\sum_{l,m,j} i^l C(j,j_z;l,m,\tfrac{1}{2},s)$$
$$\cdot Y_{lm}^*(\hat{\mathbf{p}})\begin{pmatrix}1\\-i\frac{\vec{\sigma}\cdot\nabla}{\epsilon+1}\end{pmatrix}j_l(pr)\chi_{jj_z l}(\hat{\mathbf{r}}) . \qquad (3.276)$$

For large r the spherical Bessel function behaves as [35; 56]

$$j_l(pr) \sim \frac{1}{pr} \sin(pr - \tfrac{1}{2}\pi l) . \qquad (3.277)$$

Comparison of (3.276) with the wave function (3.259) shows that

$$\psi^\pm(\mathbf{p}, \vec{\zeta}, \mathbf{r}) = 4\pi \sqrt{\frac{\epsilon+1}{2\epsilon}} \sum_{l,m,j} i^l C(j, j_z; l, m, \tfrac{1}{2}, s) Y^*_{lm}(\hat{\mathbf{p}}) e^{\pm i\eta_\kappa} \Phi_{jj_zl}(\mathbf{r}) , \qquad (3.278)$$

satisfies the boundary conditions if the asymptotic behavior of $g_\kappa(r)$ is given by (3.268). This is verified by taking the asymptotic limit of the difference $\psi^\pm(\mathbf{p}, \vec{\zeta}, \mathbf{r}) - \psi_0(\mathbf{p}, \vec{\zeta}, \mathbf{r})$. Since ψ_0 represents a plane wave, this difference has to be an outgoing or ingoing spherical wave (the potential $V(r)$ is assumed to tend to zero faster than $1/r$ for $r \to \infty$):

$$\psi^\pm(\mathbf{p}, \vec{\zeta}, \mathbf{r}) - \psi_0(\mathbf{p}, \vec{\zeta}, \mathbf{r}) \sim 4\pi \sqrt{\frac{\epsilon+1}{2\epsilon}} \sum_{l,m,j} i^l C(j, j_z; l, m, \tfrac{1}{2}, s)$$

$$\cdot Y^*_{lm}(\hat{\mathbf{p}}) \begin{pmatrix} 1 \\ -i(\epsilon+1)^{-1}\vec{\sigma}\cdot\nabla \end{pmatrix}$$

$$\cdot \left\{ e^{\pm i\eta_\kappa} \frac{\sin(pr - \tfrac{1}{2}\pi l + \eta_\kappa)}{pr} - \frac{\sin(pr - \tfrac{1}{2}\pi l)}{pr} \right\} \chi_{jj_zl} . \qquad (3.279)$$

As was required, the curly bracket has the values (according to whether the sign of η_κ is plus or minus)

$$\{\ldots\} = \frac{1}{2ipr} \cdot \left\{ \begin{matrix} e^{i(pr-\pi l/2)}(e^{2i\eta_\kappa} - 1) \\ e^{-i(pr-\pi l/2)}(1 - e^{-2i\eta_\kappa}) \end{matrix} \right\} . \qquad (3.280)$$

By means of Eq. (3.256) the double sum over l and j can be replaced by a sum over all integer values except zero of κ; besides, remembering that $j_z = m + s$, the sum over m can be replaced by a sum over j_z. This results in

$$\psi^\pm(\mathbf{p}, \vec{\zeta}, \mathbf{r}) = 4\pi \sqrt{\frac{\epsilon+1}{2\epsilon}} \sum_{\kappa,j_z} i^l C(j, j_z; l, m, \tfrac{1}{2}, s) Y^*_{lm}(\hat{\mathbf{p}}) e^{\pm i\eta_\kappa} \Phi_{\kappa j_z}(\mathbf{r}) . \qquad (3.281)$$

In Eqs. (3.278) and (3.281), $\Phi_{\kappa j_z}(\mathbf{r}) = \Phi_{jj_zl}(\mathbf{r})$ is given by either of the relations (3.259), (3.261), or (3.263).

The wave function $\psi^\pm(\mathbf{p},\vec{\zeta},\mathbf{r})$ can be transformed to an expression where the Clebsch-Gordan coefficients are absent [65]. In order to derive it we note that the angular part of the wave function obeys the relation

$$(\vec{\sigma}\cdot\mathbf{L}+\kappa+1)\chi_{\kappa j_z}(\hat{\mathbf{r}})=0\,,\tag{3.282}$$

which follows from (3.249), (3.253), and (3.255). This equation is solved by setting

$$\chi_{\kappa j_z}(\hat{\mathbf{r}})=N(\kappa-\vec{\sigma}\cdot\mathbf{L})P_l(\hat{\mathbf{p}}\cdot\hat{\mathbf{r}})v(\vec{\zeta})\,,\tag{3.283}$$

where N is a normalization factor. To show the proof we use

$$(\vec{\sigma}\cdot\mathbf{L})^2=\mathbf{L}^2+i\vec{\sigma}\cdot(\mathbf{L}\times\mathbf{L})=\mathbf{L}^2-\vec{\sigma}\cdot\mathbf{L}\,,\tag{3.284}$$

$$\mathbf{L}^2 P_l(\hat{\mathbf{p}}\cdot\hat{\mathbf{r}})=l(l+1)P_l(\hat{\mathbf{p}}\cdot\hat{\mathbf{r}})\,,\tag{3.285}$$

and (3.258) yielding

$$(\vec{\sigma}\cdot\mathbf{L}+\kappa+1)(\vec{\sigma}\cdot\mathbf{L}-\kappa)P_l(\hat{\mathbf{p}}\cdot\hat{\mathbf{r}})=\left[(\vec{\sigma}\cdot\mathbf{L})^2+\vec{\sigma}\cdot\mathbf{L}-\kappa(\kappa+1)\right]P_l(\hat{\mathbf{p}}\cdot\hat{\mathbf{r}})=0\,.\tag{3.286}$$

Inserting (3.283) into a linear combination of functions $\Phi_{\kappa j_z}(\mathbf{r})$ [Eq. (3.259)] gives

$$\psi^\pm(\mathbf{p},\vec{\zeta},\mathbf{r})=\sum_{\kappa=-\infty}^{+\infty}(1-\delta_{\kappa 0})C_\kappa N\begin{pmatrix}1\\ -i(\epsilon-V+1)^{-1}\vec{\sigma}\cdot\nabla\end{pmatrix}$$
$$\cdot g_\kappa(r)(\kappa-\vec{\sigma}\cdot\mathbf{L})P_l(\hat{\mathbf{p}}\cdot\hat{\mathbf{r}})v(\vec{\zeta})\,.\tag{3.287}$$

Here the factor $(1-\delta_{\kappa 0})$ prevents from counting the quantum number $\kappa=0$.

Proceeding in the same way as above the asymptotic form of this function is compared with the asymptotic form of the plane wave in split representation, Eqs. (3.269) and (3.270),

$$\psi_0(\mathbf{p},\vec{\zeta},\mathbf{r})=\sqrt{\frac{\epsilon+1}{2\epsilon}}\begin{pmatrix}1\\ -i\frac{\vec{\sigma}\cdot\nabla}{\epsilon+1}\end{pmatrix}e^{i\mathbf{p}\cdot\mathbf{r}}v(\vec{\zeta})$$

$$=\sqrt{\frac{\epsilon+1}{2\epsilon}}\begin{pmatrix}1\\ -i\frac{\vec{\sigma}\cdot\nabla}{\epsilon+1}\end{pmatrix}\sum_{l=0}^{\infty}i^l(2l+1)j_l(pr)P_l(\hat{\mathbf{p}}\cdot\hat{\mathbf{r}})v(\vec{\zeta})\,.$$
$$\tag{3.288}$$

On the right-hand side we transform the sum over l to a sum over κ, using (3.256):

$$\sum_{\kappa=-\infty}^{+\infty} |\kappa| i^l j_l(pr) P_l(\hat{\mathbf{p}} \cdot \hat{\mathbf{r}}) = \sum_{l=0}^{\infty} i^l j_l(pr) P_l(\hat{\mathbf{p}} \cdot \hat{\mathbf{r}}) \sum_{j=l-1/2}^{l+1/2} |\kappa|$$

$$= \sum_{l=0}^{\infty} i^l (2l+1) j_l(pr) P_l(\hat{\mathbf{p}} \cdot \hat{\mathbf{r}}). \quad (3.289)$$

This results in

$$\psi_0(\mathbf{p}, \vec{\zeta}, \mathbf{r}) = \sqrt{\frac{\epsilon+1}{2\epsilon}} \begin{pmatrix} 1 \\ -i\frac{\vec{\sigma} \cdot \nabla}{\epsilon+1} \end{pmatrix} \sum_{\kappa=-\infty}^{+\infty} |\kappa| i^l j_l(pr) P_l(\hat{\mathbf{p}} \cdot \hat{\mathbf{r}}) v(\vec{\zeta}). \quad (3.290)$$

When the asymptotic form of the plane wave (3.290) is compared with the asymptotic form of (3.287) one finds

$$C_\kappa N = \sqrt{\frac{\epsilon+1}{2\epsilon}} i^l e^{\pm i\eta_\kappa} \frac{\kappa}{|\kappa|} \quad (3.291)$$

with the upper and lower signs referring to asymptotically outgoing and ingoing spherical waves, respectively. Finally the term $\mathbf{L} P_l(\hat{\mathbf{p}} \cdot \hat{\mathbf{r}}) = \mathbf{L} P_l(\cos\theta)$ can be written as

$$\mathbf{L} P_l(\hat{\mathbf{p}} \cdot \hat{\mathbf{r}}) = -i\,\mathbf{r} \times \nabla P_l(\hat{\mathbf{p}} \cdot \hat{\mathbf{r}}) = -i\,\hat{\mathbf{r}} \times \hat{\mathbf{p}}\, \frac{dP_l}{d(\cos\theta)}$$

$$= i\frac{\mathbf{p} \times \mathbf{r}}{|\mathbf{p} \times \mathbf{r}|} \sin\theta \frac{dP_l}{d(\cos\theta)} = i P_{l1}(\hat{\mathbf{p}} \cdot \hat{\mathbf{r}})\,\mathbf{n}, \quad (3.292)$$

where $\mathbf{n} = \mathbf{p} \times \mathbf{r}/|\mathbf{p} \times \mathbf{r}|$ denotes the unit vector perpendicular to $\mathbf{r}$ and $\mathbf{p}$, and $P_{lm}(\cos\theta)$ are the associated Legendre polynomials. Substitution of (3.291) and (3.292) in (3.287) yields Darwin's solution [50] of the Dirac equation with potential $V(r)$,

$$\psi^{\pm}(\mathbf{p}, \vec{\zeta}, \mathbf{r}) = \sqrt{\frac{\epsilon+1}{2\epsilon}} \sum_{\kappa=-\infty}^{+\infty} (1 - \delta_{\kappa 0}) i^l e^{\pm i\eta_\kappa} \begin{pmatrix} 1 \\ -i(\epsilon - V + 1)^{-1} \vec{\sigma} \cdot \nabla \end{pmatrix} g_\kappa(r)$$

$$\cdot \left[|\kappa| P_l(\hat{\mathbf{p}} \cdot \hat{\mathbf{r}}) - i\frac{\kappa}{|\kappa|} P_{l1}(\hat{\mathbf{p}} \cdot \hat{\mathbf{r}})\, (\vec{\sigma} \cdot \mathbf{n}) \right] v(\vec{\zeta}). \quad (3.293)$$

3.5.2 Calculations with a relativistic self-consistent-field potential

When bremsstrahlung production takes place in the electromagnetic field of an atom one has to take into account screening of the nuclear Coulomb potential by using exact atomic-field wave functions for the incoming and outgoing electrons. This is equivalent to describing bremsstrahlung as the lowest-order process in the Furry picture of quantum electrodynamics (cf. Sec. 3.3). The atomic field is treated as a static screened Coulomb field. Hence the only effect of the target electrons on the scattering process is screening of the electron-nucleus interaction through their joint electrostatic potential (independent-particle approximation)*. This is a good approximation especially at high energies where the polarizability of the atomic shell is small during the rapid collision process. The most important effect not included in the independent-particle approximation is the exchange interaction between the projectile and the target electrons.

As described in the preceding section the wave function is separated into angular and radial parts. One obtains numerical solutions of specified energy ϵ of the radial equations, labelled by the quantum number κ. The full partial-wave solutions of the Dirac equation are computed numerically up to an order in the quantum number $\kappa = \pm 1, \pm 2, \ldots$ which is needed for convergence of the matrix element. According to Eq. (3.281) a linear combination of eigenstates $\Phi_{\kappa j_z}$ is taken which obeys the appropriate boundary condition (plane wave plus incoming or outgoing spherical wave). The photon interaction operator of the matrix element (3.21), $\vec{\alpha} \cdot \mathbf{e}^* e^{-i\mathbf{k}\mathbf{r}}$, is expanded into a multipole series [cf. Eq. (3.270)],

$$e^{-i\mathbf{k}\cdot\mathbf{r}} = \sum_{l=0}^{\infty}(-i)^l(2l+1)P_l(\hat{\mathbf{k}}\cdot\hat{\mathbf{r}})j_l(kr). \quad (3.294)$$

According to the addition theorem for spherical harmonics (3.271) the exponential can be expressed in terms of spherical harmonics Y_{lm},

$$e^{-i\mathbf{k}\cdot\mathbf{r}} = 4\pi \sum_{l=0}^{\infty} \sum_{m=-l}^{+l} i^{-l} j_l(kr) Y_{lm}^*(\hat{\mathbf{k}}) Y_{lm}(\hat{\mathbf{r}}). \quad (3.295)$$

---

*In the relativistic independent-particle model one-electron orbitals and binding energies are obtained by solving the Dirac equation for a central potential $V(r)$ giving the average interaction energy of an atomic electron at a distance r from the nucleus with the nuclear charge Ze and with the other $Z - 1$ electrons (for a neutral atom).

The calculation of the bremsstrahlung matrix element becomes computationally very extensive. Inserting the electron wave functions $\psi^\pm(\mathbf{p}, \vec{\zeta}, \mathbf{r})$ and the expansion (3.295) into the matrix element (3.21), the angular integrations can be performed analytically. The radial matrix elements

$$R_1 = \int_0^\infty r^2 j_l(kr) g_{\kappa_1}(r) f_{\kappa_2}(r) \, dr \, , \tag{3.296}$$

$$R_2 = \int_0^\infty r^2 j_l(kr) g_{\kappa_2}(r) f_{\kappa_1}(r) \, dr \tag{3.297}$$

are the basic integrals to be obtained numerically. These integrals over rapidly oscillating functions depend on p_1, p_2, k, and Z and converge slowly. In addition the evaluation of the matrix element requires summations over the photon quantum numbers l, m and the electron quantum numbers κ_1, κ_2, j_{1z}, and j_{2z}. The angular integrals limit the summation over l since they are only different from zero if $|l_2 - l_1| \leq l \leq l_2 + l_1$, where $l_{1,2}$ are the quantum numbers belonging to $\kappa_{1,2}$ according to (3.257). The main difficulty of the calculation is to ensure convergence of the multiple sum. The convergence properties of the doubly infinite sum over κ_1 and κ_2 are rather delicate which may lead to grossly incorrect results. The number of radial matrix elements that have to be retained, grows rapidly with increasing electron energies. This imposes a computer storage and precision limitation on the energy range that can be handled, the upper limit being of the order of some hundreds of keV for the triply differential cross section and a few MeV for the integrated cross section[†]. On the other hand, the radial integrals (3.296) and (3.297) require going out to a larger radius numerically as the energy decreases. Thus numerical partial-wave calculations are unavailable for initial electron energies lower than 1 keV. Likewise, the precision becomes unsatisfactory at low photon energies. This limits the calculations, as the soft photon end of the spectrum is approached, to cases where $h\nu \geq E_0/4$ [17].

Instead of directly calculating the absolute square of the matrix element

---

[†] In the partial-wave formalism the cross section of the elementary process of bremsstrahlung is most complicated whereas the cross section integrated over the angles of the scattered electron and the emitted photon is simpler because the orthogonality of the Legendre functions leaves only the $l = 0$ term. Besides, partial waves or multipoles in the sums over angular momentum states do not interfere, thus improving convergence. This is why computations of the doubly differential cross section and of the photon spectra [14; 66] were achieved much earlier than computations of the triply differential cross section.

another procedure is possible, namely to first evaluate the full matrix element. The terms of the matrix element are summed numerically and then squared the sum, thereby reducing the number of summations to three (a single summation over incident and outgoing electron partial waves and a single summation over photon angular momenta). However, this approach has the disadvantage that sums over quantities not observed (such as polarization) have to be performed numerically rather than analytically [15].

The triply differential cross section including all polarization correlations can be written as [19]

$$\frac{d^3\sigma^{pol}}{dk\,d\Omega_k\,d\Omega_{p_2}} = \frac{d^3\sigma^{unpol}}{dk\,d\Omega_k\,d\Omega_{p_2}} \left[\frac{1}{4} \sum_{l,m,n=0}^{3} C_{lmn}\zeta_l\xi_m\zeta_n \right]. \qquad (3.298)$$

Here $d^3\sigma^{unpol}/dk\,d\Omega_k\,d\Omega_{p_2} \equiv d^3\sigma$ is the triply differential cross section averaged over incident electron spin directions (unpolarized beam) and summed over final electron spins and photon polarizations; ζ_l and ζ_n characterize the polarization states of, respectively, the incident and final electron (with $\zeta_0 = 1$), and ξ_m the polarization state of the photon (with $\xi_0 = 1$). The real numbers C_{lmn}, functions of energies and angles, are the general polarization coefficients. They satisfy $C_{000} = 1$ and $|C_{lmn}| \leq 1$, since cross sections cannot be negative, and describe the polarization correlations between the initial and final electron and the emitted photon.

The photon polarization is described in terms of the three real Stokes parameters ξ_i satisfying $\sum \xi_i^2 = 1$. Photons linearly polarized parallel or perpendicular to the bremsstrahlung production plane $(\mathbf{p}_1, \mathbf{k})$ are characterized by a non-vanishing ξ_3, circularly polarized photons by non-vanishing ξ_2. Photons linearly polarized at $\pm 45°$ to the production plane imply a non-vanishing ξ_1 (see Problem 3.10). Similarly, the polarization state of the electrons is described in terms of the direction of the spins in their rest systems. For the incident electron, ζ_3 is taken along the direction $\hat{\mathbf{p}}_1$, ζ_1 in the production plane, and ζ_2 perpendicular to this plane so that $(\zeta_1, \zeta_2, \zeta_3)$ form a right-handed set.

While $d^3\sigma^{pol}/dk\,d\Omega_k\,d\Omega_{p_2}$, the most general observable, has not been measured so far, there exist experiments on the polarization coefficients C_{030} and C_{200}, as discussed in Sections 4.3.1 and 4.3.2. In most cases, however, only the unpolarized triply differential cross section $d^3\sigma$ is computed. The results can be used to check the accuracy of the simpler theories [16].

Whereas these comparisons show good agreement for low atomic numbers[‡] of the target atoms, there are large discrepancies at high atomic numbers $Z \geq 47$ and larger momentum transfers q (for the Elwert-Haug theory [13] the region of large momentum transfer may be roughly defined by $q > a$). However, the results obtained by using the SM wave functions perform fairly well for small momentum transfers, even for $Z = 79$, improving with increasing energies of the incident and outgoing electrons. In comparison with the limited experimental data the agreement is improved in some cases but not systematically (see Chapter 4).

Nonrelativistic calculations of the triply differential cross section in the partial-wave formalism were performed by Tseng [67]. Comparisons with experimental results at the initial electron energy 140 keV show large discrepancies which is not surprising since relativistic effects are significant at that energy. No experimental results are available for nonrelativistic energies.

Electron screening is usually taken into account by using the Thomas-Fermi potential, the Kohn-Sham potential [68], or the modified Hartree-Fock-Slater potential. In most works scattering from neutral atoms is considered although the theory may likewise be applied to ionic species (for ions the bremsstrahlung cross section depends on one more variable, the ionic charge Z_i). Differences between the various screening models amount to about 1%; it appears that the sensitivity to the employed potential increases at very low energies. Adding an atomic exchange potential, dependent on the kinetic energy of the projectile electron, to the model potential affects the cross section at 300 keV impact energy by up to 2% [17]. As one would expect the effect is strongest at the hard end of the photon spectrum where the outgoing electron is relatively slow. Incorporating details of atomic shell structure into the potential V yields larger differences with respect to the results of a reference standard potential, in particular for $h\nu < \frac{1}{2}E_0$. The discrepancies occur predominantly near the minima of the cross section. Effects of taking into account the finite nuclear size are marginal.

Screening effects are particularly important for target atoms with large atomic numbers Z and for high and low projectile energies. For a given

---

[‡]The deviations from Born-approximation results for $Z = 1$ at small and large photon angles suggest that a still higher number of partial waves is required to achieve agreement.

energy of the incident electron the effects are more significant near the long-wavelength end of the bremsstrahlung spectrum, as is expected from the semiclassical considerations of Sec. 3.3.1. Here the total cross section multiplied by the photon energy remains finite in contrast to the logarithmic divergence of the Coulomb case.

At the short-wavelength limit the screened cross section vanishes, unlike the finite point-Coulomb value, but rapidly rises in the first 5 to 50 eV of the spectrum [69].

3.5.3 Calculation for a pure Coulomb potential

The partial-wave method applied to the point-Coulomb potential is an important special case, since an exact solution of the Dirac equation is available. Hence part of the numerical work can be performed analytically. For $V(r) = -a/r$ the solution of the radial wave equation (3.254) has the form [64]

$$g_\kappa(r) = (2pr)^{\gamma-1} e^{\pi a_1/2} \frac{|\Gamma(\gamma + ia_1)|}{\Gamma(2\gamma + 1)} \left[\Phi(r) + \Phi^*(r)\right], \quad (3.299)$$

where

$$\Phi(r) = e^{-ipr+i\eta}(\gamma + ia_1) F(\gamma + 1 + ia_1; 2\gamma + 1; 2ipr), \quad (3.300)$$

$$\gamma = \sqrt{\kappa^2 - a^2}, \ a_1 = a\epsilon/p, \quad (3.301)$$

and the real phase η is defined by

$$e^{2i\eta} = -\frac{\kappa - ia_1/\epsilon}{\gamma + ia_1} \quad (3.302)$$

(see Problem 3.11). The phase η is real since $|\kappa - ia_1/\epsilon| = |\gamma + ia_1|$. The normalization of g_κ is such that, for large values of r,

$$g_\kappa \sim (1/pr) \sin\left(pr - \tfrac{1}{2}\pi l + \eta_\kappa\right), \quad (3.303)$$

where the phase shift η_κ depends on p and r (see Problem 3.12).

The minor component of the radial wave function pertinent to (3.299) is given by

$$f_\kappa(r) = i\sqrt{\frac{\epsilon - 1}{\epsilon + 1}} (2pr)^{\gamma-1} e^{\pi a_1/2} \frac{|\Gamma(\gamma + ia_1)|}{\Gamma(2\gamma + 1)} \left[\Phi(r) - \Phi^*(r)\right]. \quad (3.304)$$

The continuum-state solution of the Dirac equation for a Coulomb potential (3.25), as given by Darwin [50], is obtained if the radial wave function $g_\kappa(r)$ of Eq. (3.299) is inserted into the series expansion (3.293). There is no solution available in closed form. An alternative and more convenient form where Dirac matrices are used and the spin dependence is expressed by the plane wave spinor $u(\mathbf{p})$ was developed by Johnson and Deck [70]:

$$\psi_1(\mathbf{r}) = \left[N_1 + iaM_1\vec{\alpha}\cdot(\hat{\mathbf{p}}_1 - \hat{\mathbf{r}}) + L_1\vec{\alpha}\cdot(\hat{\mathbf{p}}_1 - \hat{\mathbf{r}})(\vec{\alpha}\cdot\hat{\mathbf{p}}_1)\right]u(\mathbf{p}_1), \quad (3.305)$$

where

$$N_1 = 2\sum_{n=1}^{\infty}(-1)^n n \frac{\Gamma(\gamma - ia_1)}{\Gamma(2\gamma + 1)} e^{\pi a_1/2}(-2ip_1 r)^{\gamma - 1} e^{ip_1 r}$$
$$\cdot F(\gamma - ia_1; 2\gamma + 1; -2ip_1 r)(P'_{n-1} - P'_n), \quad (3.306)$$

$$M_1 = \sum_{n=1}^{\infty}(-1)^n \frac{\Gamma(\gamma - ia_1)}{\Gamma(2\gamma + 1)} e^{\pi a_1/2}(-2ip_1 r)^{\gamma - 1} e^{ip_1 r}$$
$$\cdot F(\gamma - ia_1; 2\gamma + 1; -2ip_1 r)(P'_{n-1} + P'_n), \quad (3.307)$$

$$L_1 = \sum_{n=1}^{\infty}(-1)^n \frac{\Gamma(\gamma - ia_1)}{\Gamma(2\gamma + 1)} e^{\pi a_1/2}(-2ip_1 r)^{\gamma - 1} e^{ip_1 r}$$
$$\cdot \left\{[(\gamma - n)F(\gamma - ia_1; 2\gamma + 1; -2ip_1 r) - 2\gamma F(\gamma - ia_1; 2\gamma; -2ip_1 r)]P'_{n-1}\right.$$
$$\left.+ [(\gamma + n)F(\gamma - ia_1; 2\gamma + 1; -2ip_1 r) - 2\gamma F(\gamma - ia_1; 2\gamma; -2ip_1 r)]P'_n\right\}, \quad (3.308)$$

and

$$\gamma = \sqrt{n^2 - a^2}, \quad a_1 = (\epsilon_1/p_1)a. \quad (3.309)$$

P'_n denotes the derivation with respect to the argument of the Legendre function $P_n = P_n(\hat{\mathbf{p}}_1 \cdot \hat{\mathbf{r}})$ and F the confluent hypergeometric function.

Considered in terms of an expansion in the Coulomb parameter $a = \alpha Z$, the coefficients N_1 and M_1 are of order unity, while the coefficient L_1 is of order a^2. The latter can be demonstrated by setting $\gamma = n$; then $L_1 = 0$ (see Problem 3.14). It follows that the three terms on the right-hand side of (3.305) are, respectively, of order unity, a, and a^2.

The function $\psi_1(\mathbf{r})$ has the asymptotic form of a plane wave plus outgoing spherical wave. A function with the asymptotic character associated with a final electron state, i.e., plane wave plus ingoing spherical wave, can be derived from (3.305) via time reversal. Then the adjoint function $\psi_2^\dagger(\mathbf{r})$

can be written as

$$\psi_2^\dagger(\mathbf{r}) = u^\dagger(\mathbf{p}_2)\left[N_2 - iaM_2\vec{\alpha}\cdot(\hat{\mathbf{p}}_2+\hat{\mathbf{r}}) + L_2(\vec{\alpha}\cdot\hat{\mathbf{p}}_2)\,\vec{\alpha}\cdot(\hat{\mathbf{p}}_2+\hat{\mathbf{r}})\right], \quad (3.310)$$

where

$$N_2 = 2\sum_{n=1}^{\infty} n\,\frac{\Gamma(\gamma - ia_2)}{\Gamma(2\gamma + 1)}\,e^{\pi a_2/2}(-2ip_2r)^{\gamma-1}e^{ip_2r}$$
$$\cdot F(\gamma - ia_2; 2\gamma + 1; -2ip_2r)\,(P'_{n-1} + P'_n)\,, \qquad (3.311)$$

$$M_2 = \sum_{n=1}^{\infty} \frac{\Gamma(\gamma - ia_2)}{\Gamma(2\gamma + 1)}\,e^{\pi a_2/2}(-2ip_2r)^{\gamma-1}e^{ip_2r}$$
$$\cdot F(\gamma - ia_2; 2\gamma + 1; -2ip_2r)(P'_{n-1} - P'_n)\,, \qquad (3.312)$$

$$L_2 = \sum_{n=1}^{\infty} \frac{\Gamma(\gamma - ia_2)}{\Gamma(2\gamma + 1)}\,e^{\pi a_2/2}(-2ip_2r)^{\gamma-1}e^{ip_2r}$$
$$\cdot\Big\{\left[(\gamma - n)F(\gamma - ia_2; 2\gamma + 1; -2ip_2r) - 2\gamma F(\gamma - ia_2; 2\gamma; -2ip_2r)\right]P'_{n-1}$$
$$- \left[(\gamma + n)F(\gamma - ia_2; 2\gamma + 1; -2ip_2r) - 2\gamma F(\gamma - ia_2; 2\gamma; -2ip_2r)\right]P'_n\Big\}\,.$$
$$(3.313)$$

Here $P_n = P_n(\hat{\mathbf{p}}_2\cdot\hat{\mathbf{r}})$, $a_2 = (\epsilon_2/p_2)a$.

The Sommerfeld-Maue functions $\psi_1(\mathbf{r})$ and $\psi_2^\dagger(\mathbf{r})$, given in Eqs. (3.129) and (3.134), follow directly from (3.305) and (3.310) as a result of approximating γ by n in the latter functions (as stated above, the coefficients $L_{1,2}$ vanish in this approximation). The differences between the exact point-Coulomb wave function and the SM wave function for small values of n or $|\kappa|$ have been investigated by Fink and Pratt [60]. For high atomic numbers Z, the SM wave function differs from the exact wave function in phase as well as in magnitude. For $Z = 13$, there is better agreement for $n = 1$. As n increases, for a given Z and energy, agreement improves. The main error in the SM wave function for the small-r region ($r < 1$) is due to approximating γ by n in the multiplicative factor $(2pr)^\gamma$. As r increases, the error of setting $\gamma = n$ in the phase and in the hypergeometric functions becomes larger and thus comprises a larger percentage of the error in the wave function.

The wave functions $\psi_1(\mathbf{r})$ and $\psi_2^\dagger(\mathbf{r})$ are inserted in the bremsstrahlung matrix element (3.21). Besides, the photon interaction operator, $\vec{\alpha}\cdot\mathbf{e}^*\,e^{-i\mathbf{kr}}$, is expanded into the multipole series (3.295). Then some of the summations and the integrations can be performed in closed form yielding, however, very complicated expressions containing generalized hypergeometric functions (Appell functions) with 10 arguments which are difficult to evaluate numerically. The method benefits from the fact that the integrations over the angles of the outgoing electrons and photons can be carried out analytically using the orthonormality conditions for the states [71].

The cross sections calculated in this way are exact regarding the interaction with the pure Coulomb field and are of first order referring to the interaction with the radiation field. The radiative corrections to be expected are of relative order $\alpha \approx 1/137$ (see Sec. 3.8).

The results of these calculations can be contrasted with those obtained by means of the Born approximation to evaluate the Coulomb correction. In particular, one can explain why the cross section remains finite at the short-wavelength limit. For $p_2 \to 0$, the point-Coulomb radial wave functions $\bar{g}_\kappa$ and $\bar{f}_\kappa$ take the form (see Problem 3.13)

$$\bar{g}_\kappa = \sqrt{\frac{\pi}{2ap_2}} \left[\tfrac{1}{2} \rho \, J_{2\gamma-1}(\rho) - (\kappa + \gamma) J_{2\gamma}(\rho) \right], \tag{3.314}$$

$$\bar{f}_\kappa = -\sqrt{\frac{\pi a}{2p_2}} J_{2\gamma}(\rho), \tag{3.315}$$

where J_ν denotes the Bessel function of order ν and $\rho = 2\sqrt{2ar}$.

Hence in the expression $p_2 |M|^2$ the factor p_2 cancels. The divergence of the exact wave function for small values of p_2 is a consequence of the long-range nature of the point-Coulomb potential. If the potential is cut off at some distance the wave function will be finite as $p_2 \to 0$. Thus, for the more realistic case of a screened potential, the cross section will again vanish at the short-wavelength limit, as in the Born approximation. However, numerical computations indicate that the constant rather than $p_2^{-1/2}$ behavior of the wave function occurs at quite low kinetic energies of the outgoing electrons so that the cross section rises rapidly in the first 5 to 50 eV from the tip [69]. Since the energy resolution of x-ray detectors is worse, it is sufficient to use the point-Coulomb value for the cross section at the short-wavelength limit.

Physically, the finite bremsstrahlung cross section at the tip of the spectrum is related with the fact that the photon energy $h\nu = E_0$ is a limit only for the free-free transitions. If the electron is captured into a bound state it can emit a photon of energy $h\nu > E_0$. Highly excited bound (Rydberg) states are similar to free states at low kinetic energies. Hence there is a relationship§ between the high-frequency endpoint region of bremsstrahlung and recombination radiation [72] (which remains valid even when the lowest-order radiative corrections are included [73]): The matrix element for radiative recombination, involving a final negative binding energy state ($E_b < 0$) may be analytically continued to the corresponding bremsstrahlung matrix element involving a positive kinetic energy final state ($E_2 > 0$). However, in the case of a repulsing Coulomb field (e.g., in positron-nucleus scattering), where no bound states exist, the short-wavelength limit is a true limit and the cross section tends to zero exponentially, as will be shown in Section 7.2.

3.6 Spin-dependent cross section and bremsstrahlung asymmetry

The angular distribution of the bremsstrahlung photons shows an azimuthal asymmetry around the direction of the incident electrons if these are transversely polarized perpendicular to the production plane ($\mathbf{p}_1, \mathbf{k}$). This is a consequence of spin-orbit interaction in analogy to Mott scattering. Since calculations in first-order Born approximation give zero photon asymmetry more sophisticated theories are required. One possibility is the extension of the first Born approximation to one higher order in the Coulomb parameter $a = \alpha Z$, as was performed by Johnson and Rozics [74] and Sobolak and Stehle [29]. It is evident that this procedure yields a linear dependence on the atomic number Z of the resulting photon asymmetry A. Thus the values of $|A|$ may exceed 100% even for relatively low atomic numbers Z, and it can be concluded that the accuracy of the corresponding formula for A is sufficient only for very low values of Z. The exact numerical calculation in partial-wave formulation using screened potentials (Sec. 3.5) can be considered the best available method. However, in the case of the elementary process it requires a prohibitive number of sums to be computed [15; 18]. Therefore a practicable alternative is the calculation of the photon asymmetry by means of the Sommerfeld-Maue (SM) wave functions. It is true that the cross section calculated in this way is exact only to lowest or-

---

§The correspondence between bremsstrahlung and radiative recombination applies if the outgoing low-energy electron in the bremsstrahlung process is not observed since the elementary process of radiative recombination is described by the doubly differential cross section.

der in a. Moreover, the radiative corrections, which are of relative order α, are neglected as well. So it is not to be expected at first sight that it yields reliable values of the asymmetry. But it turns out that the results of the second Born approximation are reproduced fairly well for low atomic numbers, and what is more, the values of $|A|$ do not exceed 100%, as it should be [75]. The accuracy of the results can be checked by comparisons with the few available data of partial-wave calculations and with experimental determinations of the bremsstrahlung asymmetry (Sec. 4.3.2).

According to Eq. (3.150) the matrix element for bremsstrahlung in Sommerfeld-Maue approximation (neglecting the term including I_4) has the form

$$M = N_1 N_2^* \{ u^\dagger(\mathbf{p}_2) \, (\vec{\alpha} \cdot \mathbf{e}^*) \, u(\mathbf{p}_1) I_1 + u^\dagger(\mathbf{p}_2) \, (\vec{\alpha} \cdot \mathbf{e}^*)(\vec{\alpha} \cdot \mathbf{I}_2) \, u(\mathbf{p}_1)$$
$$+ u^\dagger(\mathbf{p}_2) \, (\vec{\alpha} \cdot \mathbf{I}_3)(\vec{\alpha} \cdot \mathbf{e}^*) \, u(\mathbf{p}_1) \}. \tag{3.316}$$

The electron polarization is most conveniently introduced by the split spinor representation. The procedure is simplified by the fact that the Sommerfeld-Maue wave function contains the plane-wave spinor $u(\mathbf{p})$. Using Eqs. (3.6), (3.143), and (3.144) the first term of the matrix element (3.316), expressed by the Pauli spinors $v_1 = v(\mathbf{p}_1)$ and $v_2 = v(\mathbf{p}_2)$, can be written as

$$N_1 N_2^* \, u^\dagger(\mathbf{p}_2) \, (\vec{\alpha} \cdot \mathbf{e}^*) \, u(\mathbf{p}_1) \, I_1$$

$$= N_1 N_2^* N_1' N_2' v_2^\dagger \left(1 + \frac{\vec{\sigma} \cdot \mathbf{p}_2}{\epsilon_2 + 1}\right) \begin{pmatrix} 0 & \vec{\sigma} \cdot \mathbf{e}^* \\ \vec{\sigma} \cdot \mathbf{e}^* & 0 \end{pmatrix} \begin{pmatrix} 1 \\ \frac{\vec{\sigma} \cdot \mathbf{p}_1}{\epsilon_1 + 1} \end{pmatrix} v_1 I_1$$

$$= N_1 N_2^* \sqrt{\frac{(\epsilon_1 + 1)(\epsilon_2 + 1)}{4\epsilon_1 \epsilon_2}}$$

$$\cdot \left(v_2^\dagger \left\{ \frac{(\vec{\sigma} \cdot \mathbf{p}_2)}{\epsilon_2 + 1} (\vec{\sigma} \cdot \mathbf{e}^*) + (\vec{\sigma} \cdot \mathbf{e}^*) \frac{(\vec{\sigma} \cdot \mathbf{p}_1)}{\epsilon_1 + 1} \right\} v_1 \right) I_1. \tag{3.317}$$

Correspondingly we get for the second and third term of M

$$N_1 N_2^* \, u^\dagger(\mathbf{p}_2) \, (\vec{\alpha} \cdot \mathbf{e}^*)(\vec{\alpha} \cdot \mathbf{I}_2) \, u(\mathbf{p}_1)$$
$$= N_1 N_2^* N_1' N_2' v_2^\dagger \left(1 + \frac{\vec{\sigma} \cdot \mathbf{p}_2}{\epsilon_2 + 1}\right) \begin{pmatrix} 0 & \vec{\sigma} \cdot \mathbf{e}^* \\ \vec{\sigma} \cdot \mathbf{e}^* & 0 \end{pmatrix}$$
$$\cdot \begin{pmatrix} 0 & \vec{\sigma} \cdot \mathbf{I}_2 \\ \vec{\sigma} \cdot \mathbf{I}_2 & 0 \end{pmatrix} \begin{pmatrix} 1 \\ \frac{\vec{\sigma} \cdot \mathbf{p}_1}{\epsilon_1 + 1} \end{pmatrix} v_1$$

$$= N_1 N_2^* \sqrt{\frac{(\epsilon_1+1)(\epsilon_2+1)}{4\epsilon_1\epsilon_2}} \left(v_2^\dagger \left\{ (\vec{\sigma}\cdot\mathbf{e}^*)(\vec{\sigma}\cdot\mathbf{I}_2) \right. \right.$$
$$\left. \left. + \frac{(\vec{\sigma}\cdot\mathbf{p}_2)(\vec{\sigma}\cdot\mathbf{e}^*)(\vec{\sigma}\cdot\mathbf{I}_2)(\vec{\sigma}\cdot\mathbf{p}_1)}{(\epsilon_1+1)(\epsilon_2+1)} \right\} v_1 \right) \quad (3.318)$$

and

$$N_1 N_2^* \, u^\dagger(\mathbf{p}_2)\,(\vec{\alpha}\cdot\mathbf{I}_3)(\vec{\alpha}\cdot\mathbf{e}^*)\, u(\mathbf{p}_1)$$
$$= N_1 N_2^* \sqrt{\frac{(\epsilon_1+1)(\epsilon_2+1)}{4\epsilon_1\epsilon_2}} \left(v_2^\dagger \left\{ (\vec{\sigma}\cdot\mathbf{I}_3)(\vec{\sigma}\cdot\mathbf{e}^*) \right. \right.$$
$$\left. \left. + \frac{(\vec{\sigma}\cdot\mathbf{p}_2)(\vec{\sigma}\cdot\mathbf{I}_3)(\vec{\sigma}\cdot\mathbf{e}^*)(\vec{\sigma}\cdot\mathbf{p}_1)}{(\epsilon_1+1)(\epsilon_2+1)} \right\} v_1 \right). \quad (3.319)$$

Together, this yields the matrix element

$$M = N_1 N_2^* \sqrt{\frac{(\epsilon_1+1)(\epsilon_2+1)}{4\epsilon_1\epsilon_2}} \left(v_2^\dagger \left\{ \left[\frac{(\vec{\sigma}\cdot\mathbf{p}_2)(\vec{\sigma}\cdot\mathbf{e}^*)}{\epsilon_2+1} + \frac{(\vec{\sigma}\cdot\mathbf{e}^*)(\vec{\sigma}\cdot\mathbf{p}_1)}{\epsilon_1+1} \right] I_1 \right. \right.$$
$$+ (\vec{\sigma}\cdot\mathbf{e}^*)(\vec{\sigma}\cdot\mathbf{I}_2) + \frac{(\vec{\sigma}\cdot\mathbf{p}_2)(\vec{\sigma}\cdot\mathbf{e}^*)(\vec{\sigma}\cdot\mathbf{I}_2)(\vec{\sigma}\cdot\mathbf{p}_1)}{(\epsilon_1+1)(\epsilon_2+1)} + (\vec{\sigma}\cdot\mathbf{I}_3)(\vec{\sigma}\cdot\mathbf{e}^*)$$
$$\left. \left. + \frac{(\vec{\sigma}\cdot\mathbf{p}_2)(\vec{\sigma}\cdot\mathbf{I}_3)(\vec{\sigma}\cdot\mathbf{e}^*)(\vec{\sigma}\cdot\mathbf{p}_1)}{(\epsilon_1+1)(\epsilon_2+1)} \right\} v_1 \right) \quad (3.320)$$

Due to the very useful relation (3.243) no higher power of the Pauli spin matrices other than the first need to occur in the formalism. Applying (3.243) repeatedly the matrix element takes the form

$$M = N_1 N_2^* \sqrt{\frac{(\epsilon_1+1)(\epsilon_2+1)}{4\epsilon_1\epsilon_2}} \left(v_2^\dagger \left\{ \left(\frac{\mathbf{p}_1\cdot\mathbf{e}^*}{\epsilon_1+1} + \frac{\mathbf{p}_2\cdot\mathbf{e}^*}{\epsilon_2+1} \right) I_1 + (\mathbf{I}_2+\mathbf{I}_3)\cdot\mathbf{e}^* \right. \right.$$
$$+ \frac{(\mathbf{e}^*\cdot\mathbf{p}_1)[(\mathbf{I}_3-\mathbf{I}_2)\cdot\mathbf{p}_2]}{(\epsilon_1+1)(\epsilon_2+1)} + \frac{(\mathbf{e}^*\cdot\mathbf{p}_2)[(\mathbf{I}_2-\mathbf{I}_3)\cdot\mathbf{p}_1]}{(\epsilon_1+1)(\epsilon_2+1)}$$
$$+ \frac{(\mathbf{p}_1\cdot\mathbf{p}_2)[\mathbf{e}^*\cdot(\mathbf{I}_2+\mathbf{I}_3)]}{(\epsilon_1+1)(\epsilon_2+1)} + \frac{i}{(\epsilon_1+1)(\epsilon_2+1)}\,\vec{\sigma}\cdot\Big[\{(\epsilon_1+1)(\mathbf{p}_2\times\mathbf{e}^*)$$
$$- (\epsilon_2+1)(\mathbf{p}_1\times\mathbf{e}^*)\}I_1 + (\epsilon_1+1)(\epsilon_2+1)\{\mathbf{e}^*\times(\mathbf{I}_2-\mathbf{I}_3)\}$$
$$+ (\mathbf{p}_1\cdot\mathbf{I}_2)(\mathbf{p}_2\times\mathbf{e}^*) - (\mathbf{p}_2\cdot\mathbf{I}_3)(\mathbf{p}_1\times\mathbf{e}^*) + (\mathbf{e}^*\cdot\mathbf{p}_1)(\mathbf{p}_2\times\mathbf{I}_3)$$
$$- (\mathbf{e}^*\cdot\mathbf{p}_2)(\mathbf{p}_1\times\mathbf{I}_2) - \{(\mathbf{p}_1\times\mathbf{p}_2)\cdot(\mathbf{I}_2+\mathbf{I}_3)\}\mathbf{e}^*$$
$$\left. \left. + \{(\mathbf{I}_2\times\mathbf{p}_1)\cdot\mathbf{e}^*\}\mathbf{p}_2 - \{(\mathbf{I}_3\times\mathbf{p}_2)\cdot\mathbf{e}^*\}\mathbf{p}_1 \Big] \right\} v_1 \right). \quad (3.321)$$

By means of the notations

$$\mathbf{I} = \left(\frac{\mathbf{p}_1}{\epsilon_1+1} + \frac{\mathbf{p}_2}{\epsilon_2+1}\right)I_1 + \mathbf{I}_2 + \mathbf{I}_3 + \frac{1}{(\epsilon_1+1)(\epsilon_2+1)}\{[(\mathbf{I}_3-\mathbf{I}_2)\cdot\mathbf{p}_2]\mathbf{p}_1$$
$$+ [(\mathbf{I}_2-\mathbf{I}_3)\cdot\mathbf{p}_1]\mathbf{p}_2 + (\mathbf{p}_1\cdot\mathbf{p}_2)(\mathbf{I}_2+\mathbf{I}_3)\}, \qquad (3.322)$$

$$\mathbf{J} = \left(\frac{\mathbf{p}_1}{\epsilon_1+1} - \frac{\mathbf{p}_2}{\epsilon_2+1}\right)I_1 + \mathbf{I}_2 - \mathbf{I}_3 + \frac{(\mathbf{I}_3\cdot\mathbf{p}_2)\mathbf{p}_1 - (\mathbf{I}_2\cdot\mathbf{p}_1)\mathbf{p}_2}{(\epsilon_1+1)(\epsilon_2+1)}, \qquad (3.323)$$

$$\mathbf{T} = \frac{\mathbf{p}_2\times\mathbf{I}_3}{(\epsilon_1+1)(\epsilon_2+1)}, \quad \mathbf{U} = \frac{\mathbf{I}_2\times\mathbf{p}_1}{(\epsilon_1+1)(\epsilon_2+1)}, \qquad (3.324)$$

$$R = \frac{(\mathbf{I}_2+\mathbf{I}_3)\cdot(\mathbf{p}_1\times\mathbf{p}_2)}{(\epsilon_1+1)(\epsilon_2+1)} = \mathbf{T}\cdot\mathbf{p}_1 + \mathbf{U}\cdot\mathbf{p}_2, \qquad (3.325)$$

the matrix element can be written as

$$M = N_1 N_2^* [(\epsilon_1+1)(\epsilon_2+1)/(4\epsilon_1\epsilon_2)]^{1/2} \Big(v_2^\dagger \{(\mathbf{e}^*\cdot\mathbf{I}) + i\vec{\sigma}\cdot(\mathbf{e}^*\times\mathbf{J})$$
$$- i(\vec{\sigma}\cdot\mathbf{e}^*)R + i(\vec{\sigma}\cdot\mathbf{p}_1)(\mathbf{e}^*\cdot\mathbf{T}) + i(\vec{\sigma}\cdot\mathbf{p}_2)(\mathbf{e}^*\cdot\mathbf{U})$$
$$+ i(\mathbf{e}^*\cdot\mathbf{p}_1)(\vec{\sigma}\cdot\mathbf{T}) + i(\mathbf{e}^*\cdot\mathbf{p}_2)(\vec{\sigma}\cdot\mathbf{U})\} v_1\Big). \qquad (3.326)$$

Combining all the terms multiplied by $\vec{\sigma}$ to the vector

$$\mathbf{X} = \mathbf{e}^*\times\mathbf{J} - R\mathbf{e}^* + (\mathbf{e}^*\cdot\mathbf{T})\mathbf{p}_1 + (\mathbf{e}^*\cdot\mathbf{U})\mathbf{p}_2$$
$$+ (\mathbf{e}^*\cdot\mathbf{p}_1)\mathbf{T} + (\mathbf{e}^*\cdot\mathbf{p}_2)\mathbf{U}, \qquad (3.327)$$

we arrive at

$$M = N_1 N_2^* [(\epsilon_1+1)(\epsilon_2+1)/(4\epsilon_1\epsilon_2)]^{1/2} \big(v_2^\dagger \{\mathbf{e}^*\cdot\mathbf{I} + i\vec{\sigma}\cdot\mathbf{X}\} v_1\big). \qquad (3.328)$$

This formula determines completely the cross section with all possible polarization effects.

The cross section for arbitrary directions of the electron spins, $\vec{\zeta}_1$ and $\vec{\zeta}_2$, is proportional to the absolute square of the matrix element. Usually the final electron spins are not observed. The summation over $\vec{\zeta}_2$ can be easily performed using

$$\sum_{\vec{\zeta}_2} v_2 v_2^\dagger = 1 \qquad (3.329)$$

which follows from (3.148). The remaining spinors v_1 can be eliminated with the aid of the normalization $v^\dagger v = 1$ and the diagonal matrix element

(3.149), whereby $|M|^2$ is expressed in terms of the spin vector $\vec{\zeta}_1$. Applying again (3.243), we get immediately

$$\sum_{\vec{\zeta}_2} |M|^2 = |N_1 N_2|^2 \frac{(\epsilon_1 + 1)(\epsilon_2 + 1)}{4\epsilon_1 \epsilon_2} \left(v_1^\dagger \{ |\mathbf{e}^* \cdot \mathbf{I}|^2 + 2\,\mathrm{Im}[(\mathbf{e}^* \cdot \mathbf{I})(\vec{\sigma} \cdot \mathbf{X}^*)] \right.$$

$$\left. + |\mathbf{X}|^2 + i\vec{\sigma} \cdot (\mathbf{X}^* \times \mathbf{X}) \} v_1 \right)$$

$$= |N_1 N_2|^2 \frac{(\epsilon_1 + 1)(\epsilon_2 + 1)}{4\epsilon_1 \epsilon_2} \{ |\mathbf{e}^* \cdot \mathbf{I}|^2 + |\mathbf{X}|^2$$

$$+ [2\,\mathrm{Im}\{(\mathbf{e}^* \cdot \mathbf{I})\mathbf{X}^*\} + i(\mathbf{X}^* \times \mathbf{X})] \cdot \vec{\zeta}_1 \}. \quad (3.330)$$

Inserting the expression (3.327) for $\mathbf{X}$ and noting that, by the definitions (3.324), $\mathbf{p}_1 \cdot \mathbf{U} = \mathbf{p}_2 \cdot \mathbf{T} = 0$, the matrix element squared takes the form

$$\sum_{\vec{\zeta}_2} |M|^2 = |N_1 N_2|^2 \frac{(\epsilon_1 + 1)(\epsilon_2 + 1)}{4\epsilon_1 \epsilon_2} \{ |\mathbf{e}^* \cdot \mathbf{I}|^2 + |\mathbf{e}^* \times \mathbf{J}|^2 + |R|^2$$

$$+ p_1^2 |\mathbf{e}^* \cdot \mathbf{T}|^2 + p_2^2 |\mathbf{e}^* \cdot \mathbf{U}|^2 + |\mathbf{e} \cdot \mathbf{p}_1|^2 |\mathbf{T}|^2 + |\mathbf{e} \cdot \mathbf{p}_2|^2 |\mathbf{U}|^2$$

$$+ 2\,\mathrm{Re}[\{ (\mathbf{e} \times \mathbf{e}^*) \cdot \mathbf{J}^* - (\mathbf{e} \cdot \mathbf{p}_1)(\mathbf{e}^* \cdot \mathbf{T}^*)$$

$$- (\mathbf{e} \cdot \mathbf{p}_2)(\mathbf{e}^* \cdot \mathbf{U}^*) \} R + (\mathbf{e} \times \mathbf{J}^*) \cdot \{ (\mathbf{e}^* \cdot \mathbf{T}) \mathbf{p}_1$$

$$+ (\mathbf{e}^* \cdot \mathbf{U}) \mathbf{p}_2 + (\mathbf{e}^* \cdot \mathbf{p}_1)\mathbf{T} + (\mathbf{e}^* \cdot \mathbf{p}_2)\mathbf{U} \}$$

$$+ (\mathbf{p}_1 \cdot \mathbf{p}_2)(\mathbf{e} \cdot \mathbf{T}^*)(\mathbf{e}^* \cdot \mathbf{U}) - (\mathbf{e}^* \cdot \mathbf{p}_1)(\mathbf{e} \cdot \mathbf{T}^*)(\mathbf{U} \cdot \mathbf{p}_2)$$

$$- (\mathbf{e}^* \cdot \mathbf{p}_2)(\mathbf{e} \cdot \mathbf{U}^*)(\mathbf{T} \cdot \mathbf{p}_1) + (\mathbf{e} \cdot \mathbf{p}_1)(\mathbf{e}^* \cdot \mathbf{p}_2)(\mathbf{T}^* \cdot \mathbf{U})]$$

$$+ \vec{\zeta}_1 \cdot \mathbf{Q}_1 \} \quad (3.331)$$

with

$$\mathbf{Q}_1 = i\{ [(\mathbf{J}^* \times \mathbf{J}) \cdot \mathbf{e}] \mathbf{e}^* + [(\mathbf{e} \times \mathbf{e}^*) \cdot \mathbf{J}^*] \mathbf{J} + (\mathbf{e} \times \mathbf{e}^*)|R|^2$$

$$+ |\mathbf{e} \cdot \mathbf{p}_1|^2 (\mathbf{T}^* \times \mathbf{T}) + |\mathbf{e} \cdot \mathbf{p}_2|^2 (\mathbf{U}^* \times \mathbf{U}) \}$$

$$+ 2\,\mathrm{Im}\{ (\mathbf{e}^* \cdot \mathbf{I})[\mathbf{e} \times \mathbf{J}^* + (\mathbf{e} \cdot \mathbf{T}^*)\mathbf{p}_1 + (\mathbf{e} \cdot \mathbf{U}^*)\mathbf{p}_2 + (\mathbf{e} \cdot \mathbf{p}_1)\mathbf{T}^*$$

$$+ (\mathbf{e} \cdot \mathbf{p}_2)\mathbf{U}^*] + (\mathbf{e} \cdot \mathbf{T}^*)[(\mathbf{e}^* \cdot \mathbf{p}_1)\mathbf{J} - (\mathbf{p}_1 \cdot \mathbf{J})\mathbf{e}^*$$

$$+ (\mathbf{e}^* \cdot \mathbf{p}_1)(\mathbf{T} \times \mathbf{p}_1) + (\mathbf{e}^* \cdot \mathbf{p}_2)(\mathbf{U} \times \mathbf{p}_1) - (\mathbf{e}^* \cdot \mathbf{U})(\mathbf{p}_1 \times \mathbf{p}_2)]$$

$$+ (\mathbf{e} \cdot \mathbf{U}^*)[(\mathbf{e}^* \cdot \mathbf{p}_2)\mathbf{J} - (\mathbf{p}_2 \cdot \mathbf{J})\mathbf{e}^* + (\mathbf{e}^* \cdot \mathbf{p}_1)(\mathbf{T} \times \mathbf{p}_2)$$

$$+ (\mathbf{e}^* \cdot \mathbf{p}_2)(\mathbf{U} \times \mathbf{p}_2)] + (\mathbf{e}^* \cdot \mathbf{p}_1)[(\mathbf{T} \cdot \mathbf{J}^*)\mathbf{e} - (\mathbf{e} \cdot \mathbf{T})\mathbf{J}^*]$$

$$+ (\mathbf{e}^* \cdot \mathbf{p}_2)[(\mathbf{U} \cdot \mathbf{J}^*)\mathbf{e} - (\mathbf{e} \cdot \mathbf{U})\mathbf{J}^*] + (\mathbf{e} \cdot \mathbf{p}_1)(\mathbf{e}^* \cdot \mathbf{p}_2)(\mathbf{U} \times \mathbf{T}^*)$$

$$- R^*[(\mathbf{e}^* \cdot \mathbf{I})\mathbf{e} + \mathbf{J} - (\mathbf{e} \cdot \mathbf{J})\mathbf{e}^* + (\mathbf{e}^* \cdot \mathbf{T})(\mathbf{p}_1 \times \mathbf{e})$$

$$+ (\mathbf{e}^* \cdot \mathbf{U})(\mathbf{p}_2 \times \mathbf{e}) + (\mathbf{e}^* \cdot \mathbf{p}_1)(\mathbf{T} \times \mathbf{e}) + (\mathbf{e}^* \cdot \mathbf{p}_2)(\mathbf{U} \times \mathbf{e})]\} \, .$$
(3.332)

This formula gives the correlation between the spin of the incoming electron and the photon polarization. The structure of the term $\mathbf{Q}_1$, representing the spin-dependent part of the cross section, gives information about an important feature of the polarization correlation. In order that $\mathbf{Q}_1$ be different from zero, a non-vanishing imaginary part of either $\mathbf{e}$ or of the quantities $\mathbf{I}, \mathbf{J}, R, \mathbf{T}, \mathbf{U}$ has to exist.

In Born approximation the integrals $\mathbf{I}_1, \mathbf{I}_2$, and $\mathbf{I}_3$ are given by Eqs. (3.177) to (3.179). Hence, apart from the common factor K_1, the quantities $\mathbf{I}, \mathbf{J}, R, \mathbf{T}$, and $\mathbf{U}$ are real so that $\mathbf{Q}_1$ vanishes for real $\mathbf{e}$. Therefore the linear polarization is no longer dependent on the spin state $\vec{\zeta}_1$ of the initial electron, and the linear polarization in Born approximation can be calculated most simply from Eq. (3.76) as will be done in Section 3.7.2. On the other hand, the spin-independent part of (3.331) does not depend on the circular polarization. Hence in Born approximation the bremsstrahlung is circularly polarized only when the incoming electrons are spin-polarized (see Sec. 3.7.1), unless the polarization of the outgoing electron is recorded in coincidence.

Finally, the summation over the photon polarization vectors $\mathbf{e}$ is carried out by means of the relation (3.79) yielding

$$\sum_{\vec{\zeta}_2, \mathbf{e}} |M|^2 = |N_1 N_2|^2 \frac{(\epsilon_1 + 1)(\epsilon_2 + 1)}{4\epsilon_1 \epsilon_2} \Big\{ |\mathbf{I} \times \hat{\mathbf{k}}|^2 + |\mathbf{J}|^2 + |\mathbf{J} \cdot \hat{\mathbf{k}}|^2$$
$$+ p_1^2 |\mathbf{T} \times \hat{\mathbf{k}}|^2 + p_2^2 |\mathbf{U} \times \hat{\mathbf{k}}|^2 + (\mathbf{p}_1 \times \hat{\mathbf{k}})^2 |\mathbf{T}|^2 + (\mathbf{p}_2 \times \hat{\mathbf{k}})^2 |\mathbf{U}|^2$$
$$+ 2\,\mathrm{Re}\big[\{(\mathbf{p}_1 \cdot \hat{\mathbf{k}})(\mathbf{T}^* \cdot \hat{\mathbf{k}}) + (\mathbf{p}_2 \cdot \hat{\mathbf{k}})(\mathbf{U}^* \cdot \hat{\mathbf{k}})\} R$$
$$- (\mathbf{T} \cdot \hat{\mathbf{k}})(\mathbf{J}^* \times \mathbf{p}_1) \cdot \hat{\mathbf{k}} - (\mathbf{U} \cdot \hat{\mathbf{k}})(\mathbf{J}^* \times \mathbf{p}_2) \cdot \hat{\mathbf{k}}$$
$$- (\mathbf{p}_1 \cdot \hat{\mathbf{k}})(\mathbf{J}^* \times \mathbf{T}) \cdot \hat{\mathbf{k}} - (\mathbf{p}_2 \cdot \hat{\mathbf{k}})(\mathbf{J}^* \times \mathbf{U}) \cdot \hat{\mathbf{k}}$$
$$+ 2(\mathbf{p}_1 \cdot \mathbf{p}_2)(\mathbf{T}^* \cdot \mathbf{U}) - 2(\mathbf{T}^* \cdot \mathbf{p}_1)(\mathbf{U} \cdot \mathbf{p}_2)$$
$$- (\mathbf{p}_1 \cdot \mathbf{p}_2)(\mathbf{T}^* \cdot \hat{\mathbf{k}})(\mathbf{U} \cdot \hat{\mathbf{k}}) + (\mathbf{p}_1 \cdot \hat{\mathbf{k}})(\mathbf{T}^* \cdot \hat{\mathbf{k}})(\mathbf{U} \cdot \mathbf{p}_2)$$
$$+ (\mathbf{p}_2 \cdot \hat{\mathbf{k}})(\mathbf{U}^* \cdot \hat{\mathbf{k}})(\mathbf{T} \cdot \mathbf{p}_1) - (\mathbf{p}_1 \cdot \hat{\mathbf{k}})(\mathbf{p}_2 \cdot \hat{\mathbf{k}})(\mathbf{T}^* \cdot \mathbf{U})\big]$$
$$+ \vec{\zeta}_1 \cdot \mathbf{Q}_2 \Big\} \, , \qquad (3.333)$$

where

$$\begin{aligned}
\mathbf{Q}_2 = i\{&\mathbf{J}^* \times \mathbf{J} - [(\mathbf{J}^* \times \mathbf{J}) \cdot \hat{\mathbf{k}}]\hat{\mathbf{k}} + (\mathbf{p}_1 \times \hat{\mathbf{k}})^2(\mathbf{T}^* \times \mathbf{T}) \\
&+ (\mathbf{p}_2 \times \hat{\mathbf{k}})^2(\mathbf{U}^* \times \mathbf{U})\} + 2\operatorname{Im}\{R^*[\mathbf{J} - (\mathbf{J} \cdot \hat{\mathbf{k}})\hat{\mathbf{k}} - \mathbf{I} + (\mathbf{I} \cdot \hat{\mathbf{k}})\hat{\mathbf{k}} \\
&+ (\mathbf{T} \cdot \hat{\mathbf{k}})(\mathbf{p}_1 \times \hat{\mathbf{k}}) + (\mathbf{U} \cdot \hat{\mathbf{k}})(\mathbf{p}_2 \times \hat{\mathbf{k}}) + (\mathbf{p}_1 \cdot \hat{\mathbf{k}})(\mathbf{T} \times \hat{\mathbf{k}}) \\
&+ (\mathbf{p}_2 \cdot \hat{\mathbf{k}})(\mathbf{U} \times \hat{\mathbf{k}})] + \mathbf{I} \times \mathbf{J}^* + (\mathbf{I} \cdot \hat{\mathbf{k}})(\mathbf{J}^* \times \hat{\mathbf{k}}) \\
&+ \mathbf{p}_1[(\mathbf{I} - \mathbf{J}) \cdot \mathbf{T}^* - (\mathbf{I} \cdot \hat{\mathbf{k}})(\mathbf{T}^* \cdot \hat{\mathbf{k}})] + \mathbf{p}_2[(\mathbf{I} - \mathbf{J}) \cdot \mathbf{U}^* \\
&- (\mathbf{I} \cdot \hat{\mathbf{k}})(\mathbf{U}^* \cdot \hat{\mathbf{k}})] + (\mathbf{p}_1 \times \mathbf{p}_2)[(\mathbf{T} \cdot \mathbf{U}^*) - (\mathbf{T} \cdot \hat{\mathbf{k}})(\mathbf{U}^* \cdot \hat{\mathbf{k}})] \\
&+ \hat{\mathbf{k}}[(\mathbf{p}_1 \cdot \mathbf{J})(\mathbf{T}^* \cdot \hat{\mathbf{k}}) + (\mathbf{p}_2 \cdot \mathbf{J})(\mathbf{U}^* \cdot \hat{\mathbf{k}}) + (\mathbf{p}_1 \cdot \hat{\mathbf{k}})(\mathbf{T}^* \cdot \mathbf{J}) \\
&+ (\mathbf{p}_2 \cdot \hat{\mathbf{k}})(\mathbf{U}^* \cdot \mathbf{J})] + \mathbf{T}^*[(\mathbf{I} - \mathbf{J}) \cdot \mathbf{p}_1 - (\mathbf{I} \cdot \hat{\mathbf{k}})(\mathbf{p}_1 \cdot \hat{\mathbf{k}})] \\
&+ \mathbf{U}^*[(\mathbf{I} - \mathbf{J}) \cdot \mathbf{p}_2 - (\mathbf{I} \cdot \hat{\mathbf{k}})(\mathbf{p}_2 \cdot \hat{\mathbf{k}})] \\
&+ (\mathbf{p}_1 \cdot \hat{\mathbf{k}})[(\mathbf{T}^* \cdot \hat{\mathbf{k}})(\mathbf{p}_1 \times \mathbf{T}) + (\mathbf{U}^* \cdot \hat{\mathbf{k}})(\mathbf{p}_2 \times \mathbf{T}) - 2(\mathbf{T}^* \cdot \hat{\mathbf{k}})\mathbf{J}] \\
&+ (\mathbf{p}_2 \cdot \hat{\mathbf{k}})[(\mathbf{T}^* \cdot \hat{\mathbf{k}})(\mathbf{p}_1 \times \mathbf{U}) + (\mathbf{U}^* \cdot \hat{\mathbf{k}})(\mathbf{p}_2 \times \mathbf{U}) - 2(\mathbf{U}^* \cdot \hat{\mathbf{k}})\mathbf{J}] \\
&+ (\mathbf{T}^* \cdot \mathbf{p}_1)(\mathbf{T} \times \mathbf{p}_1) + (\mathbf{U}^* \cdot \mathbf{p}_2)(\mathbf{U} \times \mathbf{p}_2) \\
&+ [\mathbf{p}_1 \cdot \mathbf{p}_2 - (\mathbf{p}_1 \cdot \hat{\mathbf{k}})(\mathbf{p}_2 \cdot \hat{\mathbf{k}})](\mathbf{T} \times \mathbf{U}^*)\} \quad (3.334)
\end{aligned}$$

is the spin-dependent part of the squared matrix element. In experimental investigations of the photon asymmetry the incoming electrons are usually polarized perpendicular to the bremsstrahlung production plane, i.e.

$$\vec{\zeta}_1 = \frac{\mathbf{k} \times \mathbf{p}_1}{|\mathbf{k} \times \mathbf{p}_1|} \,. \quad (3.335)$$

Then a number of terms in the expression (3.334) for $\mathbf{Q}_2$, that is, those proportional to $\mathbf{k}$ and $\mathbf{p}_1$, vanish identically.

The triply differential cross section is given by Eq. (3.24),

$$\begin{aligned}
\frac{d^3\sigma(\mathbf{p}_1, \mathbf{p}_2, \mathbf{k}, \vec{\zeta}_1)}{dk\, d\Omega_k\, d\Omega_{p_2}} &= \frac{\alpha}{(2\pi)^4} \frac{\epsilon_1 \epsilon_2 p_2 k}{p_1} \left(\frac{\hbar}{mc}\right)^2 \sum_{\vec{\zeta}_2, \mathbf{e}} |M|^2 \\
&= \tfrac{1}{2}\{C(\mathbf{p}_1, \mathbf{p}_2, \mathbf{k}) + \vec{\zeta}_1 \cdot \mathbf{Q}(\mathbf{p}_1, \mathbf{p}_2, \mathbf{k})\}\,. \quad (3.336)
\end{aligned}$$

If this cross section is summed over two opposite spin directions $\vec{\zeta}_1$ of the incoming electrons, the second term containing $\mathbf{Q}$ cancels so that $C(\mathbf{p}_1, \mathbf{p}_2, \mathbf{k})$ is equal to the cross section (3.184).

For incident electrons polarized perpendicularly to the plane of emission [Eq. (3.335)], the photon asymmetry, according to the notation (3.298), is given by the polarization coefficient C_{200}. Since in Born approximation C_{200} vanishes, the entire polarization effect reflects the deviation from Born approximation.

Generally, the photon asymmetry is defined as the ratio of the spin-dependent part of the cross section to the spin-independent part,

$$A(\mathbf{p}_1, \mathbf{p}_2, \mathbf{k}, \vec{\zeta}_1) = \frac{\mathbf{Q} \cdot \vec{\zeta}_1}{C} \,. \tag{3.337}$$

Since the cross section has to be a non-negative quantity, $|\mathbf{Q}| \leq C$, i.e., $|A|$ cannot be larger than unity.

Within the scope of the theory using SM wave functions, atomic-electron screening can only be taken into account by means of a form-factor approach which requires the validity of the Born approximation. However, investigations of polarization correlations with the partial-wave method [18] show that the effect of screening on the photon asymmetry is small even for relatively high atomic numbers, justifying neglect of screening in the SM wave function used. In the case of form-factor screening this result is easily understood since both numerator and denominator of (3.337) are multiplied by the same factor.

3.7 Bremsstrahlung polarization

3.7.1 General polarization correlation

We calculate the bremsstrahlung polarization for arbitrary spin polarization of the incoming electrons by means of the Sommerfeld-Maue wave functions. In order to include the spin state of the initial electrons in the matrix element we employ the split spinor representation of Sec. 3.4.2. Since there is not a single experiment where the polarization of the final electron is detected we shall start from the squared matrix element (3.331) which is summed over the spin states of the outgoing electron. To demonstrate the polarization properties of the radiation we split the expressions (3.331) and (3.332) into two parts according to the identity

$$2(\mathbf{a} \cdot \mathbf{e})(\mathbf{b} \cdot \mathbf{e}^*) = \{(\mathbf{a} \cdot \mathbf{e})(\mathbf{b} \cdot \mathbf{e}^*) + (\mathbf{a} \cdot \mathbf{e}^*)(\mathbf{b} \cdot \mathbf{e})\} + (\mathbf{a} \times \mathbf{b}) \cdot (\mathbf{e} \times \mathbf{e}^*) \,. \tag{3.338}$$

The first part in curly brackets is invariant under the operation $\mathbf{e} \leftrightarrow \mathbf{e}^*$ and describes the linear polarization. The last term changes sign under the operation $\mathbf{e} \leftrightarrow \mathbf{e}^*$ and describes the circular polarization. Applying (3.338) to the expressions (3.331) and (3.332) we get

$$\sum_{\vec{\zeta_2}} |M|^2 = |N_1 N_2|^2 \frac{(\epsilon_1 + 1)(\epsilon_2 + 1)}{8\epsilon_1 \epsilon_2} \{H_l + \vec{\zeta_1} \cdot \mathbf{G}_l + [\mathbf{H}_c + \mathbf{G}_c(\vec{\zeta_1})] \, i(\mathbf{e} \times \mathbf{e}^*)\}$$

(3.339)

with the notations

$$\begin{aligned}
H_l &= |\mathbf{e} \cdot \mathbf{I}|^2 + |\mathbf{e}^* \cdot \mathbf{I}|^2 + 2|\mathbf{J}|^2 - |\mathbf{e} \cdot \mathbf{J}|^2 - |\mathbf{e}^* \cdot \mathbf{J}|^2 + 2|R|^2 \\
&\quad + p_1^2(|\mathbf{e} \cdot \mathbf{T}|^2 + |\mathbf{e}^* \cdot \mathbf{T}|^2) + p_2^2(|\mathbf{e} \cdot \mathbf{U}|^2 + |\mathbf{e}^* \cdot \mathbf{U}|^2) \\
&\quad + 2|\mathbf{e} \cdot \mathbf{p}_1|^2 |\mathbf{T}|^2 + 2|\mathbf{e} \cdot \mathbf{p}_2|^2 |\mathbf{U}|^2 - 2\operatorname{Re}\{[(\mathbf{e} \cdot \mathbf{p}_1)(\mathbf{e}^* \cdot \mathbf{T}) \\
&\quad + (\mathbf{e}^* \cdot \mathbf{p}_1)(\mathbf{e} \cdot \mathbf{T}) + (\mathbf{e} \cdot \mathbf{p}_2)(\mathbf{e}^* \cdot \mathbf{U}) + (\mathbf{e}^* \cdot \mathbf{p}_2)(\mathbf{e} \cdot \mathbf{U})] R^*\} \\
&\quad + 2\operatorname{Re}\{(\mathbf{T} \cdot \mathbf{e})(\mathbf{J}^* \times \mathbf{p}_1) \cdot \mathbf{e}^* + (\mathbf{T} \cdot \mathbf{e}^*)(\mathbf{J}^* \times \mathbf{p}_1) \cdot \mathbf{e} \\
&\quad + (\mathbf{U} \cdot \mathbf{e})(\mathbf{J}^* \times \mathbf{p}_2) \cdot \mathbf{e}^* + (\mathbf{U} \cdot \mathbf{e}^*)(\mathbf{J}^* \times \mathbf{p}_2) \cdot \mathbf{e} \\
&\quad + (\mathbf{p}_1 \cdot \mathbf{e})(\mathbf{J}^* \times \mathbf{T}) \cdot \mathbf{e}^* + (\mathbf{p}_1 \cdot \mathbf{e}^*)(\mathbf{J}^* \times \mathbf{T}) \cdot \mathbf{e} \\
&\quad + (\mathbf{p}_2 \cdot \mathbf{e})(\mathbf{J}^* \times \mathbf{U}) \cdot \mathbf{e}^* + (\mathbf{p}_2 \cdot \mathbf{e}^*)(\mathbf{J}^* \times \mathbf{U}) \cdot \mathbf{e} \\
&\quad + (\mathbf{p}_1 \cdot \mathbf{p}_2)[(\mathbf{T}^* \cdot \mathbf{e})(\mathbf{U} \cdot \mathbf{e}^*) + (\mathbf{T}^* \cdot \mathbf{e}^*)(\mathbf{U} \cdot \mathbf{e})] \\
&\quad - (\mathbf{U} \cdot \mathbf{p}_2)[(\mathbf{p}_1 \cdot \mathbf{e}^*)(\mathbf{T}^* \cdot \mathbf{e}) + (\mathbf{p}_1 \cdot \mathbf{e})(\mathbf{T}^* \cdot \mathbf{e}^*)] \\
&\quad - (\mathbf{T} \cdot \mathbf{p}_1)[(\mathbf{p}_2 \cdot \mathbf{e}^*)(\mathbf{U}^* \cdot \mathbf{e}) + (\mathbf{p}_2 \cdot \mathbf{e})(\mathbf{U}^* \cdot \mathbf{e}^*)] \\
&\quad + (\mathbf{T}^* \cdot \mathbf{U})[(\mathbf{e} \cdot \mathbf{p}_1)(\mathbf{e}^* \cdot \mathbf{p}_2) + (\mathbf{e}^* \cdot \mathbf{p}_1)(\mathbf{e} \cdot \mathbf{p}_2)]\} \,, \quad (3.340) \\
\mathbf{H}_c &= i\big[\mathbf{I} \times \mathbf{I}^* - \mathbf{J} \times \mathbf{J}^* + p_1^2(\mathbf{T} \times \mathbf{T}^*) + p_2^2(\mathbf{U} \times \mathbf{U}^*)\big] \\
&\quad + 2\operatorname{Im}\{[\mathbf{p}_1 \times \mathbf{T} + \mathbf{p}_2 \times \mathbf{U} - 2\mathbf{J}]R^* - 2(\mathbf{T} \cdot \mathbf{p}_1)\mathbf{J}^* \\
&\quad + (\mathbf{T} \cdot \mathbf{J}^*)\mathbf{p}_1 + (\mathbf{J}^* \cdot \mathbf{p}_1)\mathbf{T} - 2(\mathbf{U} \cdot \mathbf{p}_2)\mathbf{J}^* + (\mathbf{U} \cdot \mathbf{J}^*)\mathbf{p}_2 \\
&\quad + (\mathbf{J}^* \cdot \mathbf{p}_2)\mathbf{U} + (\mathbf{p}_1 \cdot \mathbf{p}_2)(\mathbf{T}^* \times \mathbf{U}) + (\mathbf{T}^* \cdot \mathbf{U})(\mathbf{p}_1 \times \mathbf{p}_2) \\
&\quad - (\mathbf{T}^* \cdot \mathbf{p}_1)(\mathbf{p}_2 \times \mathbf{U}) + (\mathbf{U} \cdot \mathbf{p}_2)(\mathbf{p}_1 \times \mathbf{T}^*)\} \,, \quad (3.341) \\
\mathbf{G}_l &= 2i\{|\mathbf{e} \cdot \mathbf{p}_1|^2 (\mathbf{T}^* \times \mathbf{T}) + |\mathbf{e} \cdot \mathbf{p}_2|^2 (\mathbf{U}^* \times \mathbf{U})\} \\
&\quad + 2\operatorname{Im}\{R[2\mathbf{J}^* + \{\mathbf{e} \cdot (\mathbf{I}^* - \mathbf{J}^*)\}\mathbf{e}^* + \{\mathbf{e}^* \cdot (\mathbf{I}^* - \mathbf{J}^*)\}\mathbf{e} \\
&\quad + (\mathbf{T}^* \cdot \mathbf{e})(\mathbf{p}_1 \times \mathbf{e}^*) + (\mathbf{T}^* \cdot \mathbf{e}^*)(\mathbf{p}_1 \times \mathbf{e}) + (\mathbf{U}^* \cdot \mathbf{e})(\mathbf{p}_2 \times \mathbf{e}^*) \\
&\quad + (\mathbf{U}^* \cdot \mathbf{e}^*)(\mathbf{p}_2 \times \mathbf{e}) + (\mathbf{p}_1 \cdot \mathbf{e})(\mathbf{T}^* \times \mathbf{e}^*) + (\mathbf{p}_1 \cdot \mathbf{e}^*)(\mathbf{T}^* \times \mathbf{e}) \\
&\quad + (\mathbf{p}_2 \cdot \mathbf{e})(\mathbf{U}^* \times \mathbf{e}^*) + (\mathbf{p}_2 \cdot \mathbf{e}^*)(\mathbf{U}^* \times \mathbf{e})] + [(\mathbf{J} \times \mathbf{J}^*) \cdot \mathbf{e}]\mathbf{e}^* \\
&\quad + (\mathbf{I} \cdot \mathbf{e}^*)[\mathbf{e} \times \mathbf{J}^* + (\mathbf{T}^* \cdot \mathbf{e})\mathbf{p}_1 + (\mathbf{U}^* \cdot \mathbf{e})\mathbf{p}_2 + (\mathbf{p}_1 \cdot \mathbf{e})\mathbf{T}^* \\
&\quad + (\mathbf{p}_2 \cdot \mathbf{e})\mathbf{U}^*] + (\mathbf{I} \cdot \mathbf{e})[(\mathbf{e}^* \times \mathbf{J}^*) + (\mathbf{T}^* \cdot \mathbf{e}^*)\mathbf{p}_1 + (\mathbf{U}^* \cdot \mathbf{e}^*)\mathbf{p}_2 \\
&\quad + (\mathbf{p}_1 \cdot \mathbf{e}^*)\mathbf{T}^* + (\mathbf{p}_2 \cdot \mathbf{e}^*)\mathbf{U}^*] + 2\big[(\mathbf{T}^* \cdot \mathbf{e})(\mathbf{p}_1 \cdot \mathbf{e}^*)
\end{aligned}$$

$$+ (\mathbf{T}^* \cdot \mathbf{e}^*)(\mathbf{p}_1 \cdot \mathbf{e}) + (\mathbf{U}^* \cdot \mathbf{e})(\mathbf{p}_2 \cdot \mathbf{e}^*) + (\mathbf{U}^* \cdot \mathbf{e}^*)(\mathbf{p}_2 \cdot \mathbf{e})]\mathbf{J}$$
$$- (\mathbf{p}_1 \cdot \mathbf{J})[(\mathbf{T}^* \cdot \mathbf{e})\mathbf{e}^* + (\mathbf{T}^* \cdot \mathbf{e}^*)\mathbf{e}]$$
$$- (\mathbf{p}_2 \cdot \mathbf{J})[(\mathbf{U}^* \cdot \mathbf{e})\mathbf{e}^* + (\mathbf{U}^* \cdot \mathbf{e}^*)\mathbf{e}]$$
$$- (\mathbf{T}^* \cdot \mathbf{J})[(\mathbf{e} \cdot \mathbf{p}_1)\mathbf{e}^* + (\mathbf{e}^* \cdot \mathbf{p}_1)\mathbf{e}]$$
$$- (\mathbf{U}^* \cdot \mathbf{J})[(\mathbf{e} \cdot \mathbf{p}_2)\mathbf{e}^* + (\mathbf{e}^* \cdot \mathbf{p}_2)\mathbf{e}]$$
$$+ [(\mathbf{p}_1 \cdot \mathbf{e}^*)(\mathbf{T}^* \cdot \mathbf{e}) + (\mathbf{p}_1 \cdot \mathbf{e})(\mathbf{T}^* \cdot \mathbf{e}^*)](\mathbf{T} \times \mathbf{p}_1)$$
$$+ [(\mathbf{p}_2 \cdot \mathbf{e}^*)(\mathbf{T}^* \cdot \mathbf{e}) + (\mathbf{p}_2 \cdot \mathbf{e})(\mathbf{T}^* \cdot \mathbf{e}^*)](\mathbf{U} \times \mathbf{p}_1)$$
$$- [(\mathbf{T}^* \cdot \mathbf{e})(\mathbf{U} \cdot \mathbf{e}^*) + (\mathbf{T}^* \cdot \mathbf{e}^*)(\mathbf{U} \cdot \mathbf{e})](\mathbf{p}_1 \times \mathbf{p}_2)$$
$$+ [(\mathbf{p}_1 \cdot \mathbf{e}^*)(\mathbf{U}^* \cdot \mathbf{e}) + (\mathbf{p}_1 \cdot \mathbf{e})(\mathbf{U}^* \cdot \mathbf{e}^*)](\mathbf{T} \times \mathbf{p}_2)$$
$$+ [(\mathbf{p}_2 \cdot \mathbf{e}^*)(\mathbf{U}^* \cdot \mathbf{e}) + (\mathbf{p}_2 \cdot \mathbf{e})(\mathbf{U}^* \cdot \mathbf{e}^*)](\mathbf{U} \times \mathbf{p}_2)$$
$$+ [(\mathbf{e} \cdot \mathbf{p}_1)(\mathbf{e}^* \cdot \mathbf{p}_2) + (\mathbf{e}^* \cdot \mathbf{p}_1)(\mathbf{e} \cdot \mathbf{p}_2)](\mathbf{U} \times \mathbf{T}^*)\}, \qquad (3.342)$$

$$\mathbf{G}_c = 2\operatorname{Re}\{[(\mathbf{I} + \mathbf{J}) \cdot \vec{\zeta}_1]\mathbf{J}^* + R^*[\vec{\zeta}_1 \times (\mathbf{I} + \mathbf{J})]$$
$$- [\mathbf{I} \cdot \mathbf{J}^* + 2(\mathbf{T} \cdot \mathbf{U}^*)(\mathbf{p}_1 \cdot \mathbf{p}_2) - 2(\mathbf{T} \cdot \mathbf{p}_1)(\mathbf{U}^* \cdot \mathbf{p}_2)]\vec{\zeta}_1$$
$$+ (\mathbf{p}_1 \cdot \vec{\zeta}_1)(\mathbf{I} \times \mathbf{T}^*) + (\mathbf{T}^* \cdot \vec{\zeta}_1)(\mathbf{I} \times \mathbf{p}_1) + (\mathbf{p}_1 \cdot \mathbf{J})(\mathbf{T}^* \times \vec{\zeta}_1)$$
$$+ (\mathbf{T}^* \cdot \mathbf{J})(\mathbf{p}_1 \times \vec{\zeta}_1) + (\mathbf{p}_2 \cdot \vec{\zeta}_1)(\mathbf{I} \times \mathbf{U}^*) + (\mathbf{U}^* \cdot \vec{\zeta}_1)(\mathbf{I} \times \mathbf{p}_2)$$
$$+ (\mathbf{p}_2 \cdot \mathbf{J})(\mathbf{U}^* \times \vec{\zeta}_1) + (\mathbf{U}^* \cdot \mathbf{J})(\mathbf{p}_2 \times \vec{\zeta}_1)$$
$$+ [(\mathbf{U}^* \cdot \vec{\zeta}_1)(\mathbf{p}_1 \cdot \mathbf{p}_2) - (\mathbf{U}^* \cdot \mathbf{p}_2)(\mathbf{p}_1 \cdot \vec{\zeta}_1)]\mathbf{T}$$
$$+ [(\mathbf{T}^* \cdot \vec{\zeta}_1)(\mathbf{p}_1 \cdot \mathbf{p}_2) - (\mathbf{T}^* \cdot \mathbf{p}_1)(\mathbf{p}_2 \cdot \vec{\zeta}_1)]\mathbf{U}$$
$$+ [(\mathbf{T} \cdot \mathbf{U}^*)(\mathbf{p}_2 \cdot \vec{\zeta}_1) - (\mathbf{T} \cdot \vec{\zeta}_1)(\mathbf{U}^* \cdot \mathbf{p}_2)]\mathbf{p}_1$$
$$+ [(\mathbf{T} \cdot \mathbf{U}^*)(\mathbf{p}_1 \cdot \vec{\zeta}_1) - (\mathbf{T} \cdot \mathbf{p}_1)(\mathbf{U}^* \cdot \vec{\zeta}_1)]\mathbf{p}_2$$
$$+ p_1^2(\mathbf{T} \cdot \vec{\zeta}_1)\mathbf{T}^* + p_2^2(\mathbf{U} \cdot \vec{\zeta}_1)\mathbf{U}^*\} + 2|\mathbf{T}|^2\{(\mathbf{p}_1 \cdot \vec{\zeta}_1)\mathbf{p}_1 - p_1^2\vec{\zeta}_1\}$$
$$+ 2|\mathbf{U}|^2\{(\mathbf{p}_2 \cdot \vec{\zeta}_1)\mathbf{p}_2 - p_2^2\vec{\zeta}_1\}. \qquad (3.343)$$

In the expression (3.343) the scalar triple products were transformed by means of the formula (3.78).

The degree of linear polarization (real **e**) referring to the plane of emission is defined as

$$P_l = \frac{\sigma_\perp - \sigma_\|}{\sigma_\perp + \sigma_\|}, \qquad (3.344)$$

where $\sigma_\perp$ and $\sigma_\|$ are the cross sections (3.24), proportional to $\sum |M|^2$, for the polarization directions

$$\mathbf{e}_\perp = \frac{\mathbf{p}_1 \times \mathbf{k}}{|\mathbf{p}_1 \times \mathbf{k}|} \tag{3.345}$$

and

$$\mathbf{e}_\parallel = \frac{\mathbf{k} \times (\mathbf{p}_1 \times \mathbf{k})}{k|\mathbf{p}_1 \times \mathbf{k}|} = \frac{k^2 \mathbf{p}_1 - (\mathbf{k} \cdot \mathbf{p}_1)\mathbf{k}}{k|\mathbf{p}_1 \times \mathbf{k}|}, \tag{3.346}$$

respectively. Since $\mathbf{e} \times \mathbf{e}^* = 0$ for real $\mathbf{e}$, only the terms H_l and $\mathbf{G}_l$ of (3.339) contribute to P_l. As pointed out in the last section, $\mathbf{G}_l = 0$ in Born approximation, so that the linear bremsstrahlung polarization is independent of the initial electron spin $\vec{\zeta}_1$. In terms of the polarization coefficients of Eq. (3.298), the degree of linear polarization is $P_l = C_{030}$.

The probability of circular polarization is obtained by setting

$$\mathbf{e} = \frac{1}{\sqrt{2}} (\mathbf{e}_\perp + i\delta \mathbf{e}_\parallel), \tag{3.347}$$

where $\delta = +1$ for right-handed and $\delta = -1$ for left-handed photon polarization ($i[\mathbf{e} \times \mathbf{e}^*] = \delta \hat{\mathbf{k}}$). Designating the pertaining cross sections σ_+ and σ_-, respectively, the degree of circular polarization is given by

$$P_c = \frac{\sigma_+ - \sigma_-}{\sigma_+ + \sigma_-}. \tag{3.348}$$

P_c corresponds to the polarization coefficient $C_{\ell 20}$ of the classification scheme of Eq. (3.298), where the index ℓ depends on the spin polarization of the incident electron and the second index 2 refers to the Stokes parameter ξ_2 designating circular photon polarization. For instance, the polarization coefficient C_{320} is a transmitter of helicity, producing circularly polarized photons from longitudinally polarized electrons.

In the case of circular polarization the terms containing H_l and $\mathbf{G}_l$ in the numerator of (3.348) vanish so that only the terms $\mathbf{H}_c$ and $\mathbf{G}_c$ contribute to the numerator of P_c. In Born approximation the spin-independent term $\mathbf{H}_c = 0$ since, apart from the common factor K_1, the quantities $\mathbf{I}$, $\mathbf{J}$, R, $\mathbf{T}$, and $\mathbf{U}$ are real. Hence there is only non-vanishing circular polarization if the incoming electrons are polarized. Accordingly the circular polarization from unpolarized electrons, as calculated in more sophisticated theories, is weak.

If we average over the spins of the initial electrons the terms $\mathbf{G}_l \cdot \vec{\zeta}$ and $\mathbf{G}_c(\vec{\zeta})$ of (3.339) are eliminated, since $-\vec{\zeta}_1$ represents a state opposite to $\vec{\zeta}_1$, and we get

$$\tfrac{1}{2}\sum_{\vec{\zeta}_1,\vec{\zeta}_2}|M|^2 = |N_1 N_2|^2 \frac{(\epsilon_1+1)(\epsilon_2+1)}{8\epsilon_1\epsilon_2}\{H_l + i(\mathbf{e}\times\mathbf{e}^*)\cdot\mathbf{H}_c\}. \quad (3.349)$$

3.7.2 Linear polarization in Born approximation

The bremsstrahlung produced by unpolarized electrons is, in general, partially linearly polarized. We shall calculate the degree of linear polarization in the simplest way, i.e., in Born approximation [76]. As we have seen in the last section, the linear bremsstrahlung polarization in Born approximation is independent of the initial electron spin state.

According to Eqs. (3.24) and (3.76) the bremsstrahlung cross section in Born approximation including the polarization vector $\mathbf{e}$ (assumed to be real) has the form

$$\begin{aligned}\frac{d^3\sigma_B}{dk\,d\Omega_k\,d\Omega_{p_2}} &= \frac{\alpha Z^2 r_0^2}{\pi^2}\frac{kp_2}{p_1 q^4}\bigg\{\frac{4\epsilon_2^2-q^2}{D_1^2}(\mathbf{e}\cdot\mathbf{p}_1)^2 + \frac{4\epsilon_1^2-q^2}{D_2^2}(\mathbf{e}\cdot\mathbf{p}_2)^2 \\ &\quad -2\frac{4\epsilon_1\epsilon_2-q^2}{D_1 D_2}(\mathbf{e}\cdot\mathbf{p}_1)(\mathbf{e}\cdot\mathbf{p}_2) + \frac{(\mathbf{k}\times\mathbf{q})^2}{D_1 D_2}\bigg\} \\ &= \frac{\alpha Z^2 r_0^2}{\pi^2}\frac{kp_2}{p_1 q^4}\bigg\{4\Big(\frac{\epsilon_2}{D_1}\mathbf{e}\cdot\mathbf{p}_1 - \frac{\epsilon_1}{D_2}\mathbf{e}\cdot\mathbf{p}_2\Big)^2 \\ &\quad -q^2\Big(\frac{\mathbf{e}\cdot\mathbf{p}_1}{D_1} - \frac{\mathbf{e}\cdot\mathbf{p}_2}{D_2}\Big)^2 + \frac{(\mathbf{k}\times\mathbf{q})^2}{D_1 D_2}\bigg\}. \quad (3.350)\end{aligned}$$

The last expression shows that the Bethe-Heitler cross section and the linear polarization depend on the three quantities $(\epsilon_2/D_1)\mathbf{e}\cdot\mathbf{p}_1 - (\epsilon_1/D_2)\mathbf{e}\cdot\mathbf{p}_2$, $(\mathbf{e}\cdot\mathbf{p}_1/D_1 - \mathbf{e}\cdot\mathbf{p}_2/D_2)q$, and $\mathbf{k}\times\mathbf{q}/\sqrt{D_1 D_2}$.

Using the system of polar coordinates (3.86) where the photon is emitted along the z axis, we get

$$\begin{aligned}\frac{d^3\sigma_B}{dk\,d\Omega_k\,d\Omega_{p_2}} &= \frac{\alpha Z^2 r_0^2}{4\pi^2}\frac{p_2}{kp_1 q^4}\bigg\{\frac{4\epsilon_2^2-q^2}{(\epsilon_1-p_1\cos\theta_1)^2}(\mathbf{e}\cdot\mathbf{p}_1)^2 \\ &\quad +\frac{4\epsilon_1^2-q^2}{(\epsilon_2-p_1\cos\theta_2)^2}(\mathbf{e}\cdot\mathbf{p}_2)^2 - 2\frac{(4\epsilon_1\epsilon_2-q^2)(\mathbf{e}\cdot\mathbf{p}_1)(\mathbf{e}\cdot\mathbf{p}_2)}{(\epsilon_1-p_1\cos\theta_1)(\epsilon_2-p_2\cos\theta_2)} \\ &\quad +k^2\frac{p_1^2\sin^2\theta_1 + p_2^2\sin^2\theta_2 - 2p_1 p_2\sin\theta_1\sin\theta_2\cos\varphi}{(\epsilon_1-p_1\cos\theta_1)(\epsilon_2-p_2\cos\theta_2)}\bigg\}. \quad (3.351)\end{aligned}$$

Introducing the azimuth angle ϕ for the direction $\mathbf{e}$ measured in a plane perpendicular to **k**,

$$\mathbf{e} = \{\cos\phi, \sin\phi, 0\}, \tag{3.352}$$

the cross section can be written as

$$\begin{aligned}\frac{d^3\sigma_B}{dk\,d\Omega_k\,d\Omega_{p_2}} &= A + B\cos(2\phi) + C\sin(2\phi) \\ &= A + D\cos(2\phi - 2\phi_0) \\ &= A - D + 2D\cos^2(\phi - \phi_0),\end{aligned} \tag{3.353}$$

where

$$\begin{aligned}A = F_1 \bigg\{ &(4\epsilon_2^2 - q^2)\frac{p_1^2 \sin^2\theta_1}{(\epsilon_1 - p_1\cos\theta_1)^2} + (4\epsilon_1^2 - q^2)\frac{p_2^2 \sin^2\theta_2}{(\epsilon_2 - p_2\cos\theta_2)^2} \\ &- (4\epsilon_1\epsilon_2 - q^2)\frac{2p_1 p_2 \sin\theta_1 \sin\theta_2 \cos\varphi}{(\epsilon_1 - p_1\cos\theta_1)(\epsilon_2 - p_2\cos\theta_2)} \\ &+ 2k^2 \frac{p_1^2\sin^2\theta_1 + p_2^2\sin^2\theta_2 - 2p_1p_2\sin\theta_1\sin\theta_2\cos\varphi}{(\epsilon_1 - p_1\cos\theta_1)(\epsilon_2 - p_2\cos\theta_2)} \bigg\}, \end{aligned} \tag{3.354}$$

$$\begin{aligned}B = F_1 \bigg\{ &(4\epsilon_2^2 - q^2)\frac{p_1^2 \sin^2\theta_1}{(\epsilon_1 - p_1\cos\theta_1)^2} + (4\epsilon_1^2 - q^2)\frac{p_2^2 \sin^2\theta_2 \cos(2\varphi)}{(\epsilon_2 - p_2\cos\theta_2)^2} \\ &- (4\epsilon_1\epsilon_2 - q^2)\frac{2p_1 p_2 \sin\theta_1 \sin\theta_2 \cos\varphi}{(\epsilon_1 - p_1\cos\theta_1)(\epsilon_2 - p_2\cos\theta_2)} \bigg\}, \end{aligned} \tag{3.355}$$

$$\begin{aligned}C = F_1 \bigg\{ &(4\epsilon_1^2 - q^2)\frac{p_2^2 \sin^2\theta_2 \sin(2\varphi)}{(\epsilon_2 - p_2\cos\theta_2)^2} \\ &- (4\epsilon_1\epsilon_2 - q^2)\frac{2p_1 p_2 \sin\theta_1 \sin\theta_1 \sin\varphi}{(\epsilon_1 - p_1\cos\theta_1)(\epsilon_2 - p_2\cos\theta_2)} \bigg\}, \end{aligned} \tag{3.356}$$

$$D = \sqrt{B^2 + C^2}, \quad \tan(2\phi_0) = C/B, \tag{3.357}$$

and

$$F_1 = \frac{\alpha Z^2 r_0^2}{8\pi^2} \frac{p_2}{kp_1 q^4}. \tag{3.358}$$

The intensity variation is such as would be obtained if the radiation consisted of a superposition of unpolarized radiation of relative intensity $A - D$ and linearly polarized radiation of relative intensity $2D$. The direction of the electric vector of the linearly polarized component is $\phi = \phi_0$.

The degree of linear polarization referring to the plane of emission is defined by (3.344) where $\sigma_\perp$ and $\sigma_\parallel$ are the cross sections (3.353) for the angles $\phi = 90°$ and $\phi = 0°$, respectively. This gives

$$P_l = -B/A \,. \tag{3.359}$$

The maximum degree of polarization is obtained if we set $\phi = \phi_0 + 90°$ and $\phi = \phi_0$, respectively, yielding

$$P_{l,\max} = -D/A \,. \tag{3.360}$$

For coplanar geometry ($\varphi = 0$, $C = 0$) the radiation is polarized in the plane of emission ($\phi = 0$) and the two polarizations (3.359) and (3.360) are identical. The degree of linear polarization is particularly large when the component of the recoil momentum perpendicular to $\mathbf{k}$, $q_\perp = |\mathbf{q} \times \hat{\mathbf{k}}|$, is small, since $A = B + 8k^2 F_1 (\mathbf{q} \times \mathbf{k})^2 / D_1 D_2$ for $\varphi = 0$. In the special case that the recoil momentum lies in the plane of emission, $\mathbf{q} \times \mathbf{k} = 0$ or $p_1 \sin\theta_1 = p_2 \sin\theta_2$, the radiation is completely polarized, $|P_l| = 100\%$ (this holds only in Born approximation!).

Since C is proportional to p_2^2, it vanishes at the short-wavelength limit even if the Coulomb correction factor (3.90) is applied to the cross section. Therefore $\phi_0 = 0$, i.e., the linear polarization with respect to the plane of emission is maximum at the short-wavelength limit.

Inserting the polarization vectors (3.345) and (3.346) into the cross section (3.350) results in

$$\begin{aligned}
\sigma_\perp - \sigma_\parallel = \frac{\alpha Z^2 r_0^2}{\pi^2} \frac{k p_2}{p_1 q^4} &\left\{ -\frac{4\epsilon_2^2 - q^2}{D_1^2} (\mathbf{p}_1 \times \hat{\mathbf{k}})^2 \right. \\
&+ \frac{4\epsilon_1^2 - q^2}{D_2^2} \left[p_2^2 - \frac{\{(\mathbf{k} \cdot \mathbf{p}_2)\mathbf{p}_1 - (\mathbf{p}_1 \cdot \mathbf{p}_2)\mathbf{k}\}^2}{(\mathbf{p}_1 \times \mathbf{k})^2} \right. \\
&\left. - \frac{\{\mathbf{p}_1 \cdot \mathbf{p}_2 - (\hat{\mathbf{k}} \cdot \mathbf{p}_1)(\hat{\mathbf{k}} \cdot \mathbf{p}_2)\}^2}{(\mathbf{p}_1 \times \hat{\mathbf{k}})^2} \right] \\
&\left. + 2 \frac{4\epsilon_1 \epsilon_2 - q^2}{D_1 D_2} [\mathbf{p}_1 \cdot \mathbf{p}_2 - (\hat{\mathbf{k}} \cdot \mathbf{p}_1)(\hat{\mathbf{k}} \cdot \mathbf{p}_2)] \right\} . \tag{3.361}
\end{aligned}$$

The denominator of (3.344), $\sigma_\perp + \sigma_\parallel$, is given by (3.81). Employing the coordinate system (3.196), Eq. (3.361) can be written as

$$\sigma_\perp - \sigma_\parallel = \frac{\alpha Z^2 r_0^2}{\pi^2} \frac{k p_2}{p_1 q^4} \left\{ \frac{4\epsilon_1^2 - q^2}{D_2^2} p_2^2 [\sin^2 \theta_e \sin^2 \varphi \right.$$

$$-(\sin\theta_k\cos\theta_e - \cos\theta_k\sin\theta_e\cos\varphi)^2] - \frac{4\epsilon_2^2 - q^2}{D_1^2}p_1^2\sin^2\theta_k$$

$$+ 2\frac{4\epsilon_1\epsilon_2 - q^2}{D_1 D_2}p_1 p_2 \sin\theta_k(\sin\theta_k\cos\theta_e - \cos\theta_k\sin\theta_e\cos\varphi)\Big\} \quad (3.362)$$

and

$$\sigma_\perp + \sigma_\parallel = \frac{\alpha Z^2 r_0^2}{\pi^2}\frac{kp_2}{p_1 q^4}\Big\{\left(\frac{4\epsilon_2^2 - q^2}{D_1^2} + \frac{2k^2}{D_1 D_2}\right)p_1^2\sin^2\theta_k$$

$$+ \left(\frac{4\epsilon_1^2 - q^2}{D_2^2} + \frac{2k^2}{D_1 D_2}\right)p_2^2[1 - (\cos\theta_k\cos\theta_e + \sin\theta_k\sin\theta_e\cos\varphi)^2]$$

$$- 2\frac{4\epsilon_1\epsilon_2 - q^2 + 2k^2}{D_1 D_2}p_1 p_2 \sin\theta_k(\sin\theta_k\cos\theta_e - \cos\theta_k\sin\theta_e\cos\varphi)\Big\}. \quad (3.363)$$

For coplanar geometry ($\varphi = \varphi_e - \varphi_k = 0$) the degree of linear polarization takes the form

$$P_l = -\big\{4[\epsilon_2 p_1 \sin\theta_k/D_1 - \epsilon_1 p_2 \sin\vartheta/D_2]^2$$

$$- q^2[p_1 \sin\theta_k/D_1 - p_2 \sin\vartheta/D_2]^2\big\}/\mathcal{N}_1, \quad (3.364)$$

where

$$\mathcal{N}_1 = 4[\epsilon_2 p_1 \sin\theta_k/D_1 - \epsilon_1 p_2 \sin\vartheta/D_2]^2 - q^2[p_1 \sin\theta_k/D_1 - p_2 \sin\vartheta/D_2]^2$$

$$+ 2k^2[p_1 \sin\theta_k - p_2 \sin\vartheta]^2/(D_1 D_2) \quad (3.365)$$

and $\vartheta = \theta_k - \theta_e$.

If the angle θ_e of the scattered electron is fixed this equation allows to calculate the variation of the linear polarization with the photon angle θ_k relative to the incident electron. The degree of linear polarization is maximum ($P_l = -1$) for $\mathbf{q} \times \mathbf{k} = 0$ or $p_1 \sin\theta_k = p_2 \sin(\theta_k - \theta_e)$ yielding

$$\tan\theta_k = -\frac{p_2 \sin\theta_e}{p_1 - p_2 \cos\theta_e}. \quad (3.366)$$

The photon angular distribution is easily obtained from the denominator (3.365).

3.7.3 Circular polarization from polarized electrons in Born approximation

Polarized electrons can produce circularly polarized bremsstrahlung, as was first noticed by Zel'dovich [77]. In order to derive the degree of polarization in Born approximation we start from the matrix element squared given in (3.331) and (3.332) and apply the approximations (3.177) to (3.179) for the integrals $\mathbf{I}_1$, $\mathbf{I}_2$, and $\mathbf{I}_3$ occurring in Eqs. (3.322) to (3.325). Besides, the factor (3.176) takes the value $F \approx 1$. Setting

$$\mathbf{e} = (\mathbf{e}_\perp + i\delta\mathbf{e}_\parallel)/\sqrt{2} \qquad (3.367)$$

with $\delta = \pm 1$, one notes that all the terms in Eq. (3.331) except $\mathbf{Q}_1$ become independent of δ, whereas all the terms of $\mathbf{Q}_1$ [Eq. (3.332)] are proportional to δ. Therefore the triply differential cross section in Born approximation takes the form

$$\frac{d^3\sigma}{dk\,d\Omega_k\,d\Omega_{p_2}} = \frac{\alpha Z^2 r_0^2}{4\pi^2 |K_1|^2}(\epsilon_1 + 1)(\epsilon_2 + 1)\frac{p_2 k}{p_1}\Big\{\tfrac{1}{2}\big[|\mathbf{J}|^2 + |\mathbf{J}\cdot\hat{\mathbf{k}}|^2$$
$$+ |\mathbf{I}\times\hat{\mathbf{k}}|^2 + p_1^2|\mathbf{T}\times\hat{\mathbf{k}}|^2 + p_2^2|\mathbf{U}\times\hat{\mathbf{k}}|^2 + (\mathbf{p}_1\times\hat{\mathbf{k}})^2|\mathbf{T}|^2$$
$$+ (\mathbf{p}_2\times\hat{\mathbf{k}})^2|\mathbf{U}|^2\big] + \big[(\mathbf{p}_1\cdot\hat{\mathbf{k}})(\mathbf{T}\times\mathbf{J}^*) + (\mathbf{p}_2\cdot\hat{\mathbf{k}})(\mathbf{U}\times\mathbf{J}^*)\big]\cdot\hat{\mathbf{k}}$$
$$- \big[(\mathbf{T}\cdot\hat{\mathbf{k}})(\mathbf{p}_1\times\hat{\mathbf{k}}) + (\mathbf{U}\cdot\hat{\mathbf{k}})(\mathbf{p}_2\times\hat{\mathbf{k}})\big]\cdot\mathbf{J}^* + (\mathbf{T}\cdot\mathbf{U}^*)\big[2(\mathbf{p}_1\cdot\mathbf{p}_2)$$
$$- (\mathbf{p}_1\cdot\hat{\mathbf{k}})(\mathbf{p}_2\cdot\hat{\mathbf{k}})\big] - (\mathbf{p}_1\cdot\mathbf{p}_1)(\mathbf{T}\cdot\hat{\mathbf{k}})(\mathbf{U}^*\cdot\hat{\mathbf{k}})$$
$$+ (\mathbf{p}_1\cdot\hat{\mathbf{k}})(\mathbf{T}^*\cdot\hat{\mathbf{k}})\big[(\mathbf{p}_1\cdot\mathbf{T}) + 2(\mathbf{p}_2\cdot\mathbf{U})\big]$$
$$+ (\mathbf{p}_2\cdot\hat{\mathbf{k}})(\mathbf{U}^*\cdot\hat{\mathbf{k}})\big[(\mathbf{p}_2\cdot\mathbf{U}) + 2(\mathbf{p}_1\cdot\mathbf{T})\big]$$
$$- 2(\mathbf{p}_1\cdot\mathbf{T})(\mathbf{p}_2\cdot\mathbf{U}^*) + \delta(\vec{\zeta}_1\cdot\mathbf{Q}_1)\Big\} \qquad (3.368)$$

with

$$\mathbf{Q}_1 = (\mathbf{J}\cdot\hat{\mathbf{k}})\mathbf{J}^* + |R|^2\hat{\mathbf{k}} + R^*\big\{[(\mathbf{I}+\mathbf{J})\cdot\mathbf{e}_\parallel]\mathbf{e}_\perp - [(\mathbf{I}+\mathbf{J})\cdot\mathbf{e}_\perp]\mathbf{e}_\parallel$$
$$+ (\mathbf{e}_\parallel\cdot\mathbf{T})(\mathbf{p}_1\times\mathbf{e}_\perp) - (\mathbf{e}_\perp\cdot\mathbf{T})(\mathbf{p}_1\times\mathbf{e}_\parallel) + (\mathbf{e}_\parallel\cdot\mathbf{U})(\mathbf{p}_2\times\mathbf{e}_\perp)$$
$$- (\mathbf{e}_\perp\cdot\mathbf{U})(\mathbf{p}_2\times\mathbf{e}_\parallel) + (\mathbf{e}_\parallel\cdot\mathbf{p}_1)(\mathbf{T}\times\mathbf{e}_\perp) + (\mathbf{e}_\parallel\cdot\mathbf{p}_2)(\mathbf{U}\times\mathbf{e}_\perp)$$
$$- (\mathbf{e}_\perp\cdot\mathbf{p}_2)(\mathbf{U}\times\mathbf{e}_\parallel)\big\} + (\mathbf{e}_\perp\cdot\mathbf{I})(\mathbf{e}_\parallel\times\mathbf{J}^*) - (\mathbf{e}_\parallel\cdot\mathbf{I})(\mathbf{e}_\perp\times\mathbf{J}^*)$$
$$+ \big\{(\mathbf{e}_\perp\cdot\mathbf{I})(\mathbf{e}_\parallel\cdot\mathbf{T}^*) - (\mathbf{e}_\parallel\cdot\mathbf{I})(\mathbf{e}_\perp\cdot\mathbf{T}^*)\big\}\mathbf{p}_1$$
$$+ \big\{(\mathbf{e}_\perp\cdot\mathbf{I})(\mathbf{e}_\parallel\cdot\mathbf{U}^*) - (\mathbf{e}_\parallel\cdot\mathbf{I})(\mathbf{e}_\perp\cdot\mathbf{U}^*)\big\}\mathbf{p}_2 + (\mathbf{e}_\parallel\cdot\mathbf{p}_1)(\mathbf{e}_\perp\cdot\mathbf{I})\mathbf{T}^*$$
$$+ \big\{(\mathbf{e}_\parallel\cdot\mathbf{p}_2)(\mathbf{e}_\perp\cdot\mathbf{I}) - (\mathbf{e}_\perp\cdot\mathbf{p}_2)(\mathbf{e}_\parallel\cdot\mathbf{I})\big\}\mathbf{U}^*$$
$$+ (\mathbf{p}_1\cdot\mathbf{J}^*)\big\{(\mathbf{e}_\perp\cdot\mathbf{T})\mathbf{e}_\parallel - (\mathbf{e}_\parallel\cdot\mathbf{T})\mathbf{e}_\perp\big\}$$

$$+ (\mathbf{p}_2 \cdot \mathbf{J}^*)\{(\mathbf{e}_\perp \cdot \mathbf{U})\mathbf{e}_\| - (\mathbf{e}_\| \cdot \mathbf{U})\mathbf{e}_\perp\} - (\mathbf{e}_\| \cdot \mathbf{p}_1)(\mathbf{T} \cdot \mathbf{J}^*)\mathbf{e}_\perp$$
$$+ (\mathbf{U} \cdot \mathbf{J}^*)\{(\mathbf{e}_\perp \cdot \mathbf{p}_2)\mathbf{e}_\| - (\mathbf{e}_\| \cdot \mathbf{p}_2)\mathbf{e}_\perp\} - (\mathbf{e}_\| \cdot \mathbf{p}_1)(\mathbf{e}_\perp \cdot \mathbf{T})(\mathbf{T}^* \times \mathbf{p}_1)$$
$$+ \{(\mathbf{e}_\perp \cdot \mathbf{p}_2)(\mathbf{e}_\| \cdot \mathbf{T}) - (\mathbf{e}_\| \cdot \mathbf{p}_2)(\mathbf{e}_\perp \cdot \mathbf{T})\}(\mathbf{U}^* \times \mathbf{p}_1)$$
$$+ \{(\mathbf{e}_\perp \cdot \mathbf{T})(\mathbf{e}_\| \cdot \mathbf{U}^*) - (\mathbf{e}_\| \cdot \mathbf{T})(\mathbf{e}_\perp \cdot \mathbf{U}^*)\}(\mathbf{p}_1 \times \mathbf{p}_2)$$
$$+ (\mathbf{e}_\| \cdot \mathbf{p}_1)(\mathbf{e}_\perp \cdot \mathbf{U}^*)(\mathbf{p}_2 \times \mathbf{T}) + \{(\mathbf{e}_\| \cdot \mathbf{p}_2)(\mathbf{e}_\perp \cdot \mathbf{U})$$
$$- (\mathbf{e}_\perp \cdot \mathbf{p}_2)(\mathbf{e}_\| \cdot \mathbf{U})\}(\mathbf{p}_2 \times \mathbf{U}^*) + (\mathbf{e}_\| \cdot \mathbf{p}_1)(\mathbf{e}_\perp \cdot \mathbf{p}_2)(\mathbf{U}^* \times \mathbf{T}). \quad (3.369)$$

After substituting the quantities (3.322) to (3.325), (3.345), and (3.346), and some rearrangement this can be written as [78]

$$\frac{d^3\sigma}{dk\, d\Omega_k\, d\Omega_{p_2}} = \frac{1}{2}\left(\frac{d^3\sigma_B}{dk\, d\Omega_k\, d\Omega_{p_2}} + \delta \frac{d^3\sigma_c}{dk\, d\Omega_k\, d\Omega_{p_2}}\right), \quad (3.370)$$

where $d^3\sigma_B/(dk\, d\Omega_k\, d\Omega_{p_2})$ is given by (3.81), and

$$\frac{d^3\sigma_c}{dk\, d\Omega_k\, d\Omega_{p_2}} = \frac{\alpha Z^2 r_0^2}{\pi^2} \frac{kp_2}{p_1 q^4} \Bigg\{ (\mathbf{p}_1 \cdot \vec{\zeta}_1)\left[\frac{4\epsilon_2^2 - q^2}{D_1^2}\left(k - \frac{\mathbf{p}_1 \cdot \mathbf{k}}{\epsilon_1 + 1}\right)\right.$$
$$- \frac{4\epsilon_1^2 - q^2}{D_2^2}\left(k + \frac{\mathbf{p}_1 \cdot \mathbf{k}}{\epsilon_1 + 1}\right) + 2\frac{4\epsilon_1\epsilon_2 + q^2}{D_1 D_2}\frac{\mathbf{p}_1 \cdot \mathbf{k}}{\epsilon_1 + 1} + 2(\epsilon_1 + \epsilon_2)\left(\frac{1}{D_2} - \frac{1}{D_1}\right)$$
$$+ \frac{4}{D_1 D_2^2}[(\mathbf{p}_1 \times \mathbf{k})^2 - (\mathbf{p}_2 \times \mathbf{k})^2]\left(\epsilon_1 + \frac{\mathbf{p}_1 \cdot (\mathbf{p}_2 + \mathbf{k})}{\epsilon_1 + 1}\right)\bigg]$$
$$- (\mathbf{k} \cdot \vec{\zeta}_1)\left[\frac{4\epsilon_2^2 - q^2}{D_1^2} + \frac{4\epsilon_1^2 - q^2}{D_2^2} - 2\frac{4\epsilon_1\epsilon_2 + q^2}{D_1 D_2}\right]$$
$$+ (\mathbf{k} + \mathbf{p}_2) \cdot \vec{\zeta}_1 \frac{4}{D_1 D_2^2}[(\mathbf{p}_1 \times \mathbf{k})^2 - (\mathbf{p}_2 \times \mathbf{k})^2]\Bigg\} \quad (3.371)$$

is the spin-dependent part of the cross section. Then, according to Eq. (3.348), the degree of circular polarization is given by

$$P_c = \frac{d^3\sigma_c}{dk\, d\Omega_k\, d\Omega_{p_2}} \bigg/ \frac{d^3\sigma_B}{dk\, d\Omega_k\, d\Omega_{p_2}}. \quad (3.372)$$

This holds for completely polarized incident electrons. Otherwise, P_c has to be multiplied by the degree of electron polarization, P_e. Hence, unpolarized electrons cannot emit circularly polarized bremsstrahlung (in more sophisticated theories the term $\mathbf{H}_c$ of Eq. (3.339) gives a small contribution). Due to the factor $1/k$ in the Born approximation cross section the photon circular polarization vanishes at the low-energy end of the bremsstrahlung spectrum. In general, P_c increases with increasing photon energy.

We will consider the particular cases of the spin parallel to the direction of the incident electron (longitudinal spin) and of the spin perpendicular to this direction (transversal spin).

(a) For longitudinal spin ($\vec{\zeta}_1 = \hat{\mathbf{p}}_1$) the cross section (3.371) can be transformed to

$$\frac{d^3\sigma_c}{dk\,d\Omega_k\,d\Omega_{p_2}} = \frac{2\alpha Z^2 r_0^2}{\pi^2} \frac{p_2}{p_1^2 q^4} \left\{ \epsilon_1 (\mathbf{p}_1 + \mathbf{p}_2) \cdot \hat{\mathbf{k}} \left[\frac{\mathbf{p}_1 \times \mathbf{k}}{D_1} - \frac{\mathbf{p}_2 \times \mathbf{k}}{D_2} \right]^2 \right.$$
$$+ (1 - \epsilon_1 k) \left[\frac{(\mathbf{p}_1 \times \mathbf{k})^2}{D_1} + \frac{(\mathbf{p}_2 \times \mathbf{k})^2}{D_2} \right] \left(\frac{1}{D_1} - \frac{1}{D_2} \right)$$
$$\left. + \left(\frac{1}{D_2^2} - \frac{1}{D_1^2} \right)(\mathbf{p}_1 \times \mathbf{k}) \cdot (\mathbf{p}_2 \times \mathbf{k}) + 2k^2 \frac{(\mathbf{p}_2 \times \mathbf{k})^2 - (\mathbf{p}_1 \times \mathbf{k})^2}{D_1^2 D_2} \right\}. \tag{3.373}$$

At high energies of the initial electrons, $\epsilon_1 \gg 1$, the circular polarization is high, approaching $P_c = 100\%$ at the short-wavelength limit. This characteristic is utilized for the detection of electron spin polarization in radioactive beta decay [79]. At decreasing energies and towards the long-wavelength end of the bremsstrahlung spectrum the degree of polarization drops to values $|P_c| \ll 1$.

(b) If the electron spin is transversal perpendicular to the production plane, $\vec{\zeta}_1 = (\mathbf{k} \times \mathbf{p}_1)/|\mathbf{k} \times \mathbf{p}_1|$, the spin-dependent cross section is

$$\frac{d^3\sigma_c}{dk\,d\Omega_k\,d\Omega_{p_2}} = \frac{4\alpha Z^2 r_0^2}{\pi^2} \frac{kp_2}{p_1 q^4 D_1 D_2^2} \frac{\mathbf{p}_2 \cdot (\mathbf{k} \times \mathbf{p}_1)}{|\mathbf{k} \times \mathbf{p}_1|} \left[(\mathbf{p}_1 \times \mathbf{k})^2 - (\mathbf{p}_2 \times \mathbf{k})^2 \right]. \tag{3.374}$$

In this case only the term proportional to $\mathbf{p}_2 \cdot \vec{\zeta}_1$ in the curly bracket of (3.371) is left. Hence the degree of circular polarization vanishes at the short-wavelength limit and for coplanar geometry. It is generally small compared to unity. The pertinent polarization coefficient according to (3.298) is C_{220}.

(c) For the spin directed transversally in the production plane,

$$\vec{\zeta}_1 = \frac{\mathbf{p}_1 \times (\mathbf{k} \times \mathbf{p}_1)}{p_1 |\mathbf{k} \times \mathbf{p}_1|} = \frac{p_1^2 \mathbf{k} - (\mathbf{k} \cdot \mathbf{p}_1)\mathbf{p}_1}{p_1 |\mathbf{k} \times \mathbf{p}_1|}, \tag{3.375}$$

the spin-dependent cross section takes the form

$$\frac{d^3\sigma_c}{dk\,d\Omega_k\,d\Omega_{p_2}} = \frac{\alpha Z^2 r_0^2}{\pi^2} \frac{kp_2}{p_1^2 q^4} \left\{ |\mathbf{p}_1 \times \mathbf{k}| \left[2 \frac{4\epsilon_1 \epsilon_2 + q^2}{D_1 D_2} - \frac{4\epsilon_2^2 - q^2}{D_1^2} - \frac{4\epsilon_1^2 - q^2}{D_2^2} \right] \right.$$

$$+ \frac{4}{D_1 D_2^2} \left[|\mathbf{p}_1 \times \mathbf{k}| + \frac{(\mathbf{p}_1 \times \mathbf{p}_2) \cdot (\mathbf{p}_1 \times \mathbf{k})}{|\mathbf{p}_1 \times \mathbf{k}|} \right] \left[(\mathbf{p}_1 \times \mathbf{k})^2 - (\mathbf{p}_2 \times \mathbf{k})^2 \right] \right\}.$$
(3.376)

The circular photon polarization from electrons which are transversely polarized in the production plane is described by the polarization coefficient C_{120}. As in case (b) the degree of polarization is generally small.

At the short-wavelength limit, $p_2 = 0$, a number of terms in (3.371) vanishes. Furthermore, we have $q^2 = D_1$ and

$$\frac{(\mathbf{p}_1 \times \mathbf{k})^2}{q^4} = \frac{\epsilon_1 k}{q^2} - \frac{k^2}{q^4} - \frac{1}{4}.$$
(3.377)

Hence the spin-dependent cross section (3.371) reduces to

$$\frac{d^3 \sigma_c}{dk \, d\Omega_k \, d\Omega_{p_2}} = \frac{\alpha Z^2 r_0^2}{\pi^2} \frac{p_2}{k p_1^3} \frac{(\mathbf{p}_1 \times \mathbf{k})^2}{q^8} \{ k^2 (q^2 - 4)(\mathbf{p}_1 \cdot \vec{\zeta}_1) + 4 p_1^2 (\mathbf{k} \cdot \vec{\zeta}_1) \}$$
(3.378)

so that, using (3.213), the degree of circular polarization takes the simple form

$$P_c = \frac{(k^2/p_1^2)(q^2 - 4)(\mathbf{p}_1 \cdot \vec{\zeta}_1) + 4(\mathbf{k} \cdot \vec{\zeta}_1)}{4 + (k-1)q^2}, \quad p_2 = 0.$$
(3.379)

3.8 Radiative corrections to bremsstrahlung

The electromagnetic interaction of the electron (and positron) is given accurately by quantum electrodynamics (QED). Since the interaction, characterized by the fine-structure constant $\alpha \approx 1/137$, is relatively weak, a perturbation treatment is appropriate. Most applications of QED are performed in lowest order with respect to the interaction with the radiation field. In discussing radiative corrections to a QED process, the terms 'virtual photons' and 'real photons' are often referred to.

- By real photons, we mean photons corresponding to Feynman diagrams in which only one end of the photon line is attached to some particle line such as those in Fig. 3.3.
- Virtual photons occur in diagrams in which both ends of the photon line are attached, e.g., to electron lines or to a potential such as those in Fig. 3.4.

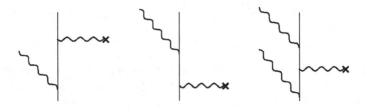

Fig. 3.3 Diagrams corresponding to the emission of real photons.

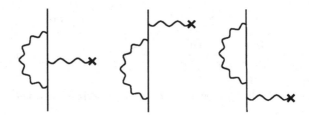

Fig. 3.4 Diagrams corresponding to the emission of virtual photons.

The effects of both real and virtual photons enter the experimentally measured cross section, which includes the radiative corrections. The important distinction from the theoretical point of view is that the square of the four-momentum of real photons is zero (real photons have zero mass), $k^2 - \mathbf{k}^2 = 0$, whereas this is not the case for virtual photons. By radiative corrections to the bremsstrahlung process we mean the corrections due to the emission and reabsorption of virtual photons and to the emission of any number of real soft photons with a total energy less than some small Δk [80]. Each of the two contributions is divergent in the limiting case that the photon energy approaches zero, $k \to 0$. This behaviour is known as the infrared catastrophe. It is found, however, that the infrared divergence occurring in the cross section of the virtual-photon radiative correction through a term $\ln \lambda$ (where λ is the fictitious photon mass) is removed by adding the real-soft-photon radiative correction (two-photon bremsstrahlung).

Radiative corrections to the bremsstrahlung cross section can be calculated with a huge amount of labour using standard theory of quantum electrodynamics. Moreover, they depend sensitively on the experimental conditions (energy and angular resolutions). The corrections are of relative order α. Calculations have been carried out in Born approximation by Fomin [81] and Mitra et al. [82]. The analytic expressions obtained are rather complicated and no numerical evaluation of these formulae has been given. Besides, the results seem to disagree in the extreme-relativistic limit [82].

In general, radiative corrections to bremsstrahlung are negligible at low energies (up to a few GeV). This may not hold near the short-wavelength limit which has been studied by McEnnan and Gavrila [73].

Chapter 4

Experiments on the elementary process of electron-nucleus bremsstrahlung

4.1 Survey of experimental devices

The schematic diagram of Fig. 4.1 shows the main components employed in coincidence experiments to study the elementary processes of electron-nucleus and electron-electron bremsstrahlung. In the following short compilation of the individual components we refer only to corresponding sections of this book where the pertinent references are given.

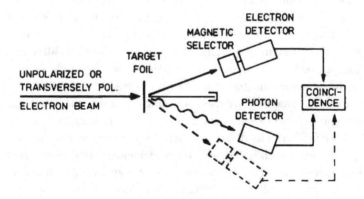

Fig. 4.1 Schematic diagram of experimental arrangements. The dashed-lined blocks mark a second magnetic selector with electron detector used for the second outgoing electron in electron-electron bremsstrahlung experiments (Section 6.2.1).

(a) Electron beams. The experiments were mostly performed with the help of electrostatic accelerators in an energy range of order 100 keV. For unpolarized beams the usual directly heated tungsten cathodes were employed. The source of the spin-polarized electron beam used the photoemission of electrons from a GaAsP crystal irradiated by circularly polarized light of a helium-neon laser (Sec. 4.3.2).

(b) Targets. In all experiments of the topic discussed here thin solid foils are used as targets with thickness mostly between 10 and 50 $\mu g/cm^2$ to keep plural scattering of electrons as small as possible. Nevertheless, there remains a finite probability that the electrons will be scattered (Mott, Møller) before and/or after the bremsstrahlung process. The influence of plural scattering was checked by using targets of different thickness or by Monte Carlo calculations. In comparison with theoretical results the environment of the target atom is usually ignored.

(c) Photon detectors. For the detection and energy analysis of bremsstrahlung these are used: Scintillation detectors (Sec. 4.2 and Sec. 4.3), lithium-drifted germanium detectors (Sec. 4.3), and intrinsic germanium detectors (Sec. 4.2.3 and Sec. 6.1.2). Those detectors sensitive to the linear polarization of the photons depend on the polarization dependence of Compton scattering (Sec. 4.3.1 and Sec. 6.2.1).

(d) Electron detectors. For the detection of the decelerated outgoing electrons scintillation detectors as well as surface barrier detectors (Sec. 4.2) and lithium-drifted silicon detectors (Sec. 6.2.1) were used. It is essential to avoid detecting elastically scattered electrons since their count rates in general are orders of magnitude higher than those of electrons after radiative scattering. In order to eliminate the elastically scattered electrons before reaching the detector a magnetic selector is inserted in front of the detector. Mostly, dispersive magnets are used, focusing electrons of a well-defined energy onto the detector (Sections 4.2, 4.3, and 6.2). Thus it is possible to use plastic scintillators as detectors with poor energy resolution but good timing properties, which is desirable for coincidence experiments. However, also non-dispersive and doubly focusing (i.e., triply focusing) magnets were employed in order to eliminate elastically scattered electrons but to transmit the inelastically scattered electrons within a broad energy range onto an energy dispersive (semiconductor) detector. With such a device two-parameter coincidence measurements in an energy-sharing experiment were performed (Sec. 6.1.2). The dashed-lined blocks in Fig. 4.1 represent a detector for the second outgoing electron of e-e bremsstrahlung composed of a non-dispersive magnet and a lithium-drifted silicon detector (Sec. 6.2.1).

4.2 Electron-photon coincidence experiments without regard to polarization variables

In this section we discuss electron-photon coincidence experiments using unpolarized incident electron beams as well as detectors not sensitive to polarization of the outgoing electrons and the emitted photons.

4.2.1 Angular distributions of photons for fixed directions of outgoing electrons

Measurements of photon angular distributions for fixed directions of the outgoing electrons were reported by Nakel, beginning in 1966 [2; 3; 4]. The incident electron energy was 300 keV, the targets consisted of gold foils. Outgoing electrons with an energy of 170 keV and scattering angles of $0°, 5°$, and $10°$ were selected (Fig. 4.2a-c). Since the experimental values were only relative, they were normalized at the maximum of the angular distributions to the theoretical curves of Elwert and Haug [13] (see Sec. 3.4) in the absence of anything better. Although the condition $\alpha Z \ll 1$ for the validity of the calculation is not satisfied for $Z = 79$, the measured angular distributions are in qualitative agreement with the Elwert-Haug theory. First measurements of the absolute cross section at a single point of the angular distribution revealed that the calculation underestimates the cross section for gold [83], whereas for aluminum ($Z = 13$) good agreement was found [84].

In 1972 an angular distribution of the absolute triply differential cross section for silver targets ($Z = 47$) was reported by Aehlig and Scheer [85]. Figure 4.3 depicts the experimental cross sections in comparison with the partial-wave results of Keller and Dreizler [17] and of Shaffer et al. [15], and the results of Elwert and Haug [13] including form-factor screening. The partial-wave results of Keller and Dreizler (and those of Tseng [18] not shown here) are very similar to the Elwert-Haug values, but they do not agree with those of Shaffer et al., in marked contrast to other cases. The disagreement between the partial-wave results is probably due to the fact that the calculation of Shaffer et al. does not include enough partial-wave terms [18]. It is interesting to note that the cross sections calculated in second Born approximation by Deck, Moroi, and Alling [62] (see Section 3.4.7) deviate even more from the experimental values [86], indicating that the Born expansion converges poorly for high atomic numbers.

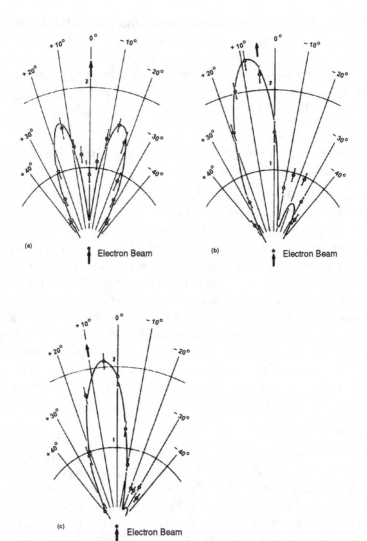

Fig. 4.2 (a)-(c) Measured (circles) and calculated (full line) photon angular distributions for fixed directions of outgoing electrons. Incident electron energy $E_0 = 300$ keV, outgoing electron directions of (a) $0°$, (b) $5°$, and (c) $10°$, and energy $E_e = 170$ keV; atomic number $Z = 79$. The theoretical curves are calculated by the theory of Elwert and Haug [13] (from Nakel [4]).

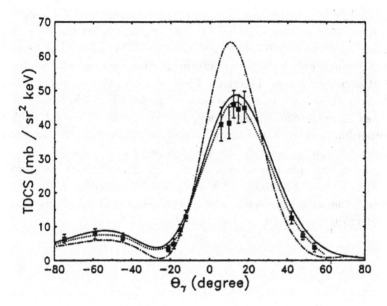

Fig. 4.3 Experimental photon angular distribution for fixed direction of outgoing electrons ($\theta_e = 30°$) in comparison with calculations of Keller and Dreizler [17] (full curve), Shaffer et al. [15] (chain curve), and Elwert and Haug [13] (dotted curve). Silver target ($Z = 47$), $E_0 = 180\,\text{keV}$, $h\nu = 80\,\text{keV}$ (from Keller and Dreizler [17]).

For carbon targets ($Z = 6$), where the results of the Elwert-Haug and first Born calculations differ very little, Nakel and Pankau [87] found good agreement with their experimental values.

In all these (coplanar) measurements it turned out that there is a strong angular correlation with the photons being predominantly emitted on the same side relative to the primary beam as the decelerated electrons. This behaviour can be understood classically by considering the radiation of the electrons along the hyperbolic orbit around the nucleus as discussed in Section 2.1.

4.2.2 Angular distributions of electrons for fixed photon directions

The measurement of an angular distribution of decelerated outgoing electrons for a fixed photon emission direction of 2° was reported by Hub and Nakel [88]. The incident electron energy was 300 keV, the target was aluminum (atomic number $Z = 13$). A magnetic spectrometer selected outgoing electrons of 220 keV, i.e., the energy of the emitted photons was relatively small (80 keV). The electron distribution (Fig. 4.4) is strongly peaked into the forward direction. Thus, photons with an energy that is far from the high-energy limit are radiated preferably by electrons that are deflected through relatively small angles in the nuclear Coulomb field. On the other hand, at the high-energy limit where all the kinetic energy of the incident electron is radiated, the electron angular distributions are more isotropic, as shown in the Figures 4.5 and 4.6. The distributions were calculated by means of Sommerfeld-Maue functions according to Eq. (3.212). Although the outgoing electrons have zero velocity, the cross sections are dependent on the direction of the final electrons.

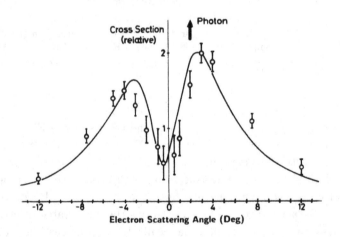

Fig. 4.4 Angular distribution of decelerated outgoing electrons for fixed photon direction $\theta_k = 2°$. The curve represents the theory of Elwert and Haug [13]. Incident electron energy 300 keV, outgoing electron energy 220 keV, and atomic number $Z = 13$ (from Hub and Nakel [88]).

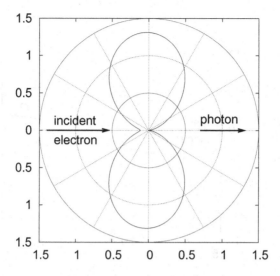

Fig. 4.5 Calculated angular distributions of decelerated outgoing electrons at the short-wavelength limit for fixed photon angle $\theta_k = 0°$ and coplanar geometry. The incident electron energy is $E_0 = 300$ keV and the target atomic number $Z = 13$.

4.2.3 Energy distributions for fixed electron and photon directions

The earliest measurements of energy distributions for fixed electron and photon directions were reported in 1973 by Faulk and Quarles [89] using aluminum and gold targets and an incident electron energy of 140 keV. The data were compared with the Elwert-Haug and with the Bethe-Heitler calculations. Both theories give generally satisfactory agreement for aluminum ($Z = 13$). For gold ($Z = 79$) neither theory agrees satisfactorily with the data even though the Elwert-Haug theory is somewhat more accurate.

In 1977 Aehlig, Metzger, and Scheer [90] reported an extensive investigation of the energy dependence of the absolute cross section for gold. The incident electron energy was $E_0 = 300$ keV. Measurements were made for four different combinations of electron scattering angle and photon emission angle over a wide range of the photon energy. An example is shown in Fig. 4.7. On the average, the deviations from the Elwert-Haug calculations (including screening) are somewhat less than those from the Bethe-Heitler formula. The partial-wave computations of Keller and Dreizler [17] are in

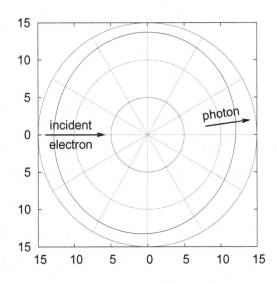

Fig. 4.6 Same as Figure 4.5 for the photon angle $\theta_k = 8°$.

agreement with those presented by Shaffer et al. [15] for hard-photon emission ($h\nu \geq 150$ keV) (note that the angles given in the caption of Fig. 1 of Ref. [15] are wrong). For the angle combination $\theta_e = 20°$, $\theta_k = -35°$, where the cross sections are much smaller, the partial-wave results of Keller and Dreizler and those of Tseng [18] agree quite well with the experimental data in the hard-photon region. However, there are large discrepancies between the theoretical curves at lower photon energies $h\nu < 150$ keV.

In measurements of Komma and Nakel [91] on gold targets a set of parameters was used where the calculations of Elwert and Haug and of Bethe and Heitler exhibit a distinctly different behaviour. Here, the measurements favour the Elwert-Haug calculation (Fig. 4.8) although the condition $\alpha Z \ll 1$ is not satisfied. In the same experiment further measurements of energy distributions were performed on carbon ($Z = 6$), copper ($Z = 29$), and silver ($Z = 47$). The results on carbon and copper are shown in Sec. 6.1.2 together with the results for electron-electron bremsstrahlung measured simultaneously (see Fig. 6.5).

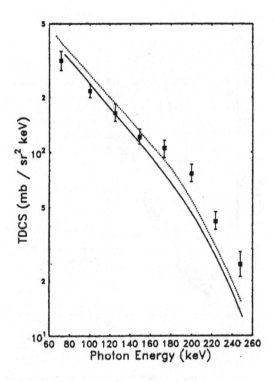

Fig. 4.7 Spectral distribution for electron scattering angle $\theta_e = 20°$ and photon emission angle $\theta_k = 10°$. Gold target ($Z = 79$), incident electron energy $E_0 = 300$ keV. Full squares with error bars: experimental results of Aehlig et al. [90]; full curve: partial-wave calculation of Keller and Dreizler; dotted curve: theory of Elwert and Haug [13] (from Keller and Dreizler [17]).

4.2.4 Wide-angle bremsstrahlung experiments at very high energies

In order to test quantum electrodynamics for time-like four-momenta, coincidence experiments on wide-angle bremsstrahlung at incident electron energies of up to 10 GeV have been performed. Bernardini et al. [92] employed 900-MeV electrons of the Frascati Electron Synchrotron impinging on a hydrogen target. The three outgoing particles were detected in coincidence. No significant deviations between the measured and the Bethe-Heitler cross section, corrected for proton recoil and proton form factor, were found to within 5% accuracy. In the experiments carried out at the

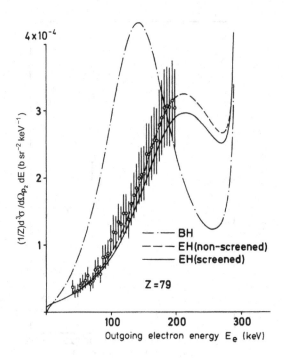

Fig. 4.8 Absolute triply differential cross section for bremsstrahlung from a gold target as a function of the outgoing electron energy; $E_0 = 300$ keV, $\theta_e = 20°$, $\theta_k = -35°$. The theoretical curves are calculated according to the Bethe-Heitler (BH) and Elwert-Haug (EH) theories (from Komma and Nakel [91]).

Cornell 10-GeV synchrotron [93; 94], electrons of 2.5 to 9.5 GeV struck carbon and copper targets. The momentum transfer q to the nucleus was kept low by requiring that the resultant transverse momentum of the final electron-photon system be zero. The experimental cross sections were compared with the predictions of the Bethe-Heitler formula corrected for nuclear form factors and radiative corrections. Again the agreement was good, corroborating the validity of quantum electrodynamics.

4.3 Electron-photon coincidence experiments including polarization variables

The most detailed independently observable quantity in the bremsstrahlung process is the triply differential cross section including all polarization correlations, corresponding to an electron-photon coincidence experiment with a spin-polarized primary beam and detectors sensitive to polarization of the photons and of the outgoing electrons [cf. Eq. (3.298) of Sec. 3.5.2]. Such a 'complete' scattering experiment has not yet been done and is not likely to be soon. The hitherto realized experiments on polarization correlations associated with the triply differential cross section yield the linear polarization of bremsstrahlung emitted by unpolarized electrons (Sec. 4.3.1) and the photon emission asymmetry of bremsstrahlung from transversely polarized electrons (Sec. 4.3.2).

4.3.1 *Linear polarization of bremsstrahlung emitted by unpolarized electrons*

Bremsstrahlung produced by a beam of unpolarized electrons in the field of an atomic nucleus is, in general, partially linearly polarized depending on the photon energy, the photon emission angle, the initial electron energy, and the atomic number of the target (see, for example, Refs. [28] and [95]).

In experiments without observation of the decelerated outgoing electrons in coincidence with the photons, the resulting polarization is produced by averaging over many elementary processes with different directions of the outgoing electrons. Thus the polarization behaviour in non-coincidence experiments is rather complex and provides only very limited information on the elementary radiation process itself.

Electron-photon coincidence experiments which involve measuring the angular dependence of the photon linear polarization for fixed directions of outgoing electrons were reported by Behncke and Nakel [96; 97] for a carbon target and by Bleier and Nakel [98] for copper and gold targets.

A schematic view of the experimental arrangement used in both experiments [97; 98] is shown in Fig. 4.9. A beam of unpolarized electrons of 300 keV impinges on a thin target. Coplanar geometry was used; the momenta of the incoming and outgoing electrons and the photon all lie in the same plane, the reaction plane. Outgoing electrons of definite energy are selected from the continuous spectrum by means of an electron detector

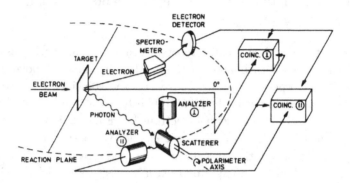

Fig. 4.9 Schematic view of the experimental arrangement. For the sake of simplicity, only two analyzers are shown instead of the four used [97].

(magnetic spectrometer and plastic scintillation detector). The photon polarization analyzer depends on the polarization sensitivity of the Compton process, and consists of a central scatterer and four analyzers [NaJ(Tl)] to measure the scattered photons in the reaction plane and perpendicular to it (in Fig. 4.9 only two analyzers are shown). In order to reduce the background, a plastic scintillation detector was used instead of a simple scatterer, employed in a previous measurement [96]. The degree of photon linear polarization is defined by the expression

$$P_l = \frac{I_\perp - I_\parallel}{I_\perp + I_\parallel} \qquad (4.1)$$

where $I_\perp$ and $I_\parallel$ are the bremsstrahlung intensity components with electric vectors perpendicular and parallel, respectively, to the reaction plane. Complete polarization with the electric vector parallel to the reaction plane thus corresponds to $P_l = -1$.

P_l corresponds to the polarization coefficient C_{030} according to the classification scheme of Pratt and Feng (cf. Sec. 3.5.2). The meaning of the three indices is as follows (from left to right): the primary electron beam is unpolarized, the linear polarization of the photon is measured, and the spin polarization of the outgoing electron is ignored.

The pulses from the electron detector and the photon detectors are fed to two conventional coincidence systems. Acceptable events were those producing a coincidence between the electron detector, the central scatterer, and either the horizontal or vertical analyzers. The quantity which is measured directly is the ratio of the counting rates of the true coincidences $N_\perp/N_\parallel$. To obtain the linear polarization from the measured data, the asymmetry ratio R of the Compton polarimeter must be known. R is the ratio of counting rates which one would obtain in the analyzer perpendicular and parallel to the reaction plane for a photon beam completely polarized in this plane. R was calculated by means of a Monte Carlo program. The polarization P_l, the ratio of the coincidence rates $N_\perp/N_\parallel$, and the asymmetry ratio R are connected through the relation

$$P_l = \frac{R+1}{R-1} \frac{1 - N_\perp/N_\parallel}{1 + N_\perp/N_\parallel}.\quad(4.2)$$

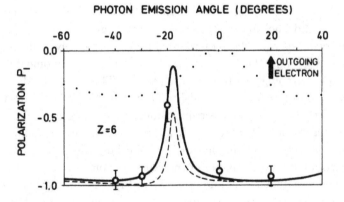

Fig. 4.10 Linear photon polarization in the elementary bremsstrahlung process as a function of the photon-emission angle. Primary electrons of 300 keV are used, incident on a carbon target; the outgoing electrons have energy 140 keV and $\theta_e = +20°$. The experimental values are shown by open circles. The theoretical curve (solid line) was calculated for the present experimental situation by means of the Elwert-Haug theory [13]. The dashed curve shows the theoretical polarization without corrections for the finite solid angles, the finite energy width of the magnetic electron spectrometer, and the electron plural scattering in the target. The correction for the contribution of electron-electron bremsstrahlung, however, was performed. For comparison, the dotted curve gives the calculated degree of polarization integrated over all directions of the outgoing electron (from Behncke and Nakel[97]).

In the experiment of Behncke and Nakel [97] the targets consisted of carbon foils (thicknesses 37–48 μg/cm^2). The measurement was performed at five different photon-emission angles for outgoing electrons of the fixed direction $\theta_e = 20°$ and energy $E_e = 140$ keV. The incident electron energy was 300 keV. The results are shown in Fig. 4.10 (open circles). The theoretical curve (solid line) was calculated by means of the Elwert-Haug theory [13]. It includes corrections for (i) the finite solid angles and energy width of the detectors, (ii) the plural scattering of the electrons in the target, (iii) a contribution of electron-electron bremsstrahlung (see Fig. 4.11).

A calculation in Born approximation (cf. Sec. 3.7.2) yields practically the same result for these parameters except for the region of lowest polarization $|P_l|$. Thus, Coulomb effects might still play no significant role for the low atomic number $Z = 6$.

The theoretical degree of polarization, calculated by means of Sommerfeld-Maue functions (cf. Sec. 3.7) and represented without taking into account corrections, is shown in Fig. 4.11a (solid line). Looking at the angular dependence of the polarization of the electron-nucleus bremsstrahlung (Fig. 4.11a, solid line) and at the corresponding cross section (Fig. 4.11b, solid line) one sees that the radiation is almost completely polarized with the electric vector in the reaction plane, containing the momenta of the incoming and outgoing electron and the photon. This is in accordance with the conception of the simple physical picture that the radiation emitted in the plane of the orbit of the electron is linearly polarized with the electric vector in this plane. However, such a simple model cannot explain the deviation from the complete polarization, especially the strong decrease of the polarization at the minimum of the cross section.

In a theoretical analysis Fano, McVoy, and Albers [99] investigated the contributions to the bremsstrahlung arising from the orbital motion of the electron and a change of spin orientation. They point out that orbital motion and spin changes are not independent and that the contributions of orbital motion and change of spin orientation yield, respectively, linearly and circularly polarized bremsstrahlung. Interference effects of both contributions yield observable features even if the incident electron beam is unpolarized, e.g., the dominant component of the electric vector may be perpendicular to the reaction plane. However, when the momenta of the incoming and outgoing electron and the photon all lie in the same plane, the entire contribution of the orbital motion is polarized in that plane. There-

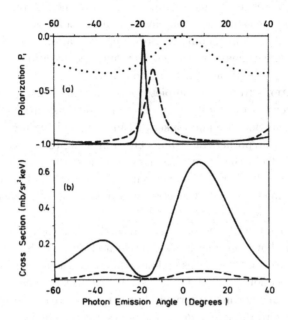

Fig. 4.11 (a) Theoretical photon linear polarization (uncorrected) in the elementary process of electron-nucleus bremsstrahlung (———) and electron-electron bremsstrahlung (- - - -) as a function of the photon-emission angle, calculated for the experimental parameters of Fig. 4.10. For comparison with the elementary process, the theoretical polarization of electron-nucleus bremsstrahlung is shown, integrated over all directions of the outgoing electron (· · · · ·). (b) Theoretical triply differential cross section of electron-nucleus bremsstrahlung (———) and electron-electron bremsstrahlung (- - - -) as a function of the photon emission angle, calculated for the experimental parameters and with corrections for the finite solid angles and the finite energy width of the magnetic electron spectrometer. The cross section of electron-electron bremsstrahlung is multiplied by a factor of 6 according to the number of electrons in the carbon atom (from Behncke and Nakel [97]).

fore, deviations from the complete polarization in the reaction plane must be caused by the influence of the spin-flip radiation. The strong decrease of the polarization in the minimum of the cross section indicates that the influence of the spin is particularly strong there. The decrease of the polarization could be confirmed by the measured value at $-20°$ (Fig. 4.10). In the further course of the polarization, as depicted in Fig. 4.11a, the degree of polarization generally does not reach $P_l = -1$, either. Only when the momentum transfer to the nucleus, **q**, is parallel to the photon-emission direction, is the bremsstrahlung completely linearly polarized, as was shown in Section 3.7.2 (this holds only in Born approximation). With the parameters of the experiment considered, maximum polarization occurs at the photon angle $\theta_k \approx -29°$ according to Eq. (3.366).

The deviations from the complete polarization in the reaction plane should not occur in the radiation process of a spinless particle. To verify this expectation, Behncke and Nakel [97] investigated the bremsstrahlung polarization of a hypothetical spinless electron using the cross-section formula following from the Duffin-Kemmer equation, the wave equation for scalar particles [33]. For coplanar geometry the linear polarization turns out to be complete with the electric vector lying in the reaction plane ($P_l = -1$).

For comparison with the polarization in the elementary process, Figs. 4.10 and 4.11 show the polarization integrated over all directions of the outgoing electrons (dotted lines). This curve was calculated in Born approximation (Section 3.7.2) and corresponds to a measurement in which the outgoing electrons are not observed. For symmetry reasons the radiation emitted in forward direction ($\theta_k = 0°$) shows a complete absence of linear polarization, whereas the polarization of the elementary process is nearly complete in this direction. The general features of the integrated polarization are discussed in terms of orbital motion and spin effects by Motz and Placious [28].

The calculations by means of Sommerfeld-Maue wave functions for the linear polarization of the elementary bremsstrahlung process (Sec. 3.7.1) show that the decrease of the polarization at the minimum of the cross section diminishes strongly with increasing target atomic number Z (Fig. 4.12), whereas in Born approximation, disregarding the distortion of the electron wave functions by the Coulomb field, the polarization is not dependent on Z. Stimulated by these findings, Bleier and Nakel [98] used targets of copper ($Z = 29$) and gold ($Z = 79$) and measured the angular dependence of the photon linear polarization in the elementary process of bremsstrahlung. The results are shown in Fig. 4.13a for copper and in Fig. 4.13b for gold (open circles). According to definition (4.1), the negative values of polarization indicate that the radiation is polarized parallel to the emission plane.

All theoretical curves were computed for the present experiment taking into account detector apertures and plural scattering; screening is neglected.

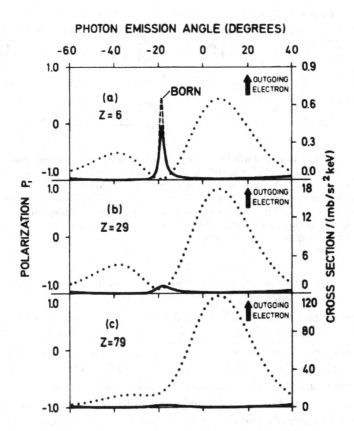

Fig. 4.12 Theoretical photon linear polarization in the elementary process of electron-nucleus bremsstrahlung as a function of photon-emission angle computed for point geometry by means of Sommerfeld-Maue functions (solid lines) for (a) carbon, (b) copper, and (c) gold. Screening is neglected. The dashed line in (a) represents a calculation in Born approximation which for the other angles is virtually coincident with the solid curve. In Born approximation, polarization is not dependent on Z. Dotted lines represent triply differential cross sections computed with the Elwert-Haug formula ([13]). Incident electron energy 300 keV; outgoing electron energy 140 keV; scattering angle 20° (from Bleier and Nakel[98]).

The Born approximation (dashed lines, calculated according to Sec. 3.7.2) assumes $\alpha Z \ll \beta_2$, where β_2 is the velocity of the outgoing electron in

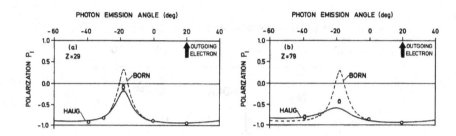

Fig. 4.13 Photon linear polarization in the elementary bremsstrahlung process as a function of the photon-emission angle for outgoing electrons of +20° and 140 keV. Primary electrons of 300 keV are used incident on (a) copper and (b) gold targets. Dashed lines represent calculations in Born approximation; solid lines represent calculations with Sommerfeld-Maue wave functions. All theoretical curves were computed for the experiment considered taking into account detector apertures and plural scattering. Screening is neglected (from Bleier and Nakel[98]).

units of the velocity of light. The calculation by means of Sommerfeld-Maue functions assumes $\alpha Z \ll 1$. Although the assumptions of the calculations are poorly satisfied for medium and large Z, the agreement with the measurements is acceptable except for the range near the minimum of the cross section where both calculations fail. However, the discrepancies are smaller for the calculation by means of SM functions than for the Born approximation. In the measurement for $Z = 6$ [97], agreement with the theories was obtained within experimental errors. Further computations of the polarization according to the partial-wave method [15; 17; 18; 100] would be interesting (cf. Sec. 3.5.2).

4.3.2 Photon emission asymmetry of bremsstrahlung from transversely polarized electrons

Bremsstrahlung emitted by transversely polarized electrons shows a left-right asymmetry in spatial distribution if the spin polarization is perpendicular to the reaction plane. Disregarding the decelerated outgoing electrons such left-right photon emission asymmetries were measured by several groups [101; 102; 103]. However, since in all these experiments only the emitted photons were observed, the results are necessarily averaged over all electron scattering angles, implying small values of the asymmetry. Therefore the test of theoretical predictions is not as strong as it could be.

To get detailed information on the elementary collision process Mergl et al. [104] performed a coincidence measurement between outgoing electrons and photons emitted by a transversely polarized beam. According to the classification scheme of Pratt and Feng [19] the pertinent polarization coefficient is denoted as C_{200}. The meaning of the three indices is as follows (from left to right): the primary electron beam is transversely polarized with the spin perpendicular to the reaction plane, the polarization of the photon is not measured, and the polarization of the outgoing electron is not measured. Calculations in the first-order Born approximation yield zero emission asymmetry (cf. Section 3.6). Measurements of the asymmetry, therefore, are a proper test for theories going beyond the first Born approximation.

A sketch of the experimental arrangement is shown in Fig. 4.14. The source of the polarized electron beam (described in detail in Ref. [105]) used the photoemission of electrons from a GaAsP crystal irradiated by circularly polarized light of a helium-neon laser. The source is contained in a high-voltage terminal of a 300-kV accelerator tube and produces a continuous transversely polarized beam with a polarization degree in the range of 35% to 40%. The bremsstrahlung photons from the target leave the vacuum chamber through a thin plastic window before entering a high-purity germanium detector. The electron detector system consisted of a magnetic spectrometer for the energy analysis combined with a plastic-scintillation detector.

There are two experimental methods of obtaining spin-dependent bremsstrahlung emission asymmetry: (1) by measuring the relative, triply differential cross sections for fixed positions of the photon and electron detectors but for two opposite orientations of the primary electron spins; (2) by measuring them for fixed spin orientation but changing the angular position of both the photon and the electron detector symmetrically to the direction of the primary beam. The first method was used since the arrangement allows to easily change the orientation of the electron spin by changing the helicity of the circularly polarized light. In this way many possible instrumental asymmetries do not enter.

The quantity measured directly is the counting rate of the true coincidences alternately for spin-up and spin-down electrons of the primary beam. The asymmetry coefficient C_{200} is obtained as the ratio $C_{200} = A/P$, where P is the polarization of the beam and, for spin-up and spin-down counting rates $n_\uparrow$ and $n_\downarrow$, respectively, $A = (n_\uparrow - n_\downarrow)/(n_\uparrow + n_\downarrow)$.

138 Experiments on the elementary process of electron-nucleus bremsstrahlung

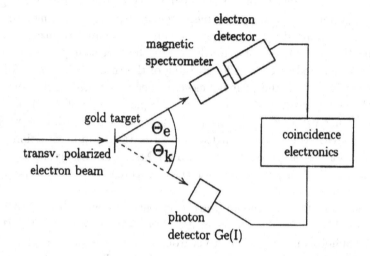

Fig. 4.14 Sketch of the coplanar electron-photon coincidence experiment. The spin direction of the primary electron beam is perpendicular to the reaction plane (from Mergl et al. [104]).

A result of the measurement of C_{200} as a function of the photon emission angle θ_k for fixed outgoing electron angle $\theta_e = 45°$ is shown in Fig. 4.15. The full curve depicts a calculation of Haug [75], and the dashed curve is the pertinent triply differential cross section for unpolarized primary electrons from the theory of Elwert and Haug [13] using Sommerfeld-Maue eigenfunctions (cf. Sec. 3.4.3). The region of validity can be expressed by $\alpha Z \ll 1$. Although the condition is not satisfied for gold ($Z = 79$), the measurements are in qualitative agreement with the calculations. The experimental data agree excellently with the partial-wave computations of Tseng [18] except for one data point. The partial-wave results also make it clear that the effect of atomic screening on C_{200} is not significant.

Comparison of the calculated triply differential cross section with the pertinent photon emission asymmetries shows that high asymmetries occur in regions of small cross sections, whereas in the region of the large lobe the asymmetry is very small. Notice in this context the open diamonds in Fig. 4.15, giving non-coincident photon asymmetries C_{20} measured simultaneously. The values of C_{20} are very small compared to C_{200} and have the opposite sign, as this particular measurement integrates over all angles of

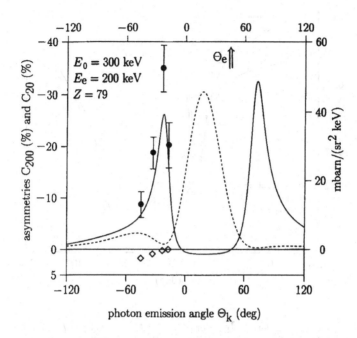

Fig. 4.15 Photon emission asymmetry C_{200} (full circles) as a function of the photon emission angle θ_k for incident electron energy 300 keV and outgoing electrons of energy $E_e = 200$ keV and scattering angle $\theta_e = 45°$. The solid line displays the emission asymmetry from the theory of Haug [75], and the broken line the pertinent triply differential cross section for unpolarized primary electrons [13]. The error bars represent the standard deviations only; the systematic error of the asymmetry scale was estimated to be ±2%. The open diamonds give the non-coincident photon asymmetry coefficients C_{20} measured simultaneously (from Mergl et al. [104]).

the outgoing electrons. Thus small asymmetries associated with large cross sections mask the large asymmetries with small cross sections. In comparison with it, the results of the coincidence experiment yield very detailed information on the elementary collision process.

Further computations of the photon asymmetry by means of the partial-wave method (cf. Sec. 3.5.2) have been performed for the same energies and $\theta_e = 20°$ [15; 17]. Figure 4.16 presents the data of Shaffer et al. [15] compared with the experimental results of Mergl et al. [104] and the theory of Haug [75]. The agreement between the two theories is quite

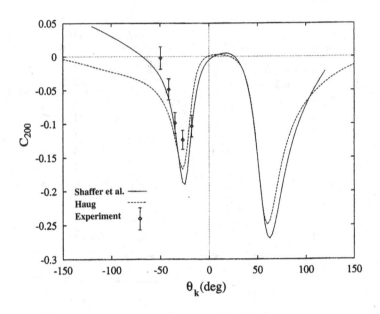

Fig. 4.16 Photon asymmetry, C_{200}, for $Z = 79$, $E_0 = 300$ keV, $E_e = 200$ keV, and $\theta_e = 20°$, as a function of photon angle. The curves represent the theoretical results of Shaffer et al. [15] (partial-wave method) and of Haug [75] (use of SM wave functions), experimental data are from Mergl et al. [104] (from Shaffer et al. [15]).

good. It appears that even for $Z = 79$ the Haug results for the polarization variable C_{200} are close to the partial-wave predictions in the cases which have been studied. The partial-wave calculations agree relatively well with the asymmetry experiments.

Finally, Mergl et al. [104] have studied the interesting special case of the electron detector being placed in the direction of the primary beam to detect outgoing decelerated electrons with deflection angle $\theta_e = 0°$. Then, for an unpolarized primary beam the corresponding photon angular distribution has to be symmetric about the beam (see, e.g., Ref. [3]). For a transversely polarized beam, however, the authors did find an emission asymmetry. The measured values of C_{200} as a function of the photon emission angle are shown with error bars in Fig. 4.17. The solid line represents the calculation of Haug [75], and the dashed curve depicts the pertinent triply differential cross section [13] for unpolarized primary electrons. There are large dis-

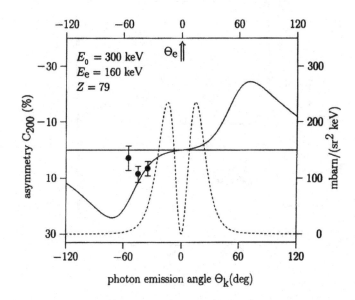

Fig. 4.17 Measured photon emission asymmetry C_{200} (full circles) as a function of the photon emission angle θ_k for gold target, incident electron energy $E_0 = 300$ keV, and outgoing electrons of scattering angle $\theta_e = 0°$ and energy $E_e = 160$ keV [104]. The error bars represent the standard deviations only; the systematic error of the asymmetry scale was estimated to be ±2%. The solid line depicts the emission asymmetry from a calculation of Haug [75], and the broken line is the pertinent triply differential cross section for unpolarized primary electrons [13] (from Mergl et al. [104]).

crepancies between the theoretical and experimental results which are not surprising since the Haug theory does not claim to be valid for high atomic numbers. These are largely removed by the partial-wave calculations of Tseng [18] which are presented in Fig. 4.18 as solid curve together with the calculation of Haug [75] (dashed curve) and the experimental data*. The dots displaying the partial-wave results for the point-dots displaying the partial-wave results for the point-Coulomb potential indicate that the effect of atomic screening on C_{200} is not significant even for high atomic numbers.

After an analysis of Sobolak and Stehle [29] the sign of the asymmetry is

*Note that the definitions of C_{200} in Figs. 4.17 and 4.18 have opposite signs.

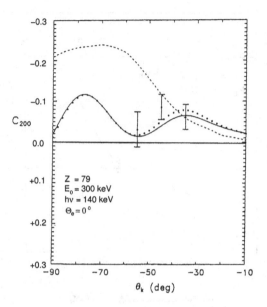

Fig. 4.18 Measured photon emission asymmetry C_{200} as a function of the photon emission angle θ_k for gold target, incident electron energy $E_0 = 300$ keV, and outgoing electrons of scattering angle $\theta_e = 0°$ and energy $E_e = 160$ keV [104]. The solid line and the dots show the emission asymmetry according to the partial-wave computations of Tseng [18] with the Kohn-Sham potential and point Coulomb potential, respectively, the dashed curve displays the theory of Haug [75] (adapted from Tseng [18]).

dependent on whether or not the electron spin flips in the bremsstrahlung emission process. If for primary electrons with spin up the radiation is more right than left — which is the case in the experiment — the spin-flip processes should be dominant. The authors also give a classical argument for the origin of the asymmetry considering the force on a magnetic dipole moving in the Coulomb field of an atomic nucleus.

Measurements of the spin asymmetry, C_{200}, as a function of the kinetic energy of the outgoing electrons for fixed angles of the outgoing electrons and the emitted photons were performed by Geisenhofer and Nakel [106].

All the asymmetry measurements show that the electron-photon coincidence technique combined with a polarized electron beam provides a very sensitive method for testing the bremsstrahlung theory.

4.3.3 Further polarization correlations

Apart from the correlations considered above where in each case only one polarization variable is involved, there are also correlations between the polarizations of the collision partners. We mention some examples, although for none of them the elementary process was measured so far.

A correlation that is well known is the production of circularly polarized bremsstrahlung by longitudinally polarized electrons (see Sec. 3.7.3). The process has been frequently used in parity non-conservation experiments for measuring the polarization of β particles [79]. A circular bremsstrahlung component, though with lower degree of polarization, is also produced by transversely polarized electrons (see Sec. 3.7.3).

Following Olsen [65], it is interesting to note that the time reversal argument shows that to the lowest order in the interaction for any process only correlations between an even number of polarizations are present. So in lowest order approximation unpolarized electrons do not produce circularly polarized bremsstrahlung unless the polarization of the final state electron is recorded in coincidence.

For a complete scattering experiment (cf. Sec. 4.3) the emitted photons are expected to be completely elliptically polarized.

4.4 Tagged photons: The application of the coincidence technique in producing quasi-monochromatic photon beams

In several fields of physics, such as atomic or nuclear physics, quasi-monochromatic photon beams are required. In the tagged photon method, the outgoing electron from the bremsstrahlung process is detected in coincidence with a particle produced by the photon-induced process to be studied, e.g., a nuclear-decay product following a photon-induced reaction.

Figure 4.19 shows a schematic diagram of the tagged photon method. An electron beam of known energy E_0 strikes a radiator (thin metallic target or a monocrystal) generating a stream of bremsstrahlung photons. The outgoing electrons are momentum analyzed in the field of a magnetic spectrometer and correlated in time with some subsequent reaction product of the bremsstrahlung photons. The energy of the tagged photon is given by $h\nu = E_0 - E_e$, where E_e is the energy of the outgoing electron.

The tagged photon method using thin metallic foils is described, e.g., in

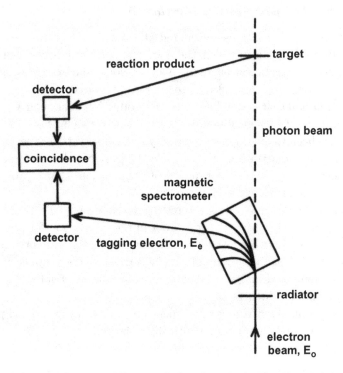

Fig. 4.19 Schematic diagram of the tagged photon method. The energy of the tagged photon is given by the difference between the incoming and outgoing electron energies, $E_0 - E_e$.

Refs. [107] and [108]. A considerable progress in producing monochromatic linearly polarized photons was achieved by combining the tagged photon method with the coherent bremsstrahlung[†] production on a monocrystal, e.g., a thin diamond crystal [109; 110]. Incoherent bremsstrahlung is the sum of contributions from atoms acting independently. In the case of coherent bremsstrahlung on a monocrystal the recoil is taken up by the entire lattice (without creation or annihilation of a phonon). The regular structure of the atoms in the crystal within a coherence volume enhances the radiation of polarized bremsstrahlung at certain energies and angles because the

[†]The adjective 'coherent' in coherent bremsstrahlung does not indicate that the photons in the beam are in a coherent state, as is light from a laser. Rather it refers to the coherent effect of multiple atoms in the crystal lattice.

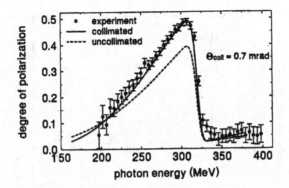

Fig. 4.20 Linear polarization of tagged photons from a crystalline radiator. The degree of polarization of the collimated coherent bremsstrahlung is measured by means of the beam asymmetry of coherent π^0 photoproduction (from Rambo et al. [109]).

lattice allows only particular directions of the momentum transfer. Bremsstrahlung from a monocrystalline target consists of both coherent and incoherent radiation. By suitable orientation of the crystal with respect to the incident electron beam coherent bremsstrahlung may dominate in a selected range showing strong linear polarization. An example is shown in Fig. 4.20.

All efforts to understand the tagging process lead naturally to a study of the elementary bremsstrahlung process itself. In a theoretical analysis Maximon et al. [111] investigated in detail the triply differential cross section of both polarized and unpolarized bremsstrahlung from electrons, for angles and energies in the range of interest for a photon tagging system (50 MeV to 1 GeV). The calculations were performed in first Born approximation (Bethe-Heitler formula). One interesting aspect of the findings is the pronounced structure, a sharp dip in the region of very small momentum transfers $\mathbf{q}$, which exists in the angular distribution of the triply differential cross section but is absent from the cross sections integrated over the angles of either the photon or the final electron. This structure is found both in the cross sections for polarized photons and in the cross section summed over photon polarizations. The reason for the dip is the strong dependence of the cross sections on the photon and electron angles at very high energies. Consider the Bethe-Heitler cross section written in the form (3.82). The

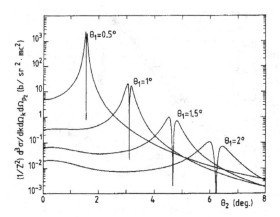

Fig. 4.21 Triply differential cross section in Born approximation for unpolarized bremsstrahlung as a function of the electron angle θ_2 for various photon angles θ_1 in coplanar geometry. Incident electron energy $E_0 = 140$ MeV, photon energy $h\nu = 95$ MeV (from Maximon et al. [111]).

two terms $\mathbf{B}^2$ and $\mathbf{C}^2$ in the curly brackets are proportional to q^2 and are therefore small. If now the first term $\mathbf{A}^2$ vanishes, the cross section falls off within a very narrow angular range (see Problem 4.1). For the unpolarized cross section, examples of the dip as a function of the electron angle θ_2 are shown in Fig. 4.21 for coplanar geometry and various photon angles. Coulomb corrections to the Born-approximation cross sections are found not to fill in the dips, in agreement with the fact that the Born approximation is accurate at very high energies.

The dips of the cross sections for photon polarization perpendicular and parallel to the production plane have a different dependence on the angles of the final electron and photon over most of the photon spectrum. This behaviour results in rather large degrees of polarization in the dips which may be utilized in experiments. Thus the linear polarization of a tagged photon beam can be greatly enhanced by means of a kinematic selection of the outgoing electrons [?]. One has, however, to bear in mind that an evaluation of the cross section for a specific set of discrete angles could give totally incorrect values for the cross section integrated over the finite solid angles of the detectors.

No experimental verification of the theoretical findings concerning the dips is known to us.

Chapter 5

Theory of the elementary process of electron-electron bremsstrahlung

5.1 Introduction

There are two main differences between bremsstrahlung produced in the static Coulomb field of a nucleus and bremsstrahlung emitted in the collision of two electrons:

- Whereas the nuclear recoil can be neglected in most cases, the recoiling electron changes the kinematics of the problem and gives rise to retardation effects.
- Exchange effects occur for electron-electron collisions.

For that reason 8 Feynman diagrams contribute to the matrix element (Fig. 5.1), and the expression for the bremsstrahlung cross section in the field of an electron is much more complicated than the corresponding Bethe-Heitler formula for the nuclear Coulomb field. The first calculation of the fully differential cross section for unpolarized electrons, exact to first-order perturbation theory, was achieved by Hodes [113]. Since the formula is extremely lengthy it is nearly impossible to evaluate it without employing computers. Therefore the common practice was to use approximations, either by considering the nonrelativistic and the high-energy limits or by neglecting some of the eight Feynman diagrams. In the first case the cross sections for the intermediate energy region had to be evaluated by interpolation, in the second case the importance of the missing terms was not well understood except for very high energies.

In electron-electron collisions exchange implies a repulsion which depends strongly on the distance of the colliding particles in the phase space.

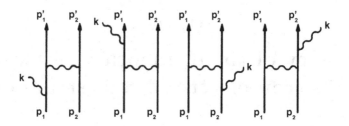

Fig. 5.1 Four of the 8 Feynman diagrams for e-e bremsstrahlung in lowest-order perturbation theory. The other 4 graphs are obtained by exchanging p_1 and p_2.

At high energies in the rest system of the target electrons the incoming electrons are scattered predominantly by very small angles, whereas the recoil particles move with low momenta at relatively large angles to the direction of incidence. Thus the two final electrons have a large distance in phase space and exchange can be neglected. At low energies, however, the momenta of the two electrons are more comparable and exchange effects are expected to be significant. Corresponding to the decrease of the phase space available the cross section is reduced.

In the photon emission accompanying the collision of two electrons the effect of one particle can no longer be replaced by an external field, contrary to electron-nucleus or electron-atom bremsstrahlung. The electric dipole moment of two particles with the charges e_1 and e_2 and the masses m_1 and m_2 in the center-of-mass system is given by

$$\mathbf{d} = \mu\left(\frac{e_1}{m_1} - \frac{e_2}{m_2}\right)\mathbf{r} \tag{5.1}$$

where $\mu = m_1 m_2/(m_1 + m_2)$ denotes the reduced mass and $\mathbf{r} = \mathbf{r}_1 - \mathbf{r}_2$ is the radius vector between the particles. Since the dipole moment of a two-electron system is equal to zero, the emission in the nonrelativistic collision of two electrons is electric quadrupole radiation.

5.2 Kinematics

In the elementary process of electron-electron (e-e) bremsstrahlung two electrons with the four-momenta $\underline{p}_1 = (\epsilon_1, \mathbf{p}_1)$ and $\underline{p}_2 = (\epsilon_2, \mathbf{p}_2)$ are colliding under the emission of a photon with the four-momentum $\underline{k} = (k, \mathbf{k})$. The outgoing electrons have the four-momenta $p'_1 = (\epsilon'_1, \mathbf{p}'_1)$ and $p'_2 = (\epsilon'_2, \mathbf{p}'_2)$ (Fig. 5.2). The kinematics of the process can be derived most simply from the energy-momentum conservation law

$$p_1 + p_2 = p'_1 + p'_2 + k \,. \tag{5.2}$$

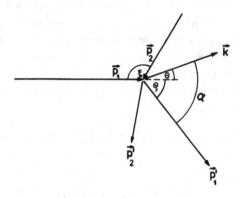

Fig. 5.2 Particle momenta and angles in the elementary process of e-e bremsstrahlung.

To this end we eliminate p'_2 by squaring the relation $p'_2 = p_1 + p_2 - p'_1 - k$, yielding*

$$(p_1 p_2) - (p_1 p'_1) - (p_2 p'_1) - (k p_1) - (k p_2) + (k p'_1) + 1 = 0 \tag{5.3}$$

or

$$\epsilon'_1(\epsilon_1 + \epsilon_2 - k) - \mathbf{p}'_1 \cdot (\mathbf{p}_1 + \mathbf{p}_2 - \mathbf{k})$$
$$= \epsilon_1(\epsilon_2 - k) - \epsilon_2 k + 1 - \mathbf{p}_1 \cdot \mathbf{p}_2 + \mathbf{k} \cdot (\mathbf{p}_1 + \mathbf{p}_2) \,. \tag{5.4}$$

Owing to (5.2) and (5.3) only 5 of the 10 possible invariant products are independent.

*The metric is such that the product of two four-vectors $a = (a_0, \mathbf{a})$ and $b = (b_0, \mathbf{b})$ is defined as $(ab) = a_0 b_0 - \mathbf{a} \cdot \mathbf{b}$. For electrons this yields $(pp) = \epsilon^2 - p^2 = 1$. Since the photon has zero rest mass we have $(kk) = k^2 - \mathbf{k}^2 = 0$.

Solving (5.3) for the photon energy k we get

$$k = \frac{(p_1 p_2) - (p_1 p'_1) - (p_2 p'_1) + 1}{t_0 - \mathbf{t} \cdot \hat{\mathbf{k}}}, \qquad (5.5)$$

where $\hat{\mathbf{k}}$ is the unit vector in photon direction and

$$t_0 = \epsilon_1 + \epsilon_2 - \epsilon'_1, \quad \mathbf{t} = \mathbf{p}_1 + \mathbf{p}_2 - \mathbf{p}'_1. \qquad (5.6)$$

Equation (5.5) represents the polar equation of an ellipse with the semilatus rectum $p = [(p_1 p_2) - (p_1 p'_1) - (p_2 p'_1) + 1]/t_0$ and the numerical eccentricity $e = |\mathbf{t}|/t_0 < 1$. Hence, if the angle ϑ between the vectors $\mathbf{k}$ and $\mathbf{t}$ varies, the photon momentum $\mathbf{k}$ moves on a rotational ellipsoid. The maximum (minimum) photon energy is reached for $\vartheta = 0$ ($\vartheta = \pi$).

Another important relation between invariant products is obtained by squaring Eq. (5.2) yielding

$$(p_1 p_2) = (p'_1 p'_2) + (p'_1 k) + (p'_2 k) = (p'_1 p'_2) + (p_1 k) + (p_2 k). \qquad (5.7)$$

Noting that

$$(p'_1 p'_2) = \epsilon'_1 \epsilon'_2 - \mathbf{p}'_1 \cdot \mathbf{p}'_2 \geq \epsilon'_1 \epsilon'_2 - p'_1 p'_2 \geq 1, \qquad (5.8)$$

we get

$$(p_1 k) + (p_2 k) \leq (p_1 p_2) - 1. \qquad (5.9)$$

This relation allows immediately to determine the maximum photon energy for arbitrary electron momenta $\mathbf{p}_1$ and $\mathbf{p}_2$:

$$k \leq k_{\max}(\mathbf{p}_1, \mathbf{p}_2) = \frac{(p_1 p_2) - 1}{\epsilon_1 + \epsilon_2 - |\mathbf{p}_1 + \mathbf{p}_2|}. \qquad (5.10)$$

The maximum photon energy occurs for $(p'_1 p'_2) = \epsilon'_1 \epsilon'_2 - \mathbf{p}'_1 \cdot \mathbf{p}'_2 = 1$ implying

$$\epsilon'_1 = \epsilon'_2 = \tfrac{1}{2}(\epsilon_1 + \epsilon_2 - k), \quad \mathbf{p}'_1 = \mathbf{p}'_2 = \tfrac{1}{2}(\mathbf{p}_1 + \mathbf{p}_2 - \mathbf{k}). \qquad (5.11)$$

That is, at the short-wavelength limit of e-e bremsstrahlung the outgoing electrons have equal momenta.

For the calculation of the cross section we will need the energy of the scattered electron[†], ϵ'_1, as a function of its direction $\hat{\mathbf{p}}'_1 = \mathbf{p}'_1/p'_1$, for fixed

---

[†]Since the two electrons are indistinguishable, the term 'scattered electron' implies either of the outgoing electrons.

photon momentum **k**. Solving Eq. (5.4) for ϵ_1' and p_1' we get

$$\epsilon_1' = \frac{(\epsilon_1 + \epsilon_2 - k)[(p_1'p_2') + 1] \pm \hat{\mathbf{p}}_1' \cdot (\mathbf{p}_1 + \mathbf{p}_2 - \mathbf{k})R}{(\epsilon_1 + \epsilon_2 - k)^2 - [\hat{\mathbf{p}}_1' \cdot (\mathbf{p}_1 + \mathbf{p}_2 - \mathbf{k})]^2} \quad (5.12)$$

and

$$p_1' = \frac{\hat{\mathbf{p}}_1' \cdot (\mathbf{p}_1 + \mathbf{p}_2 - \mathbf{k})[(p_1'p_2') + 1] \pm (\epsilon_1 + \epsilon_2 - k)R}{(\epsilon_1 + \epsilon_2 - k)^2 - [\hat{\mathbf{p}}_1' \cdot (\mathbf{p}_1 + \mathbf{p}_2 - \mathbf{k})]^2}, \quad (5.13)$$

where

$$R = \sqrt{[(p_1'p_2') + 1]^2 + [\hat{\mathbf{p}}_1' \cdot (\mathbf{p}_1 + \mathbf{p}_2 - \mathbf{k})]^2 - (\epsilon_1 + \epsilon_2 - k)^2}. \quad (5.14)$$

The choice of the signs in the numerators of (5.12) and (5.13) depends on the frame of reference considered. The product $(p_1'p_2')$ occurring in (5.12) to (5.14) can be substituted by means of the relation (5.7).

The square root R takes a simple form if the terms in (5.14) are expressed by primed quantities,

$$\begin{aligned} R &= \left[(\epsilon_1'\epsilon_2' - \mathbf{p}_1' \cdot \mathbf{p}_2' + 1)^2 + \{\hat{\mathbf{p}}_1' \cdot (\mathbf{p}_1' + \mathbf{p}_2')\}^2 - (\epsilon_1' + \epsilon_2')^2\right]^{1/2} \\ &= \left[p_1'^2 \epsilon_2'^2 + \epsilon_1'^2 (\hat{\mathbf{p}}_1' \cdot \mathbf{p}_2')^2 - 2\epsilon_1'\epsilon_2' (\mathbf{p}_1' \cdot \mathbf{p}_2')\right]^{1/2} \\ &= |\epsilon_2' p_1' - \epsilon_1' \hat{\mathbf{p}}_1' \cdot \mathbf{p}_2'|. \end{aligned} \quad (5.15)$$

At the short-wavelength limit $[\mathbf{p}_1' = \mathbf{p}_2'$, see Eq. (5.11)] this yields $R = 0$.

We will now specify the kinematics for the two most important cases, the center-of-mass system and the laboratory system, the rest system of electron 2.

5.2.1 *Center-of-mass system*

In the center-of-mass system (cms) the two incoming electrons have equal energies and opposite momenta, i.e., $\underline{p_1} = (\epsilon, \mathbf{p})$, $\underline{p_2} = (\epsilon, -\mathbf{p})$. Inserting this into (5.5) the photon energy is given by

$$k = \frac{2\epsilon(\epsilon - \epsilon_1')}{2\epsilon - \epsilon_1' + p_1' \cos\alpha}, \quad (5.16)$$

where α denotes the angle between the momenta **k** and $\mathbf{p}_1'$ (see Fig. 5.2). Clearly, the photon energy has its maximum for $\alpha = \pi$:

$$k_{\max}(\epsilon_1') = \frac{2\epsilon(\epsilon - \epsilon_1')}{2\epsilon - \epsilon_1' - p_1'}. \quad (5.17)$$

By setting the derivative $dk_{\max}(\epsilon'_1)/d\epsilon'_1$ equal to zero it follows that the absolute maximum of the photon energy occurs for $\epsilon'_1 = (\epsilon^2 + 1)/(2\epsilon)$ and $p'_1 = p^2/(2\epsilon)$, yielding

$$k_{\max} = p^2/\epsilon. \tag{5.18}$$

In this case the outgoing electrons have equal momenta [see Eq. (5.11)], and momentum conservation, $\mathbf{p}'_1 + \mathbf{p}'_2 + \mathbf{k} = 0$, requires $\mathbf{p}'_1 = \mathbf{p}'_2 = -\frac{1}{2}\mathbf{k}$. That is, the photon energy in the cms is maximum (short-wavelength limit of e-e bremsstrahlung) if the momenta of the outgoing electrons are opposite to half the photon momentum. The result (5.18) is immediately obtained by means of Eq. (5.10).

If Eqs. (5.12) to (5.14) are specialized to the cms, the energy and momentum of the scattered electron 1 at fixed photon energy and direction are respectively given by

$$\epsilon'_1 = \frac{2\epsilon(\epsilon-k)(2\epsilon-k) \mp k\cos\alpha R_c}{(2\epsilon-k)^2 - k^2\cos^2\alpha}, \tag{5.19}$$

$$p'_1 = \frac{\pm(2\epsilon-k)R_c - 2\epsilon k(\epsilon-k)\cos\alpha}{(2\epsilon-k)^2 - k^2\cos^2\alpha}, \tag{5.20}$$

where

$$R_c = \sqrt{4\epsilon^2(\epsilon-k)^2 - (2\epsilon-k)^2 + k^2\cos^2\alpha}. \tag{5.21}$$

The allowed ranges of k and α are determined from the condition that p'_1 be positive corresponding to $\epsilon'_1 > 1$. The denominator of the expressions (5.19) and (5.20) is always positive. Hence, for photon energies smaller than

$$k_0 = 2\epsilon(\epsilon-1)/(2\epsilon-1), \tag{5.22}$$

only the upper sign in (5.19) and (5.20) is a solution. For $k > k_0$ and

$$\cos\alpha < -\frac{1}{k}\sqrt{(2\epsilon-k)^2 - 4\epsilon^2(\epsilon-k)^2} \tag{5.23}$$

both signs are allowed.

5.2.2 Laboratory system

In the laboratory system (rest system of electron 2) the two initial electrons have the four-momenta $\underline{p_1} = (\epsilon_1, \mathbf{p}_1)$ and $\underline{p_2} = (1, 0)$. Introducing a coordinate system where the incident electron moves along the z axis and the photon is emitted into the x-z plane,

$$\hat{\mathbf{p}}_1 = \{0,0,1\}\,,$$
$$\hat{\mathbf{p}}'_1 = \{\sin\theta_e \cos\varphi, \sin\theta_e \sin\varphi, \cos\theta_e\}\,, \quad (5.24)$$
$$\hat{\mathbf{k}} = \{\sin\theta_k, 0, \cos\theta_k\}\,,$$

we get

$$\hat{\mathbf{p}}_1 \cdot \hat{\mathbf{k}} = \cos\theta_k\,, \quad \hat{\mathbf{p}}_1 \cdot \hat{\mathbf{p}}'_1 = \cos\theta_e\,, \quad (5.25)$$

and

$$\hat{\mathbf{p}}'_1 \cdot \hat{\mathbf{k}} = \cos\alpha = \cos\theta_k \cos\theta_e + \sin\theta_k \sin\theta_e \cos\varphi\,. \quad (5.26)$$

Then Eq. (5.4) yields

$$\epsilon'_1(\epsilon_1 - k + 1) - (\epsilon_1 + 1)(1 - k) - kp_1 \cos\theta_k = p'_1(p_1 \cos\theta_e - k \cos\alpha)\,. \quad (5.27)$$

This results in the photon energy

$$k = \frac{p_1 p'_1 \cos\theta_e - (\epsilon_1 + 1)(\epsilon'_1 - 1)}{\epsilon_1 - \epsilon'_1 + 1 - p_1 \cos\theta_k + p'_1 \cos\alpha}\,. \quad (5.28)$$

Consider now the case that the momentum $\mathbf{p}'_1$ of one of the outgoing electrons and the photon direction $\hat{\mathbf{k}} = \mathbf{k}/k$ are fixed, corresponding to the experimental arrangement that an outgoing electron and the photon are detected in coincidence. Then the kinematics of the process is completely determined since, due to the energy-momentum conservation (5.2), only 5 of the 9 scalar quantities contained in $\mathbf{p}'_1$, $\mathbf{p}'_2$, and $\mathbf{k}$ can be chosen independently. Just these 5 scalar quantities are given by $\mathbf{p}'_1$ and $\hat{\mathbf{k}}$. Thus the photon energy is unambiguously determined by (5.28). In Fig. 6.3 the relation (5.28) is plotted as a function of the energy ϵ'_1 of the outgoing electron in comparison with the linear relation $k = \epsilon_1 - \epsilon'_1$ valid for electron-atom bremsstrahlung. It is seen that one can distinguish between the photons emitted in the two processes by measuring their energy in addition to their direction.

At fixed direction of the scattered electron, the maximum photon energy is reached for coplanar geometry, $\varphi = \pi$, if the photon is emitted at the angle θ_k given by

$$\sin\theta_k = \frac{p'_1 \sin\theta_e}{|\mathbf{p}_1 - \mathbf{p}'_1|}\,, \quad \cos\theta_k = \frac{p_1 - p'_1 \cos\theta_e}{|\mathbf{p}_1 - \mathbf{p}'_1|}\,. \quad (5.29)$$

Then Eqs. (5.26) and (5.28) yield

$$k_{\max}(\mathbf{p}'_1, \hat{\mathbf{k}}) = \frac{p_1 p'_1 \cos\theta_e - (\epsilon_1 + 1)(\epsilon'_1 - 1)}{\epsilon_1 - \epsilon'_1 + 1 - |\mathbf{p}_1 - \mathbf{p}'_1|}. \qquad (5.30)$$

The denominator of the expression (5.28) is always larger than zero. Hence, for k to be positive, the scattered electron is restricted to forward angles θ_e given by $p_1 p'_1 \cos\theta_e \geq (\epsilon_1 + 1)(\epsilon'_1 - 1)$, or

$$\cos\theta_e \geq \frac{(\epsilon_1 + 1)(\epsilon'_1 - 1)}{p_1 p'_1}, \quad \sin\theta_e \leq \sqrt{\frac{2(\epsilon_1 - \epsilon'_1)}{(\epsilon_1 - 1)(\epsilon'_1 + 1)}}. \qquad (5.31)$$

These relations give the maximum energy and momentum of the electron scattered by the angle θ_e (corresponding to elastic scattering, $k = 0$),

$$\epsilon'_1 \leq \epsilon'_{1,\max}(\epsilon_1, \theta_e) = \frac{\epsilon_1(1 + \cos^2\theta_e) + \sin^2\theta_e}{\epsilon_1 \sin^2\theta_e + 1 + \cos^2\theta_e}, \qquad (5.32)$$

$$p'_1 \leq p'_{1,\max}(\epsilon_1, \theta_e) = \frac{2p_1 \cos\theta_e}{\epsilon_1 \sin^2\theta_e + 1 + \cos^2\theta_e}. \qquad (5.33)$$

The maximum photon energy at a given photon angle θ_k is obtained from Eq. (5.9), applied to the laboratory system:

$$k_{\max}(\epsilon_1, \theta_k) = \frac{\epsilon_1 - 1}{\epsilon_1 + 1 - p_1 \cos\theta_k}. \qquad (5.34)$$

The absolute maximum of k (short-wavelength limit) is reached if the photon is emitted in forward direction ($\theta_k = 0$) yielding

$$k_{\max} = \frac{\epsilon_1 - 1}{\epsilon_1 - p_1 + 1}. \qquad (5.35)$$

At nonrelativistic energies, $\epsilon_1 \approx 1 + \frac{1}{2}p_1^2$, this is approximately equal to half the kinetic energy of the incoming electron,

$$k_{\max} \approx \tfrac{1}{2}(\epsilon_1 - 1) \approx \tfrac{1}{4}p_1^2, \quad \epsilon_1 - 1 \ll 1. \qquad (5.36)$$

In the extreme relativistic limit, the maximum photon energy is approximately equal to the kinetic energy of the incoming electron,

$$k_{\max} \approx \epsilon_1 - 3/2, \quad \epsilon_1 \gg 1. \qquad (5.37)$$

From (5.12) to (5.14) we get energy and momentum of the scattered electrons,

$$\epsilon'_1 = \frac{(\epsilon_1 - k + 1)[(\epsilon_1 + 1)(1 - k) + kp_1 \cos\theta_k] \pm (p_1 \cos\theta_e - k\cos\alpha)R_l}{(\epsilon_1 - k + 1)^2 - (p_1 \cos\theta_e - k\cos\alpha)^2},$$
(5.38)

and

$$p'_1 = \frac{(p_1 \cos\theta_e - k\cos\alpha)[(\epsilon_1 + 1)(1 - k) + kp_1 \cos\theta_k] \pm (\epsilon_1 - k + 1)R_l}{(\epsilon_1 - k + 1)^2 - (p_1 \cos\theta_e - k\cos\alpha)^2},$$
(5.39)

where

$$R_l = \sqrt{[(\epsilon_1 + 1)(1 - k) + kp_1 \cos\theta_k]^2 + (p_1 \cos\theta_e - k\cos\alpha)^2 - (\epsilon_1 - k + 1)^2}.$$
(5.40)

We will now show that both signs are valid in these expressions. From energy conservation we have

$$\epsilon_1 - k + 1 = \epsilon'_1 + \epsilon'_2 \geq 2.$$
(5.41)

Besides

$$\epsilon_1 - k + 1 > (\epsilon_1 + 1)(1 - k) + kp_1 \cos\theta_k,$$
(5.42)

since $k(\epsilon_1 - p_1 \cos\theta_k) > 0$. Thus it follows from $\epsilon'_1 \geq 1$ that the left-hand side of Eq. (5.27) is positive, i.e.,

$$p_1 \cos\theta_e - k\cos\alpha > 0.$$
(5.43)

Furthermore, the denominator of the expressions (5.38) and (5.39) is positive: From

$$\begin{aligned}
0 &\leq (\mathbf{p}_1 \times \mathbf{p}'_1 - \mathbf{k} \times \mathbf{p}'_1)^2 \\
&= (\mathbf{p}_1 \times \mathbf{p}'_1)^2 + (\mathbf{k} \times \mathbf{p}'_1)^2 - 2(\mathbf{p}_1 \times \mathbf{p}'_1) \cdot (\mathbf{k} \times \mathbf{p}'_1) \\
&= p_1^2 {p'_1}^2 - (\mathbf{p}_1 \cdot \mathbf{p}'_1)^2 + k^2 {p'_1}^2 - (\mathbf{k} \cdot \mathbf{p}'_1)^2 - 2{p'_1}^2 (\mathbf{k} \cdot \mathbf{p}_1) \\
&\quad + 2(\mathbf{p}_1 \cdot \mathbf{p}'_1)(\mathbf{k} \cdot \mathbf{p}'_1) = {p'_1}^2 (\mathbf{p}_1 - \mathbf{k})^2 - (\mathbf{p}_1 \cdot \mathbf{p}'_1 - \mathbf{k} \cdot \mathbf{p}'_1)^2 \\
&= {p'_1}^2 \big[(\mathbf{p}_1 - \mathbf{k})^2 - (p_1 \cos\theta_e - k\cos\alpha)^2\big]
\end{aligned}$$
(5.44)

we get

$$(\epsilon_1 - k + 1)^2 - (p_1 \cos\theta_e - k\cos\alpha)^2 \geq (\epsilon_1 - k + 1)^2 - (\mathbf{p}_1 - \mathbf{k})^2$$
$$= 2\big[\epsilon_1 + 1 - k(\epsilon_1 + 1 - p_1 \cos\theta_k)\big] \geq 4,$$
(5.45)

where the last inequality results from the relation (5.34) for the maximum photon energy. Thus

$$\epsilon_1 - k + 1 > p_1 \cos\theta_e - k\cos\alpha > 0 \tag{5.46}$$

and from (5.34)

$$(\epsilon_1 + 1)(1 - k) + kp_1 \cos\theta_k > 0 . \tag{5.47}$$

By means of Eqs. (5.43), (5.46), and (5.47) it follows that $p_1' > 0$ with the upper sign. The lower sign in (5.39) is valid for

$$(p_1 \cos\theta_e - k\cos\alpha)\left[(\epsilon_1 + 1)(1 - k) + kp_1 \cos\theta_k\right] \geq (\epsilon_1 - k + 1)R_l \tag{5.48}$$

or, inserting (5.40) and squaring this relation,

$$\left\{(\epsilon_1 - k + 1)^2 - \left[(\epsilon_1 + 1)(1 - k) + kp_1 \cos\theta_k\right]^2\right\}$$
$$\cdot \left\{(\epsilon_1 - k + 1)^2 - (p_1 \cos\theta_e - k\cos\alpha)^2\right\} \geq 0 . \tag{5.49}$$

According to Eqs. (5.42) and (5.46) this condition is always fulfilled, i.e., both signs hold in the expressions (5.38) and (5.39). Hence, the energy of the electron scattered into the direction specified by the angles θ_e and φ (or θ_e and α) is double-valued (cf. Fig. 6.3).

The condition $R_l^2 \geq 0$ gives the range of allowed photon angles θ_k for fixed electron angles θ_e and φ. If the photon is observed at a fixed angle θ_k, $R_l^2 \geq 0$ yields

$$k\cos\alpha = k(\cos\theta_k \cos\theta_e + \sin\theta_k \sin\theta_e \cos\varphi) \leq p_1 \cos\theta_e - R_1 , \tag{5.50}$$

where

$$R_1 = \sqrt{(\epsilon_1 - k + 1)^2 - [(\epsilon_1 + 1)(1 - k) + kp_1 \cos\theta_k]^2} . \tag{5.51}$$

Then the range of allowed angles of the outgoing electrons is given by

$$0 \leq \sin\theta_e \leq \sin\theta_e^+ < 1 ,$$
$$0 < \cos\theta_e^+ \leq \cos\theta_e \leq 1 \tag{5.52}$$

for $R_1 \leq p_1 - k\cos\theta_k$,

$$0 < \sin\theta_e^- \leq \sin\theta_e \leq \sin\theta_e^+ ,$$
$$0 < \cos\theta_e^+ \leq \cos\theta_e \leq \cos\theta_e^- < 1 \tag{5.53}$$

for $R_1 \geq p_1 - k\cos\theta_k$, and

$$\cos\varphi \leq \cos\varphi_0 = \frac{(p_1 - k\cos\theta_k)\cos\theta_e - R_1}{k\sin\theta_k \sin\theta_e}, \qquad (5.54)$$

where

$$\sin\theta_e^{\pm} = \frac{k\sin\theta_k R_1 \pm (p_1 - k\cos\theta_k)\sqrt{(\mathbf{p}_1 - \mathbf{k})^2 - R_1^2}}{(\mathbf{p}_1 - \mathbf{k})^2} \qquad (5.55)$$

and

$$\cos\theta_e^{\pm} = \frac{(p_1 - k\cos\theta_k)R_1 \mp k\sin\theta_k\sqrt{(\mathbf{p}_1 - \mathbf{k})^2 - R_1^2}}{(\mathbf{p}_1 - \mathbf{k})^2}. \qquad (5.56)$$

The above formulae hold for $\sin\theta_k \sin\theta_e \neq 0$. In the case that $\sin\theta_k = 0$, Eq. (5.50) is no longer dependent on φ resulting in $R_1/(p_1 \mp k) \leq \cos\theta_e \leq 1$ for all values of φ in accordance with (5.52) and (5.53).

5.3 Cross section

In electron-electron bremsstrahlung the two colliding particles have unity charge. Therefore Coulomb effects are not as important as in ordinary bremsstrahlung with nuclei of high atomic numbers. Also, radiative corrections may be neglected with the exception of the region of very soft photons where many-photon processes become significant (which we will not consider). Hence it is sufficient (and complicated enough) to calculate the cross section to lowest order of perturbation theory.

The cross section $d\sigma$ is defined as the number of transitions per unit time and volume divided by the incident flux and by the number of target particles per unit volume. If the summations over the spin directions of the final electrons and the photon polarizations, and the average over the spin directions of the initial electrons are carried out, the fully differential cross section of e-e bremsstrahlung is given by [34]

$$d\sigma = \frac{\alpha r_0^2}{\pi^2} \frac{\delta^{(4)}(p_1 + p_2 - p_1' - p_2' - k)}{\sqrt{(p_1 p_2)^2 - 1}} A \frac{d^3k}{k} \frac{d^3p_1'}{\epsilon_1'} \frac{d^3p_2'}{\epsilon_2'}, \qquad (5.57)$$

where

$$A = \tfrac{1}{4} \sum_{\vec{\zeta}_1, \vec{\zeta}_2} \sum_{\vec{\zeta}_1', \vec{\zeta}_2'} \sum_{\mathbf{e}} |M|^2 \qquad (5.58)$$

is the square of the matrix element averaged over the spins $\vec{\zeta}_1$ and $\vec{\zeta}_2$ of the incoming electrons and summed over the spins $\vec{\zeta}'_1$ and $\vec{\zeta}'_2$ of the outgoing electrons and the photon polarizations **e**. The product of the three relativistically invariant quantities $d^3 p'_1/\epsilon'_1$, $d^3 p'_2/\epsilon'_2$, and $d^3 k/k$ represents the density of final states, and the square root originates from the flux density

$$J = \frac{\sqrt{(p_1 p_2)^2 - 1}}{\epsilon_1 \epsilon_2} \qquad (5.59)$$

of the incident electrons for unit normalization volume.

In lowest order perturbation theory (Born approximation) the matrix element

$$M = \sum_{i=1}^{8} M_i \qquad (5.60)$$

is composed of the contributions of the 8 Feynman diagrams (Fig. 5.1). Hence the matrix element squared consists of 64 terms. The formation of the spin summations requires the calculation of traces of up to twelve Dirac matrices. Fortunately only six out of the 64 terms have to be evaluated explicitly since permutation operations can be utilized [114]. After summing over the photon polarizations a great number of algebraic manipulations and simplifications result in the lengthy formula given in Appendix B (since the thesis of Hodes [113] is almost inaccessible and we do not know of any publication where the e-e bremsstrahlung cross section is specified we thought it worthwhile to write up the formula).

A complete QED lowest-order calculation of radiative corrections to cross section and other observables in the process of polarized identical fermion scattering in covariant form was performed by Shumeiko and Suarez [115]. It includes bremsstrahlung emission with all polarization variables. A detailed analysis of polarization effects is presented there.

The delta function in (5.57) guarantees the conservation of energy and momentum. It is the product of the delta functions $\delta^{(3)}(\mathbf{p}_1 + \mathbf{p}_2 - \mathbf{p}'_1 - \mathbf{p}'_2 - \mathbf{k})$ and $\delta(\epsilon_1 + \epsilon_2 - \epsilon'_1 - \epsilon'_2 - k)$. The former can be used to integrate (5.57) over $d^3 p'_2$ yielding

$$d\sigma = \frac{\alpha r_0^2}{\pi^2} \frac{\delta(\epsilon_1 + \epsilon_2 - \epsilon'_1 - \epsilon'_2 - k)}{\epsilon'_1 \epsilon'_2 k \sqrt{(p_1 p_2)^2 - 1}} A\, d^3 p'_1\, d^3 k \,, \qquad (5.61)$$

where ϵ_2' is now a function of the momenta $\mathbf{p}_1$, $\mathbf{p}_2$, $\mathbf{p}_1'$, and $\mathbf{k}$,

$$\epsilon_2' = \sqrt{\mathbf{p}_2'^2 + 1} = \sqrt{(\mathbf{p}_1 + \mathbf{p}_2 - \mathbf{p}_1' - \mathbf{k})^2 + 1} \,. \tag{5.62}$$

Substituting

$$d^3k = k^2 dk \, d\Omega_k \,, \quad d^3p_1' = p_1'^2 dp_1' \, d\Omega_{p_1'} = \epsilon_1' p_1' \, d\epsilon_1' \, d\Omega_{p_1'} \tag{5.63}$$

gives

$$d\sigma = \frac{\alpha r_0^2}{\pi^2} \frac{p_1' k}{\epsilon_2'} \frac{\delta(\epsilon_1 + \epsilon_2 - \epsilon_1' - \epsilon_2' - k)}{\sqrt{(p_1 p_2)^2 - 1}} \, A \, dk \, d\Omega_k \, d\epsilon_1' \, d\Omega_{p_1'} \,. \tag{5.64}$$

Finally, the integration over the energies ϵ_1' can be readily performed by means of the remaining delta function. If the argument of the δ function is a function $g(x)$ of the integration variable x with n single zeros x_i, then

$$\delta\bigl(g(x)\bigr) = \sum_{i=1}^{n} \frac{\delta(x - x_i)}{|dg/dx|} \,. \tag{5.65}$$

Hence the integral over a function $f(x)$ yields

$$\int f(x) \, \delta\bigl(g(x)\bigr) \, dx = \sum_{i=1}^{n} \frac{f(x_i)}{|g'(x_i)|} \,. \tag{5.66}$$

In Eq. (5.64) we have

$$g(\epsilon_1') = \epsilon_1 + \epsilon_2 - \epsilon_1' - \epsilon_2' - k = \epsilon_1 + \epsilon_2 - \epsilon_1' - k - \sqrt{(\mathbf{p}_1 + \mathbf{p}_2 - \mathbf{p}_1' - \mathbf{k})^2 + 1} \tag{5.67}$$

with the zeros (5.12). Using $p_1' dp_1' = \epsilon_1' d\epsilon_1'$ we get

$$g'(\epsilon_1') = \frac{dg}{d\epsilon_1'} = -\frac{\epsilon_2' p_1'^2 - \epsilon_1' (\mathbf{p}_1' \cdot \mathbf{p}_2')}{\epsilon_2' p_1'^2} \,. \tag{5.68}$$

The numerator of the right-hand side can be expressed by the square root (5.15),

$$|\epsilon_2' p_1'^2 - \epsilon_1' (\mathbf{p}_1' \cdot \mathbf{p}_2')| = p_1' R \,. \tag{5.69}$$

Applying the relations (5.66) to (5.69), the cross section for the elementary process of e-e bremsstrahlung takes the form

$$\frac{d^3\sigma}{dk \, d\Omega_k \, d\Omega_{p_1'}} = \frac{\alpha r_0^2}{\pi^2} \frac{k}{R\sqrt{(p_1 p_2)^2 - 1}} \sum_{\nu=1}^{2} p_{1\nu}'^2 A(p_{1\nu}') \,. \tag{5.70}$$

Here p'_{11} and p'_{12} denote the momenta (5.13) of the outgoing electrons for fixed photon momentum **k**. In the center-of-mass system and the laboratory system they are specified by the expressions (5.20) and (5.39), respectively. The cross section is given as a function of relativistically invariant four-products, hence it can be specialized to any frame of reference.

If we want to obtain the cross section differential with respect to the electron energy ϵ'_1, the expression (5.64) has to be integrated over the photon energy k, again applying (5.66) with $g'(k) = -(kp'_2)/(\epsilon'_2 k)$. Then the cross section can be represented as

$$\frac{d^3\sigma}{d\epsilon'_1 \, d\Omega_{p'_1} \, d\Omega_k} = \frac{\alpha r_0^2}{\pi^2} \frac{k^2 p'_1}{(kp'_2)} \frac{A}{\sqrt{(p_1 p_2)^2 - 1}}. \tag{5.71}$$

This form is used in comparison with experiments where an outgoing electron is detected with fixed energy and direction, such as in coincidence measurements of the elementary process of bremsstrahlung, or in the measurement of radiation corrections in e-e scatterings. The quantity (kp'_2) can be expressed by known parameters using Eq. (B.9) in Appendix B.

When the doubly differential cross section is calculated by integrating over the total solid angle $\Omega_{p'_1}$ allowed kinematically, the expression (5.70) has to be multiplied by the factor 1/2, otherwise each outgoing electron would be counted twice.

The above cross sections are exact within lowest-order perturbation theory (Born approximation). Whereas radiative corrections are expected to be small, the Coulomb correction may be important at low electron energies. Correct results can be expected only if the conditions [cf. Eq. (3.89)]

$$a_1 = \alpha/\beta_{12} \ll 1 \,, \quad a_2 = \alpha/\beta'_{12} \ll 1 \tag{5.72}$$

are satisfied where

$$\beta_{12} = \frac{\sqrt{(p_1 p_2)^2 - 1}}{(p_1 p_2)} \,, \quad \beta'_{12} = \frac{\sqrt{(p'_1 p'_2)^2 - 1}}{(p'_1 p'_2)} \tag{5.73}$$

are the relative velocities (in units of the speed of light c) of the initial and final electrons, respectively. As in the case of the bremsstrahlung process in the Coulomb field of a nucleus, the cross section of e-e bremsstrahlung derived in Born approximation can be corrected by a simple factor. Corresponding to the Elwert factor, Eq. (3.90), it is given by the ratio of probabilities for finding the two initial and final electrons, respectively, at the same position. The correction factor has the form [116]

$$F_{ee} = \frac{a_2}{a_1} \frac{e^{2\pi a_1} - 1}{e^{2\pi a_2} - 1}. \tag{5.74}$$

In contrast to the electron-nucleus case, this factor is always less than unity as a consequence of the Coulomb repulsion between the electrons. That is, the true cross section is always lower than the one given by the Born-approximation formulae. Due to the small factor $\alpha \approx 1/137$ in the quantities a_1 and a_2, however, F_{ee} is approximately unity, especially for higher energies. An important exception is the short-wavelength limit given by $(p'_1 p'_2) = 1$ (see Section 5.2). Then $a_2 \to \infty$ and $F_{ee} \to 0$. Since the cross sections (5.70) and (5.71) tend to zero for $(p'_1 p'_2) \to 1$, this behaviour is still intensified by the exponential decrease of the Coulomb-correction factor. Thus the cross section at the high-frequency limit vanishes whereas the cross section of the electron-nucleus bremsstrahlung remains finite.

Generally the factor (5.74) is dependent on the momenta $\mathbf{p}_1$, $\mathbf{p}_2$, and $\mathbf{k}$, i.e., it is different for various photon angles. In the center-of-mass system, however, $(p'_1 p'_2) = \epsilon_1^2 + p_1^2 - 2\epsilon_1 k$ [see Eq. (5.7)], so that F_{ee} is constant for a fixed photon energy.

We will now specify the cross section in special frames of reference.

5.3.1 Laboratory system (rest system of electron 2)

For $p_2 = 0$ the cross section (5.70) has the form

$$\left(\frac{d^3\sigma}{dk\, d\Omega_k\, d\Omega_{p'_1}}\right)_{\text{lab}} = \frac{\alpha r_0^2}{\pi^2} \frac{k}{p_1 R_l} \sum_{\nu=1}^{2} p'^2_{1\nu} A(p'_{1\nu}), \tag{5.75}$$

where the $p'_{1\nu}$ are given by the two signs in (5.39).

It is interesting to note that the cross section tends to infinity at the limiting angles of the outgoing particles which are determined by the condition $R_l = 0$. For fixed photon angle θ_k these angles are given by Eqs. (5.52) to (5.56). Consider the case that the cross section is measured in a coincidence experiment where the emitted photon is registered together with one of the outgoing electrons. Each detector covers a finite solid angle Ω_e. Therefore the count rate of the electron detector is proportional to the integral of the cross section over Ω_e. If the detector approaches the limiting angles, the cross section tends to infinity as $1/R_l$. However, the integral of this reciprocal square root over the solid angle is finite, and so is the count rate.

164 *Theory of the elementary process of electron-electron bremsstrahlung*

The occurrence of the factor $1/R_l$ is purely kinematical, originating from the integration over the delta function $\delta(\epsilon_1 + \epsilon_2 - \epsilon'_1 - \epsilon'_2 - k)$ in Eq. (5.64) whereas the quantity A including the physics of the process is finite.

Figure 5.3 shows an example of the cross section (5.71) for coplanar geometry and incident electron energy $E_1 = 300 \text{ keV}$. The photon distribution is calculated for fixed energy ($E'_1 = 160 \text{ keV}$) and angle ($\theta_e = 28°$) of the scattered electron. In contrast to electron-nucleus bremsstrahlung the photon energy is not constant but varies with the emission angle θ_k according to Eq. (5.28). The maximum photon energy, 91.4 keV, is reached at the angle $\theta_k \approx -40°$, whereas the minimum photon energy, 31.2 keV, occurs at $\theta_k = 140°$.

5.3.2 Center-of-mass system ($\mathbf{p}_1 + \mathbf{p}_2 = 0$)

In the cms the cross section (5.70) has the form

$$\left(\frac{d^3\sigma}{dk\, d\Omega_k\, d\Omega_{p'_1}}\right)_{\text{cms}} = \frac{\alpha r_0^2}{\pi^2} \frac{k}{2\epsilon p R_c} \sum_{\nu=1}^{2} p'^2_{1\nu} A(p'_{1\nu}) \,. \tag{5.76}$$

The momenta $p'_{1\nu}$ and the square root R_c are given by Eqs. (5.20) and (5.21), respectively; for $k \leq k_0$ [see Eq. (5.22)] there exists only one value p'_1.

The cross section for nonrelativistic energies shows beautifully the quadrupole character of e-e bremsstrahlung (Fig. 5.4). At semirelativistic energies the radiation is much more beamed than electron-nucleus bremsstrahlung due to the different kinematics. Figure 5.5 depicts a photon angular distribution for $E_1 = E_2 = 500 \text{ keV}$, photon energy $h\nu = 250 \text{ keV}$, and the electron scattering angle $\theta_1 = 0°$. The photons are predominantly emitted near the direction of the scattered electrons. The width of the radiation lobes is about $5°$. At larger scattering angles θ_1 the cross section decreases rapidly by two to three orders of magnitude; the radiation lobes become broader, and at $\theta_1 = 90°$ all four lobes have roughly the same size.

5.3.3 Center-of-mass system of the outgoing electrons

The center-of-mass system S' of the outgoing electrons, where

$$\mathbf{p}'_1 + \mathbf{p}'_2 = \mathbf{p}_1 + \mathbf{p}_2 - \mathbf{k} = 0\,, \quad \epsilon'_1 = \epsilon'_2\,, \tag{5.77}$$

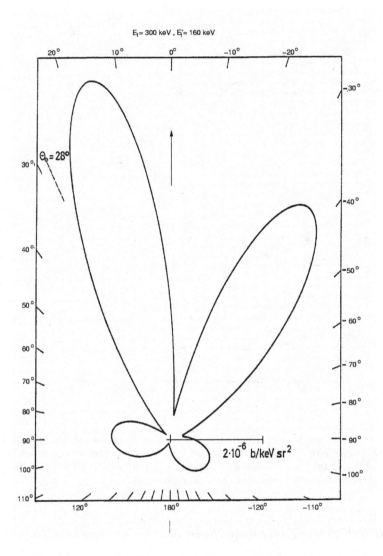

Fig. 5.3 Photon angular distribution of e-e bremsstrahlung in the laboratory system for incident electrons of 300 keV and fixed energy ($E_1' = 160\,\text{keV}$) and angle ($\theta_e = 28°$) of the outgoing electron.

is distinguished by the fact that the ambiguity in the values (5.12) and (5.13) of ϵ_1' and p_1', respectively, is removed and the square root R as given

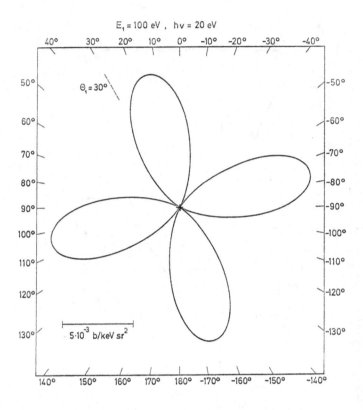

Fig. 5.4 Photon angular distribution in the elementary process of electron-electron bremsstrahlung in the center-of-mass system for coplanar geometry. Cross section (5.70) as a function of the photon angle θ for incident electron energies $E_1 = E_2 = 100$ eV, photon energy 20 eV, and electron scattering angle $\theta_1 = 30°$.

by Eq. (5.14) takes a very simple form, viz.,

$$R_{S'} = 2\epsilon'_1 p'_1 . \qquad (5.78)$$

Then the cross section (5.70) reads

$$\left(\frac{d^3\sigma}{dk\, d\Omega_k\, d\Omega_{p'_1}} \right)_{S'} = \frac{\alpha r_0^2}{\pi^2} \frac{kp'_1}{2\epsilon'_1} \frac{A}{\sqrt{(p_1 p_2)^2 - 1}} . \qquad (5.79)$$

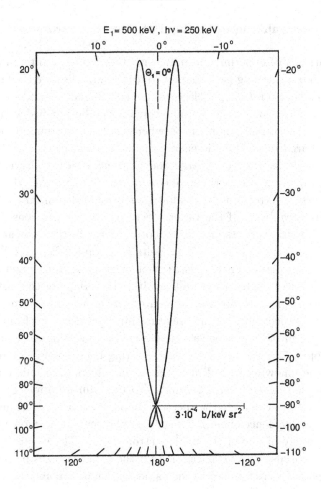

Fig. 5.5 Same as Fig. 5.4 for $E_1 = E_2 = 500$ keV, $h\nu = 250$ keV, and $\theta_1 = 0°$.

This expression is the most convenient starting point for the integration of the cross section over the solid angle $d\Omega_{p'_1}$ [117] since only the function A is dependent on the angles of the outgoing electrons.

5.4 Bremsstrahlung in the field of bound electrons

In comparisons of experimental data with theoretical results on electron-electron bremsstrahlung (see Chapter 6) it is common to use the laboratory system, i.e., the atomic target electrons are considered free and at rest. Disregarding the motion of the bound electrons can be justified if the target atoms have low atomic numbers Z where the electron binding energy is small compared to the kinetic energy of the incident electrons. Nevertheless the simple assumption of free and stationary target electrons is not strictly realized under experimental conditions. In particular, this assumption may lead to erroneous results for the fourfold differential cross section of e-e bremsstrahlung. If the target electrons are at rest, conservation of energy and momentum results in the photon energy distribution at fixed angles of the outgoing particles being a delta function [cf. Eq. (5.28)]. On the other hand, the photon energy distribution originating from target electrons bound to an atom will have a finite width. The shape of the distribution of photon energies can directly be related to the momentum distribution of the atomic electrons. Such a relationship is well known from inelastic scattering of initially monoenergetic electrons by atomic electrons [(e,2e) spectroscopy] [118] and in Compton scattering from bound electrons [119; 120]. In the following we will investigate the effects of atomic binding on the production of e-e bremsstrahlung. To this aim we apply the impulse approximation to calculate the spectral shape of e-e bremsstrahlung as observed in a coincidence experiment at fixed photon and electron angles and for fixed energy of one of the final electrons [121]. The essential assumption of the impulse approximation is that the collision takes place as if the atomic electron were free except for its possession of a momentum distribution $\chi(\mathbf{p})$, i.e., the effect of binding is limited to producing a characteristic momentum distribution of the atomic electron [56].

We consider the radiative collision of a free electron with an atomic electron which is bound to an infinitely massive nucleus. The impulse approximation can be applied to this problem supposing that the effective interaction between the incident electron and the atomic electron takes place in a region which is small as compared with the atomic dimensions. If this is the case, the energy and momentum transfers are large relative to the binding energy and the momentum of the atomic electron, respectively. We start from the cross section (5.61). It contains the flux factor $\sqrt{(p_1 p_2)^2 - 1}/(\epsilon_1 \epsilon_2)$ [cf. Eq. (5.59)] referring to colliding beams of electrons. On the other hand, the cross section for e-e bremsstrahlung from bound electrons is measured relative to a moving electron and a stationary atom. This situation corresponds to the flux factor p_1/ϵ_1. To take this into account one has to include the correction factor $\sqrt{(p_1 p_2)^2 - 1}/(\epsilon_2 p_1)$. With the aid of the relations (5.63) the cross section differential with respect to the energies of the photon and the outgoing electron 1, and the solid angles $d\Omega_k$ and $d\Omega_{p'_1}$ of the outgoing photon and electron 1, respectively, can then be written as

$$\frac{d^4\sigma}{dk\, d\Omega_k\, d\epsilon'_1\, d\Omega_{p'_1}} = \frac{\alpha r_0^2}{\pi^2} \frac{k p'_1}{\epsilon_2 \epsilon'_2 p_1} A\, \delta(\epsilon_1 + \epsilon_2 - \epsilon'_1 - \epsilon'_2 - k)\,. \quad (5.80)$$

Equation (5.80) gives the probability that in the collision of two electrons of momentum $\mathbf{p}_1$ and $\mathbf{p}_2$, respectively, one of the outgoing electrons is detected with energy ϵ'_1 (within $d\epsilon'_1$) in the direction $\mathbf{p}'_1$ (within $d\Omega_{p'_1}$) and the produced photon is detected in the direction $\mathbf{k}$ (within $d\Omega_k$). The energy of the photon is specified by the conservation of energy and momentum [see Eq. (5.5)],

$$k = \frac{(p_1 p_2) - (p_1 p'_1) - (p_2 p'_1) + 1}{\epsilon_1 + \epsilon_2 - \epsilon'_1 - \hat{\mathbf{k}} \cdot (\mathbf{p}_1 + \mathbf{p}_2 - \mathbf{p}'_1)} = K(\mathbf{p}_1, \mathbf{p}_2, \mathbf{p}'_1, \hat{\mathbf{k}})\,. \quad (5.81)$$

In the case which has been usually considered, viz., the target electrons being at rest ($\mathbf{p}_2 = 0$), the photon energy has a definite value, unambiguously determined by Eq. (5.81) for given quantities $\mathbf{p}_1$, $\mathbf{p}'_1$, and $\hat{\mathbf{k}}$. This is no longer true if the target electrons are bound in an atom. As outlined above, in the impulse approximation they can be considered free having a momentum distribution $\chi(\mathbf{p}_2)$ which is given by the Fourier transform of their wave function $\psi(\mathbf{r})$,

$$\chi(\mathbf{p}_2) = (2\pi)^{-3} \left| \int \psi(\mathbf{r})\, e^{i\mathbf{p}_2 \cdot \mathbf{r}}\, d^3 r \right|^2\,. \quad (5.82)$$

Then the photon spectrum in a certain direction $\hat{\mathbf{k}}$, observed in coincidence with electrons of energy ϵ'_1 in the direction $\hat{\mathbf{p}}'_1$, is no longer monochromatic but rather is determined by integrating the cross section (5.80), multiplied by the weight factor (5.82), over the total momentum space $d^3 p_2$:

$$\sigma_4(k) \equiv \frac{d^4\sigma}{dk\, d\Omega_k\, d\epsilon'_1\, d\Omega_{p'_1}} = \frac{\alpha r_0^2}{\pi^2} \frac{k p'_1}{p_1} \int \chi(\mathbf{p}_2) \frac{A}{\epsilon_2 \epsilon'_2}\, \delta(\epsilon_1 + \epsilon_2 - \epsilon'_1 - \epsilon'_2 - k)\, d^3 p_2\,. \quad (5.83)$$

Introducing a system of polar coordinates where the incident electrons are moving in z direction, the momentum unit vectors can be expressed by angles θ, θ_1, θ_2, φ, φ_1, φ_2 as follows:

$$\begin{aligned}
\hat{\mathbf{p}}_1 &= \{0, 0, 1\} \\
\hat{\mathbf{p}}_1' &= \{\sin\theta_1 \cos\varphi_1, \sin\theta_1 \sin\varphi_1, \cos\theta_1\} \\
\hat{\mathbf{p}}_2 &= \{\sin\theta_2 \cos\varphi_2, \sin\theta_2 \sin\varphi_2, \cos\theta_2\} \\
\hat{\mathbf{k}} &= \{\sin\theta \cos\varphi, \sin\theta \sin\varphi, \cos\theta\}
\end{aligned} \qquad (5.84)$$

Henceforth we will set $\varphi = \varphi_1 = 0$ since all the coincidence experiments have been performed in a coplanar geometry so far. Then the function ϵ_2' defined in (5.62) takes the form

$$\begin{aligned}
\epsilon_2' = \big[& p_1^2 + p_2^2 + p_1'^2 + k^2 + 1 - 2\mathbf{k}\cdot\mathbf{p}_1 - 2\mathbf{p}_1\cdot\mathbf{p}_1' + 2\mathbf{k}\cdot\mathbf{p}_1' \\
& + 2p_2 \cos\theta_2 (p_1 - p_1' \cos\theta_1 - k\cos\theta) \\
& - 2p_2 \sin\theta_2 \cos\varphi_2 (p_1' \sin\theta_1 + k\sin\theta) \big]^{1/2}.
\end{aligned} \qquad (5.85)$$

The volume element in $\mathbf{p}_2$ space is given by $d^3 p_2 = p_2^2 \, dp_2 \, d(\cos\theta_2) \, d\varphi_2$. The integration in (5.83) over φ_2 is performed by means of the δ function which according to Eq. (5.65) can be transformed into

$$\begin{aligned}
\delta(\epsilon_1 + \epsilon_2 - \epsilon_1' - \epsilon_2' - k) &= \sum_{\nu=1}^{n} \frac{\delta(\varphi_2 - \Phi_{2\nu})}{|\partial \epsilon_2'/\partial \varphi_2|} \\
&= \frac{\epsilon_2' \sum_\nu \delta(\varphi_2 - \Phi_{2\nu})}{p_2 \sin\theta_2 |\sin\varphi_2 (p_1' \sin\theta_1 + k\sin\theta)|},
\end{aligned} \qquad (5.86)$$

where the $\Phi_{2\nu}$ are n solutions of $\epsilon_1 + \epsilon_2 - \epsilon_1' - \epsilon_2' - k = 0$. Since this equation, taken as a function of $\cos\varphi_2$, has a single zero, the angles $\Phi_{2\nu}$ have the values Φ_2 and $2\pi - \Phi_2$, where Φ_2 is given by

$$\cos\Phi_2 = \frac{a + b\cos\theta_2}{c\sin\theta_2} \qquad (5.87)$$

with

$$a = \epsilon_1'(\epsilon_1 + \epsilon_2) + k(\epsilon_1 + \epsilon_2 - \epsilon_1') - \epsilon_1\epsilon_2 - 1 - p_1 p_1' \cos\theta_1$$
$$\quad - k p_1 \cos\theta + k p_1' \cos(\theta - \theta_1)$$
$$= (kp_1) - (kp_1') + (p_1 p_1') - 1 - \epsilon_2(\epsilon_1 - \epsilon_1' - k), \quad (5.88)$$
$$b = p_2(p_1 - p_1' \cos\theta_1 - k\cos\theta),$$
$$c = p_2(k\sin\theta + p_1' \sin\theta_1).$$

Then we get from (5.83)

$$\sigma_4(k) = \frac{2\alpha r_0^2}{\pi^2}\frac{kp_1'}{p_1}\int_{p_{20}}^{\infty} dp_2\, \frac{p_2^2}{\epsilon_2}\int_{\cos\theta'}^{\cos\theta''} d(\cos\theta_2)\, \frac{\chi(\mathbf{p}_2)\, A}{W}, \quad (5.89)$$

where the square root W is defined as

$$W = \sqrt{c^2 - a^2 - 2ab\cos\theta_2 - (b^2 + c^2)\cos^2\theta_2}$$
$$= \sqrt{(b^2 + c^2)(\cos\theta'' - \cos\theta_2)(\cos\theta_2 - \cos\theta')}. \quad (5.90)$$

The factor 2 in (5.89) arises from the fact that the angles Φ_2 and $2\pi - \Phi_2$ give the same contributions. The limits of the θ_2 integration are determined by the requirement that $|\cos\Phi_2| \leq 1$. Using (5.87) this condition yields $W^2 \geq 0$, or

$$\cos\theta' = -\frac{|c|\sqrt{b^2 + c^2 - a^2} + ab}{b^2 + c^2},$$
$$\cos\theta'' = \frac{|c|\sqrt{b^2 + c^2 - a^2} - ab}{b^2 + c^2}. \quad (5.91)$$

It can be easily shown that $|\cos\theta'| \leq 1$ and $|\cos\theta''| \leq 1$. The lower limit of the p_2 integration is determined by the requirement that the argument of the square root in (5.91) must not be negative. The condition $b^2 + c^2 - a^2 \geq 0$ results in

$$p_{20} = \frac{1}{2}\left||\mathbf{p}_1 - \mathbf{p}_1' - \mathbf{k}| - (\epsilon_1 - \epsilon_1' - k)\sqrt{\frac{(p_2 p_2') + 1}{(p_2 p_2') - 1}}\right|. \quad (5.92)$$

The invariant four-product $(p_2 p_2')$ can be expressed by

$$(p_2 p_2') = (kp_1) - (kp_1') + (p_1 p_1'), \quad (5.93)$$

which is obtained if one takes the square of the energy-momentum conservation $\underline{p_2} - \underline{p_2'} = \underline{k} + \underline{p_1'} - \underline{p_1}$.

By means of the relation

$$2\big[(p_2 p_2') - 1\big] = (\mathbf{p}_1 - \mathbf{p}_1' - \mathbf{k})^2 - (\epsilon_1 - \epsilon_1' - k)^2 \tag{5.94}$$

it is a simple matter to show that $p_{20} = 0$ for $k = K(\mathbf{p}_1, \mathbf{p}_2 = 0, \mathbf{p}_1', \hat{\mathbf{k}})$.

The integrations over $\cos\theta_2$ and p_2 in (5.89) have to be performed numerically. The square root W in the denominator of the integral (5.89) tends to zero at the limits of the θ_2 integration. In order to avoid these singularities we change the integration variable $\cos\theta_2$ by means of the substitution

$$dy = \frac{d(\cos\theta_2)}{\sqrt{(\cos\theta'' - \cos\theta_2)(\cos\theta_2 - \cos\theta')}} . \tag{5.95}$$

Applying the conditions $y = 0$ for $\cos\theta_2 = \cos\theta'$ and $y = \pi$ for $\cos\theta_2 = \cos\theta''$, we get

$$y = \pi - 2\arctan\sqrt{\frac{\cos\theta'' - \cos\theta_2}{\cos\theta_2 - \cos\theta'}} . \tag{5.96}$$

Then the integral over $\cos\theta_2$ in (5.89) transforms to

$$I = \int_{\cos\theta'}^{\cos\theta''} \frac{\chi(\mathbf{p}_2)\, A}{W}\, d(\cos\theta_2) = \frac{1}{\sqrt{b^2 + c^2}} \int_0^{\pi} \chi(y)\, A(y)\, dy , \tag{5.97}$$

which can be easily evaluated numerically. The final integration over p_2 is straightforward.

The quantity measured in experiments without spectral resolution of the photon detector is given by the area below the $\sigma_4(k)$ curve,

$$\int_0^{\infty} \sigma_4(k)\, dk . \tag{5.98}$$

On the other hand, if the target electrons are at rest we have $\chi(\mathbf{p}_2) = \delta(\mathbf{p}_2)$. Inserting this into (5.83), one obtains

$$\sigma_4(k) = \frac{\alpha r_0^2}{\pi^2} \frac{k p_1'}{\epsilon_2' p_1} A\, \delta(\epsilon_1 + 1 - \epsilon_1' - \epsilon_2' - k) . \tag{5.99}$$

Bearing in mind the definition (5.62) of ϵ_2' and applying the relation (5.65), we arrive at

$$\sigma_4(k) = \frac{\alpha r_0^2}{\pi^2} \frac{k^2 p_1'}{p_1} \frac{A}{(kp_2')}\, \delta\big[k - K(\mathbf{p}_1, \mathbf{p}_2 = 0, \mathbf{p}_1', \hat{\mathbf{k}})\big] . \tag{5.100}$$

where $K(\mathbf{p}_1, \mathbf{p}_2, \mathbf{p}'_1, \hat{\mathbf{k}})$ is given by (5.81). Integration over k yields the triply differential cross section for e-e bremsstrahlung in the laboratory system ($p_2 = 0$),

$$\sigma_3 = \frac{d^3\sigma}{d\epsilon'_1\, d\Omega_{p'_1}\, d\Omega_k} = \frac{\alpha r_0^2}{\pi^2} \frac{k^2 p'_1}{p_1} \frac{A}{(kp'_2)} \;,\; k = K(\mathbf{p}_1, \mathbf{p}_2=0, \mathbf{p}'_1, \hat{\mathbf{k}})\;. \quad (5.101)$$

The photon spectrum (5.89) has been computed by Haug and Keppler [121] for the parameters used in the coincidence experiments, i.e., for the incident electron energy $E_1 = 300\,\text{keV}$ and carbon targets (see Sec. 6.1.1). For the outgoing electron energy E'_1 and the angles of the outgoing photons and electrons the combinations $E'_1 = 160\,\text{keV}$, $\theta = -34°$, $\theta_1 = 28°$, and $E'_1 = 140\,\text{keV}$, $\theta = -35°$, $\theta_1 = 20°$ were chosen. Then the photon energies (5.28) for target electrons at rest ($p_2 = 0$) are $h\nu = mc^2 k \approx 91.0\,\text{keV}$ and $h\nu \approx 138.7\,\text{keV}$, respectively. The energy transfers to the target electrons, $E'_2 \approx 49.0\,\text{keV}$ and $E'_2 \approx 21.3\,\text{keV}$, respectively, are large compared to the binding energy of the K electrons of carbon ($E_b \approx 0.284\,\text{keV}$). So the suppositions for the validity of the impulse approximation are well satisfied. As was shown by comparison with exact calculations of Compton scattering from bound electrons [119], the potential in that the electron is moving has not been completely neglected in the impulse approximation. Rather the potential has cancelled out of the energy for the initial and final states. The errors of the impulse approximation are of the order $E_b^2/(E'_2 - E_2)^2$ which is negligible for the parameters used.

The width of the photon spectrum $\sigma_4(k)$ is determined by the momentum distribution $\chi(\mathbf{p}_2)$ of the atomic electrons, that is, by the absolute square of the momentum representation of the atomic wave function $\psi(\mathbf{r})$ [Eq. (5.82)]. The struck electrons can be described by using analytic Hartree-Fock wave functions [122; 123]. Since the target atoms are randomly orientated, $\chi(\mathbf{p}_2)$ is not angle dependent. Therefore the momentum distribution of the electrons has to be averaged over the magnetic substates. It is true that the outer electrons of the solid-state target foil are not accurately represented by the atomic wave functions. However, the $\sigma_4(k)$ curves for the weakly bound 2s and 2p electrons of carbon are quite narrow and are not much dependent on the exact form of the wave function.

Figure 5.6 shows the photon spectra (5.89) from a carbon target for $E_1 = 300\,\text{keV}$, $E'_1 = 140\,\text{keV}$, $\theta = -35°$, and $\theta_1 = 20°$. The angles are chosen such that the cross section σ_3 has a maximum. The curves labelled

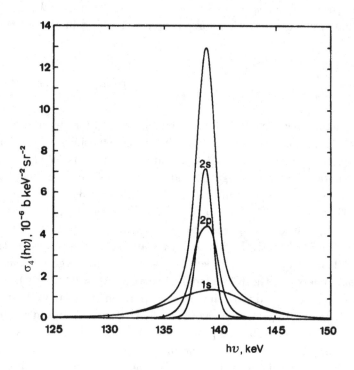

Fig. 5.6 Fourfold differential cross sections $\sigma_4(h\nu)$ from the (initially) bound electrons of a carbon target as a function of the photon energy. $E_1 = 300$ keV, $E'_1 = 140$ keV, $\theta = -35°$, and $\theta_1 = 20°$. The curves labelled 1s, 2s and 2p pertain to two electrons in each case. The upper curve represents the bremsstrahlung due to all six atomic electrons (from Haug and Keppler [121]).

1s, 2s, and 2p pertain to two electrons in each case. They are peaked around the photon energy $h\nu = 139$ keV corresponding to target electrons at rest. The spectrum originating from the most tightly bound 1s electrons is broadest with a full width at half maximum (FWHM) of about 7.9 keV; its maximum is shifted a little towards higher energies (139.3 keV) relative to the photon energy for stationary target electrons (138.7 keV). The upper curve represents the bremsstrahlung photons due to all six electrons of the carbon atom. The width of the total spectrum (FWHM $\approx$ 2.1 keV) is determined by the L electrons whereas the broader wings arise from the K electrons.

The quantity measured in experiments without spectral resolution of the photon detector is given by the area below this curve [Eq. (5.98)]. For the parameters of Fig. 5.6 we get

$$\int \sigma_4(h\nu)\,d(h\nu) \approx 3.864 \cdot 10^{-5}\,\text{b}\,\text{keV}^{-1}\,\text{sr}^{-2}\,. \tag{5.102}$$

Dividing this value by 6, the number of electrons per carbon atom, one obtains $6.44 \cdot 10^{-6}\,\text{b}\,\text{keV}^{-1}\,\text{sr}^{-2}$. In comparison, the triply differential cross section (5.101) for e-e bremsstrahlung in the laboratory system ($p_2 = 0$) amounts to $\sigma_3 \approx 6.55 \cdot 10^{-6}\,\text{b}\,\text{keV}^{-1}\,\text{sr}^{-2}$ which is close to the above result. That implies convolution of the e-e bremsstrahlung cross section with the momentum distribution $\chi(\mathbf{p})$ of the target electrons does not significantly change the value of the triply differential cross section (5.101) for low-Z target atoms. This finding gives the justification for the comparison of experimental e-e bremsstrahlung differential cross sections from atomic targets with the theoretical values calculated for target electrons at rest.

The photon spectra of Fig. 5.6 are very similar to the Compton profiles obtained by a variety of techniques: Compton scattering of x- and γ-rays [119; 120] and inelastic scattering of electrons [118; 124] or of fast ions [125]. The Compton profiles are utilized to investigate the electron momentum density of atoms, molecules and solids. In the case of e-e bremsstrahlung, however, the process involves more independent variables than inelastic electron scattering of Compton effect. The widths of the photon spectra are not broad enough to be easily measurable.

In target atoms with higher atomic number Z the binding energies E_b of K electrons increase rapidly with Z. Hence the e-e bremsstrahlung spectra become much broader than for carbon. Figure 5.7 shows the photon curve from copper K electrons ($E_b \approx 8.98\,\text{keV}$) for the parameters $E_1 = 1000\,\text{keV}$, $E_1' = 700$ keV, $\theta = -24°$, and $\theta_1 = 20°$, as computed by means of the Hartree-Fock wave functions of Clementi and Roetti [122]. Here, the photon energy for target electrons at rest is $h\nu \approx 171$ keV; the recoil energy of the struck electrons amounts to $E_2' \approx 129$ keV so that the condition $E_b \ll E_2'$ for the validity of the impulse approximation is satisfied fairly well. In this case the cross-section curve is no more bell-shaped. At the low-energy end it is rather increasing after having reached a minimum near $h\nu = 70$ keV. This behavior is a consequence of the infrared divergence of the cross section

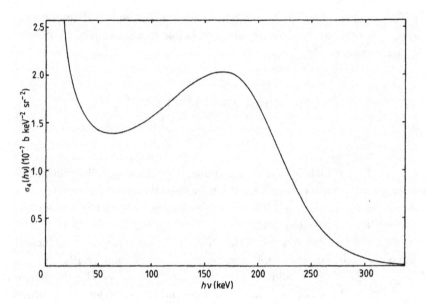

Fig. 5.7 Cross section $\sigma_4(h\nu)$ for Cu K electrons and $E_1 = 1000$ keV, $E_1' = 700$ keV, $\theta = -24°$, and $\theta_1 = 20°$ (from Haug and Keppler [121]).

(5.61), because

$$\lim_{k \to 0} A \propto k^{-2} \qquad (5.103)$$

[cf. Eq. (B.1)]. The sharp rise of the cross section $\sigma_4(k)$ near $k = 0$ would also occur in Fig. 5.6. Thus the integral (5.102), taken from the lower limit zero, will diverge, too. However, this is of no physical significance since for the comparison with experimental results only the photon regions shown in Fig. 5.6 are relevant. At the high-energy end of the spectrum, the cross-section curves are bounded by the condition $(p_2 p_2') \geq 1$ [see Eq. (5.92)]. Using (5.93) this yields for $(kp_1') > (kp_1)$ and coplanar geometry

$$k \leq k_{\max} = \frac{\epsilon_1 \epsilon_1' - p_1 p_1' \cos\theta_1 - 1}{\epsilon_1' - p_1' \cos(\theta - \theta_1) - (\epsilon_1 - p_1 \cos\theta)} . \qquad (5.104)$$

For the parameters of Fig. 5.7 the maximum photon energy amounts to $h\nu_{\max} = mc^2 k_{\max} \approx 486$ keV.

Chapter 6

Experiments on the elementary process of electron-electron bremsstrahlung

6.1 Electron-photon coincidence experiments without regard to polarization variables

Since the electron-electron system has no electric dipole moment, the non-relativistic e-e bremsstrahlung is expected to consist predominantly of electric quadrupole radiation with four lobes in the angular distribution (see Figs. 2.12, 5.3, and 5.4), and the cross section for the process is much smaller than for the dipole radiation of the electron-nucleus system. With increasing electron energy the e-e share increases but even for energies of several 100 keV the e-e bremsstrahlung in general gives such a small contribution to the total bremsstrahlung emission that it is not taken into account in most measurements of bremsstrahlung. Especially in the case of targets with high atomic number, the experiments give almost pure electron-nucleus bremsstrahlung, since its cross section is roughly proportional to Z^2, whereas for e-e bremsstrahlung at best it is proportional to the number of electrons, Z.

An important guide to discern the e-e contribution in the non-coincident bremsstrahlung spectrum is the maximum photon energy. Whereas the energy the nucleus receives can be neglected, the energy of the recoil electron may be considerable. Therefore, the e-e bremsstrahlung spectrum extends up to a maximum value less than the endpoint of the electron-nucleus bremsstrahlung spectrum. Moreover, the endpoint is angle dependent [see Eq. (5.28)]. The different upper bounds of the spectra were used to isolate the e-e contribution from the total spectrum in non-coincidence experiments [126]. However, a theoretical analysis [127] shows that the experimental results are still inconclusive. It is interesting to note that Hackl [128] did not find any evidence for a contribution of e-e bremsstrahlung using 2-MeV electrons and a lithium target ($Z = 3$). These experimental investigations show that it is difficult to isolate the e-e bremsstrahlung from the electron-nucleus bremsstrahlung when only the non-coincident photon spectrum is observed.

6.1.1 Angular distribution of photons for fixed electron direction

In the final state of the e-e bremsstrahlung process there are two outgoing electrons and one photon. The three particles have $3 \times 3 = 9$ degrees of freedom. In a kinematically complete experiment, all nine momentum components are determined. Four out of nine momentum components of the final state are determined by the initial state via momentum and energy conservation. There remain $9 - 4 = 5$ momentum components of the outgoing particles to be measured. In electron-photon coincidence experiments [5; 6; 7; 87] where the momenta of one of the outgoing electrons and of the photon (six quantities) were measured, the three-body final state is thus once overdetermined.

So it was possible to differentiate between the processes of e-e bremsstrahlung and electron-nucleus bremsstrahlung. The measuring method is based on a coincidence observation between the photon spectrum emitted in a fixed direction and outgoing electrons of definite energy and direction. As shown in Fig. 6.1 the e-e bremsstrahlung photons are well separated from the photons of electron-nucleus bremsstrahlung since the former have an energy that is reduced by the recoil energy of the second electron not observed. Figure 6.2 presents the measured photon angular distribution of e-e bremsstrahlung from a carbon target for fixed direction of outgoing electrons at $\theta_e = 20°$ (open circles). In contrast to electron-nucleus bremsstrahlung the photon energy of e-e bremsstrahlung is dependent on the photon emission angle for kinematical reasons [see Eq. (5.28)]. The measurements give the value of the cross section per carbon atom, e.g., for six bound electrons. The solid line depicts the theoretical prediction for free electron-electron collisions calculated according to Eq. (5.71) of Sec. 5.3. With regard to an estimated systematic error of $\pm 12\%$ there is agreement between experiment and theory. However, one may question the justification for the comparison. This problem is discussed in Sec. 5.4. There it is

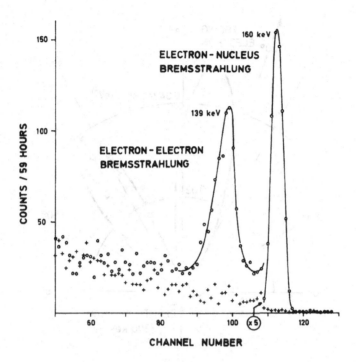

Fig. 6.1 Bremsstrahlung spectrum (pulse-height distribution) from electrons of 300 keV incident on a carbon target. Photons emitted at angle $\theta_k = -35°$ are observed in coincidence with outgoing electrons of 140 keV at scattering angle $\theta_e = 20°$ showing the coincidence peaks of e-e bremsstrahlung ($h\nu = 139$ keV) and electron-nucleus bremsstrahlung ($h\nu = 160$ keV). The crosses give the random coincidences measured simultaneously (from Nakel and Pankau [87]).

shown for the same parameters as used in the measurement presented in Fig. 6.2 that convolution of the bremsstrahlung cross section with the momentum distribution of the target electrons of carbon does not significantly change the value of the triply differential cross section. This finding gives the justification for the comparison of the experimental e-e bremsstrahlung cross sections with the theoretical values calculated for target electrons at rest. The width in the experimental coincident photon peak of e-e bremsstrahlung (Fig. 6.1) is mainly caused by the finite aperture of the electron detector and not by atomic binding of the electrons.

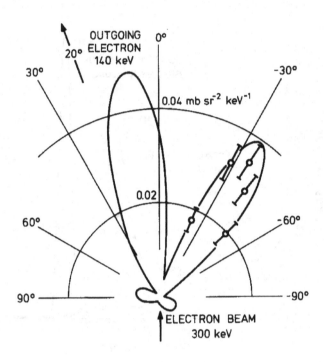

Fig. 6.2 Photon angular distribution of e-e bremsstrahlung for fixed energy ($E_e = 140$ keV) and direction ($\theta_e = 20°$) of outgoing electrons; incident electron energy $E_0 = 300$ keV, carbon target ($Z = 6$). The solid curve represents the theoretical prediction [130]. The cross section is given per atom, i.e., for six target electrons. In contrast to electron-nucleus bremsstrahlung the photon energy of e-e bremsstrahlung is dependent on the photon emission angle for kinematical reasons, e.g., from $h\nu = 139$ keV at $\theta_k = -23°$ to $h\nu = 134$ keV at $\theta_k = -47°$ (from Nakel and Pankau [87]).

6.1.2 *Energy distributions*

The processes of e-e bremsstrahlung and electron-nucleus bremsstrahlung have continuous spectra both in photon energy ($h\nu$) and outgoing electron energy (E_e), the ionization is accompanied by a characteristic x-ray line spectrum and a continuous electron spectrum (Auger electrons not considered).

In a two-parameter experiment of Komma and Nakel [91] the outgoing electrons of these processes were detected in coincidence with the emitted photons as a function of outgoing electron and photon energies for fixed detector angles. In this way it was possible to measure simultaneously the triply differential cross sections of the processes of e-e bremsstrahlung and e-nucleus bremsstrahlung as a function of the outgoing electron and photon energies (as well as the doubly differential cross section of K-shell ionization for targets of higher Z not considered here). The expected kinematical curves of the processes are delineated in Fig. 6.3. Here, the curve for the e-e bremsstrahlung is calculated for free electrons according to Eq. (5.28). For a fixed photon energy the outgoing electrons can have two energy values corresponding to the two signs in Eq. (5.38).

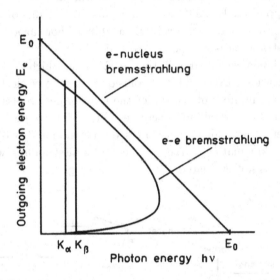

Fig. 6.3 Kinematics of the measured processes. According to energy conservation the electron-nucleus bremsstrahlung is situated along the straight line $E_e = E_0 - h\nu$ where E_0 is the incident electron energy. The curve for e-e bremsstrahlung has an angle-dependent shape and the quantum energy lies below that of e-nucleus bremsstrahlung because the second electron carries away energy. In the coincidence experiment the events of the K-shell ionization of silver and gold appear at the energies of the K_α and K_β lines (from Komma and Nakel [91]).

A schematic diagram of the experimental arrangement is shown in Fig. 6.4. The incident electron energy was 300 keV. The targets consisted of carbon, copper, silver, and gold. The photons were detected by means of a high-purity germanium detector set up at $\theta_k = -35°$ with respect to the incident beam. The outgoing electrons were observed with a surface barrier detector at the angle $\theta_e = 20°$. A non-dispersive and doubly focusing (i.e., triply focusing) magnet was inserted between the defining aperture and the surface barrier detector (cf. Sec. 4.1d) to transmit electrons between 40 and 200 keV. The timing pulses of both detectors were fed into a time-to-pulse-height converter and a window of a single-channel analyzer was set on this coincidence peak. The energy signals of both detectors were gated by the fast coincidences and fed to the two analogue-digital converters of a two-parameter multichannel analyzer. In Figs. 6.5a and 6.5b plots of two-parameter spectra of true coincidences are shown, measured for targets of carbon and copper (cf. Fig. 6.3 for the kinematics). The diagonal feature in the plots is due to electron-nucleus bremsstrahlung. For carbon (Fig. 6.5a), the e-e bremsstrahlung is well separated from the electron-nucleus bremsstrahlung. Its kinematical curve in the $h\nu$-E_e plane which depends on the angles of observation corresponds to a collision between free electrons. With increasing atomic number of the target the ratio of e-e bremsstrahlung to electron-nucleus bremsstrahlung decreases (Fig. 6.5b, copper). For silver and gold targets (plots not shown here, see [91]) the number of true coincidences on the kinematical curve for free e-e bremsstrahlung was too low and therefore was not evaluated.

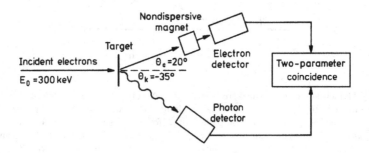

Fig. 6.4 Schematic diagram of the two-parameter arrangement (from Komma and Nakel[91]).

For comparison with theory the numbers of true coincidences of e-e bremsstrahlung as well as of electron-nucleus bremsstrahlung were pro-

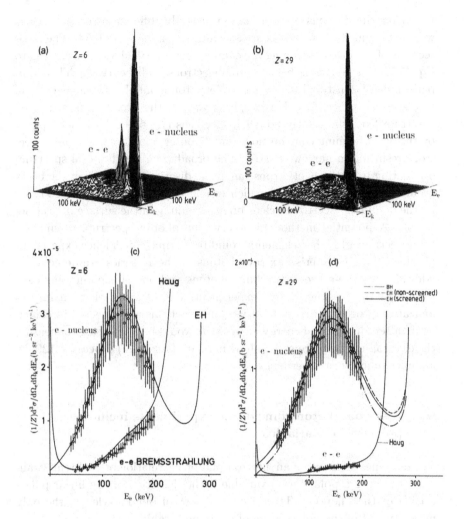

Fig. 6.5 Electron-photon coincidence spectra for (a) carbon, (b) copper, and the pertinent triply differential cross sections (c), (d) of electron-nucleus bremsstrahlung and of e-e bremsstrahlung, measured in the geometry of Fig. 6.4. In the coincidence spectra the random coincidences are subtracted. The remaining events at low energies are due to electron backscattering and Compton scattering in the detectors. The theoretical curves are calculated after formulae of Bethe-Heitler (BH), Elwert-Haug (EH), and Haug (from Komma and Nakel [91]).

184 *Experiments on the elementary process of electron-electron bremsstrahlung*

jected on the E_e axis and the absolute triply differential cross sections were determined. The results are shown in Figs. 6.5c and 6.5d. The cross sections of e-e bremsstrahlung are compared with calculations according to Eq. (5.71) for collisions between free electrons (solid curves). This comparison can be justified for targets of low atomic numbers such as carbon, as discussed in Sec. 5.4. For copper, however, the broad bremsstrahlung spectrum from the tightly bound K electrons ($E_b \approx 9$ keV) is superposed by the overwhelming contribution from 27 outer, more weakly bound, electrons resulting in narrow spectra. The broad wings of the total spectrum originating from the K electrons cannot be discriminated against the background caused by random coincidences and by a low-energy tail produced by the dominant electron-nucleus bremsstrahlung in the surface barrier detector. A potential method to observe the photon spectrum originating from electrons of high-Z elements would be a triple-coincidence experiment in which the characteristic x-ray photons of the K series are detected in addition to the e-e bremsstrahlung photons and one of the outgoing electrons. In the so-called (e,2e) experiments one observes electron impact ionization processes where both outgoing electrons are detected in coincidence after angular and energy analysis. It would be interesting to measure the electron-electron bremsstrahlung in relativistic (e,2e) processes [129] by a triple-coincidence experiment.

6.2 Electron-photon coincidence experiments including polarization variables

The only measurement of an electron-photon correlation of e-e bremsstrahlung including a polarization variable is the detection of the linear polarization of the photons. This is, to the best of our knowledge, the only measurement of the polarization of e-e bremsstrahlung.

6.2.1 *Linear polarization of electron-electron bremsstrahlung emitted by unpolarized electrons*

A calculation of the angular dependence of the degree of polarization in the elementary process of e-e bremsstrahlung is shown in Fig. 6.6a together with the corresponding triply differential cross section (Fig. 6.6b). The polarization was calculated according to a formula by Mack (unpublished) based on the theory for free electrons of Mack and Mitter [130].

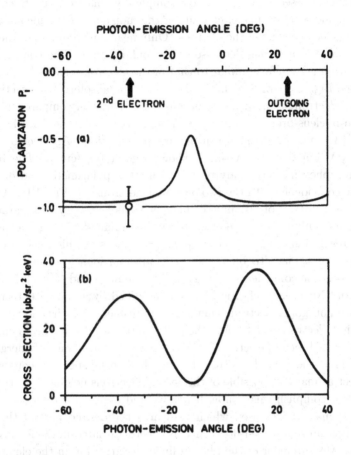

Fig. 6.6 (a) Linear photon polarization in the elementary process of e-e bremsstrahlung as a function of the photon-emission angle for outgoing electrons of energy 150 keV at $\theta_e = 25°$ (coplanar geometry). Primary electrons of 300 keV are used, incident on a carbon target. The experimental value is shown by the open circle at the photon emission angle $\theta_k = -35°$ where the photon energy is $h\nu \approx 116$ keV. The solid line is the theoretical prediction.
(b) Triply differential cross section of the elementary process of e-e bremsstrahlung calculated for the same parameters. The cross section is multiplied by a factor of 6 according to the number of electrons in the carbon atom (from Bleier and Nakel [131]).

The calculation predicts the overwhelming part of the radiation (referring to the cross section) to be almost completely polarized with the electric vector in the reaction plane, containing the momenta of the incoming and outgoing electrons and of the photons. Only near the minimum of the cross section there is a strong decrease of the polarization. The course of the polarization is therefore similar to that of the electron-nucleus bremsstrahlung (see Fig. 4.11a of Sec. 4.3.1). However, the absolute cross section for electron-electron bremsstrahlung is considerably smaller compared with the electron-nucleus bremsstrahlung (see Fig. 4.11b). Nevertheless, Bleier and Nakel [131] succeeded in measuring this polarization. An electron beam of 300 keV impinged on a thin carbon target ($20\mu g/cm^2$). The bremsstrahlung photons were analyzed in a Compton polarimeter and detected in triple coincidence with the two outgoing electrons (cf. Fig. 4.1). A triple coincidence instead of a double coincidence was necessary to separate the e-e bremsstrahlung from the electron-nucleus bremsstrahlung because the energy resolution of the Compton polarimeter was not high enough to allow the separation. To build up the triple-coincidence arrangement the electron-photon coincidence system used by Behncke and Nakel [97] and by Bleier and Nakel [98] (cf. Fig. 4.9) was modified by adding a detector for the second outgoing electron. Due to the finite energy widths and apertures of the first electron detector and the polarimeter and due to the kinematics, the associated second electrons are spread over a wide range of energy and angle. Thus the second electron detector had to be specially constructed to collect as many as possible of the second electrons belonging to the process. To accomplish this, a non-dispersive and doubly focusing (i.e. triply focusing) magnet was used which transmits the desired part of the electron spectrum to a cooled-down Si(Li) detector situated in the focus of the magnet. Measurement of the photon linear polarization in the elementary process of e-e bremsstrahlung was performed at the photon-emission angle $\theta_k = -35°$ with respect to the incident electron beam of 300 keV energy. One of the two outgoing electrons was detected at a scattering angle of 25° and energy 150 keV; then the resultant photon energy is 115.6 keV. Since for these parameters the second electron (at $-35.6°$) and the photon are emitted approximately in the same direction, the magnet of the electron detector had to be transparent to photons between the pole pieces. The binding energy of the K-shell electrons of the carbon target ($Z = 6$) is 284 eV, which is small compared with the energy ($E'_2 \approx 34.4$ keV) of the second outgoing electron.

The result of the measurement (open circle in Fig. 6.6a) shows the radiation to be completely polarized with the electric vector in the reaction plane. This finding is in best agreement with the theoretical prediction. The comparison of Figs. 6.6a and b shows that the polarization was measured at a maximum of the cross section. Nevertheless, due to the smallness of the triply differential cross section and the losses in the polarization analysis, the measuring time amounted to 420 h. The authors therefore abstained from measurements in the theoretically predicted region of the polarization decrease, occurring near the minimum cross section, as Behncke and Nakel [97] had done in the case of electron-nucleus bremsstrahlung measurements. Here, the decrease of polarization near the cross-section minimum was interpreted as being influenced by the spin-flip radiation according to a simple physical picture given by Fano et al. [99]. One can assume that this holds true for e-e bremsstrahlung, as well.

6.2.2 Further polarization correlations

To compare polarization correlations in electron-electron and electron-nucleus bremsstrahlung we follow the review article of Olsen [65] who considered the spin-polarization of the electrons, the circular photon polarization, and the momenta of the electrons and photons. Correlations between momenta and polarization are called momentum-polarization correlations, and correlations between two polarizations are called polarization-polarization correlations. An example of the first type is the photon emission asymmetry of bremsstrahlung from transversely polarized electrons (Sec. 4.3.2), an example of the latter is the helicity transfer in bremsstrahlung where the polarization of the initially polarized electron may be partially transferred to the emitted photon (Sec. 3.7.3).

Olsen shows that the momentum-polarization correlations always vanish to the lowest order in the interaction. The contributions to the momentum-polarization thus come from higher-order Coulomb interactions of an atomic nucleus taking part in the process (and from effects usually classified as radiative corrections if the process only involves photons and single charged fermions). This has the consequence that only processes in which a nucleus takes part may have large momentum-polarization correlations, of the order αZ, such as electron-nucleus bremsstrahlung, while processes involving only electrons and photons, such as electron-electron bremsstrahlung, have small correlations, of the order of α. As a consequence, e.g., the photon emission asymmetry of electron-electron bremsstrahlung from transversely polarized electrons would be very small.

Just as in the case of electron-nucleus bremsstrahlung (cf. Sec. 3.7.3) it holds for electron-electron bremsstrahlung that in lowest-order approximation unpolarized electrons do not produce circularly polarized bremsstrahlung unless the polarization of the final state electron is recorded in coincidence.

Chapter 7

Integrated cross sections and further bremsstrahlung processes

7.1 Integrated cross sections

The overwhelming part of theoretical and experimental investigations on the bremsstrahlung process deals with integrated cross sections such as doubly differential angular distributions, photon spectra, or total energy losses. Earlier theory and experiment up to 1959 were compared in the review of Koch and Motz [132]. This work is based predominantly on cross sections calculated in first Born approximation having the advantage that the integrations over the angles of the outgoing electrons and photons can be performed analytically. However, no results are presented for electron and photon polarization effects. Formulae for the linear polarization of electron-nucleus bremsstrahlung irrespective of the direction of the outgoing electron were given in Born approximation by Gluckstern and Hull [27]. Fronsdal and Überall [78] analyzed the polarization of bremsstrahlung emitted from polarized electrons, again in Born approximation taking into account screening by using a Coulomb potential with exponential cutoff. The partial-wave method (see Sec. 3.5) used by Tseng and Pratt [14] has thus far provided the best available calculations of the doubly differential cross section and associated polarization parameters. A review of the topic was given by Pratt and Feng [19] in 1985. Bremsstrahlung spectra from electrons in the kinetic energy range 1 keV to 2 MeV incident on neutral atoms with $2 \leq Z \leq 92$ were tabulated by Pratt et al. [133]. Kissel et al. [134] presented a tabulation of theoretical predictions for the shape of the photon angular distributions of atomic-field bremsstrahlung for 24 atoms with atomic numbers Z ranging from 1 to 92 for six incident electron energies from 1 to 500 keV. The tables result from interpolations in atomic number and fraction of energy radiated, $h\nu/E_0$, from a set of benchmark data computed by means of the partial-wave formalism. A similar tabulation of bremsstrahlung shape functions calculated with Sommerfeld-Maue wave functions for unscreened nuclei of low atomic numbers was given by Bernhardi et al. [135]. For the doubly differential cross section Fink and Pratt [60] find excellent agreement between numerical partial-wave computations and calculations by means of Sommerfeld-Maue wave functions for $Z \leq 13$ and the whole energy range. However, for high Z and low energies, there is no agreement in the shape of the differential cross section. A comprehensive set of bremsstrahlung cross sections, differential in the energy of the emitted photon for electrons with energies from 1 keV to 10 GeV incident on neutral atoms with atomic numbers $Z = 1$ to 100, was tabulated by Seltzer and Berger [136]. Ambrose, Altman, and Quarles [137] made comparisons between experiment and theory for the cross section differential in photon angle and energy for incident electrons of 50, 75, and 100 keV, for target elements with atomic numbers from 6 to 92, for photons emitted at $\theta_k = 90°$. Absolute doubly differential cross sections from gaseous targets have been measured for electron beam energies of 28 and 50 keV [138]. The photon spectrum was detected at 90° to the incident electron beam. The experiment provided evidence for a contribution of polarization bremsstrahlung (see Sec. 7.5).

Numerical results from a calculation of the positron bremsstrahlung spectrum calculated by means of the partial-wave formalism were presented by Feng et al. [139] both for a point-Coulomb potential and for a screened Hartree-Slater central potential. The data were compared with the corresponding electron bremsstrahlung. The photon spectrum is strongly suppressed for low-energy positrons which cannot penetrate toward the repulsive central potential. For high-energy positrons the photon spectrum approaches the one of electron bremsstrahlung.

In the case of electron-electron and electron-positron bremsstrahlung it was feasible to integrate in closed form the lengthy expressions for the triply differential cross section [see Eqs. (5.70), (5.79), (7.6), and (B.1)] over the angles of the outgoing electron or positron. The results were given in covariant form [117; 140] allowing to express the photon angular distributions in arbitrary frames of reference, e.g., center-of-mass system (cms) or laboratory system.* In the laboratory system the most conspicuous difference to the integrated photon angular distributions of electron-nucleus bremsstrahlung is that photons of higher energies, $k > (\epsilon_1 - 1)/(\epsilon_1 + p_1 + 1)$, can be emitted only into a cone with the half apex angle θ_0 given by $\cos\theta_0 = [(\epsilon_1 + 1)k - (\epsilon_1 - 1)]/(kp_1)$. As k increases, the allowed angular range is more and more restricted, in particular at relativistic energies (see [117]). In the cms the angular distributions are obviously symmetric about the emission angle $\theta = 90°$.

A further integration of the electron-electron and electron-positron cross sections over the photon angles is no longer possible in general form, so the photon spectra have to be computed numerically. However, both in the laboratory system [141][†] and in the center-of-mass system [142] of e-e bremsstrahlung a majority of the integrations has been carried out analytically. Expanding these cross sections into powers of the photon energy k up to terms of relative order $(k/k_{\max})^2$, approximate formulae for the low-energy region of the bremsstrahlung spectra have been derived. Particularly in the cms this represents a very good approximation for photon energies up to about 40% of the incident-electron kinetic energies. The approximation for the long-wavelength end of the spectrum is also valuable for the calculation of the total energy loss by bremsstrahlung proportional to

$$\Phi_{\rm rad} = \int_0^{k_{\max}} k \frac{d\sigma}{dk} dk . \qquad (7.1)$$

Since $k(d\sigma/dk)$ is logarithmically divergent at $k = 0$ (this holds for the unscreened Coulomb potential), the approximation can be used to calculate analytically the low-energy part of the integral (7.1) up to a photon energy $k \ll k_{\max}$.

For the process of electron-positron bremsstrahlung (see Sec. 7.3) the majority of integrations over the photon angles was performed for the cross section in the cms [143]. Again the approximation formula for the low-energy end of the photon spectrum was derived. This approximation is very accurate for $k < 0.3(\epsilon - 1)$, the relative error never exceeding 0.3%.

The measurement of the contribution of electron-electron bremsstrah-

---

*In Ref. [140] there is a misprint in the third line of Eq. (A1). The correct term should read as $[4 + 1/(\tau + 1)]$.

[†]Note a misprint on page 347 of Ref. [141]. In the third line the correct term should read as $(\epsilon r/w + s)$.

lung to atomic-field bremsstrahlung in non-coincidence experiments is very difficult. As discussed in Sec. 6.1, the different upper limits of the photon spectra could, in principle, be used to discriminate between the two bremsstrahlung processes. Since the relative contribution of e-e bremsstrahlung is of the order of $1/Z$, experiments with targets of low atomic number Z are most promising. The authors of experiments on beryllium ($Z = 4$) and aluminum ($Z = 13$) targets [126; 144; 145] give evidence of having discerned e-e bremsstrahlung components with spectral endpoints varying with emission angle as predicted by the kinematics [Eq. (5.34)]. A comparison with theoretical cross sections shows, however, this finding to be inconclusive since the accuracy of the measurements is not sufficient [127].

7.2 Positron-nucleus bremsstrahlung

If positrons are impinging on a nucleus or an atom under emission of radiation the process is called positron-nucleus (p-n) bremsstrahlung. The cross section of p-n bremsstrahlung can be calculated by means of the same methods as in the case of electron-nucleus bremsstrahlung (Chapter 3). For a point Coulomb field the only difference is that in the Dirac equation (3.25) the sign of the potential term a/r has to be reversed. At high energies there is rather little distinction between the presence of an electron or positron in the field of an atom (we disregard the rare events where a positron annihilates an atomic electron). But there is a substantial difference at low energies due to the opposite response to a positive nuclear charge screened by a negative charge distribution of atomic electrons. The bremsstrahlung emission is strongly suppressed for low-energy positrons which cannot penetrate toward the repulsive central potential.

In first Born approximation the cross section of positron-nucleus bremsstrahlung in the pure Coulomb field of a nucleus does not differ from that of electron-nucleus bremsstrahlung. Repeating the derivation of the matrix element in Sec. 3.3 with the potential term $+a/r$ in Eq. (3.25), the right-hand side of Eq. (3.30) gets a minus sign. Hence, $\Phi_0(\mathbf{r})$ and $\Phi_1(\mathbf{r})$ change their signs. Since the first term of the matrix element (3.54) vanishes, M changes sign, too. However, the cross section is proportional to $|M|^2$ and therefore remains unchanged. To second order in a, the matrix element can be written as $M = -aM_1 + a^2 M_2$. Thus

$$|M|^2 \approx a^2 \{|M_1|^2 - 2a\,\mathrm{Re}(M_1^* M_2)\}\,, \qquad (7.2)$$

i.e., the correction term changes sign compared to electron bremsstrahlung.

The Born-approximation results can be corrected for the Coulomb distortion of the wave functions if one multiplies them by the Elwert factor. The pertinent factor for positrons is obtained from the expression (3.90) by reversing the signs of the Coulomb parameters a_1 and a_2, resulting in

$$F_E^+ = \frac{a_2}{a_1} \frac{e^{2\pi a_1} - 1}{e^{2\pi a_2} - 1}. \tag{7.3}$$

In Born-Elwert approximation the ratio of positron to electron bremsstrahlung cross sections is $F_E^+/F_E = \exp\{2\pi(a_1 - a_2)\} < 1$. One can see that the cross section of positron-nucleus bremsstrahlung at the short-wavelength limit ($a_2 \to \infty$) now vanishes exponentially as a consequence of the repulsion by the central potential. The same result is obtained in the exact calculation by means of the partial-wave method (cf. Sec. 3.5.3).

Taking into account nuclear screening by the atomic electrons, the cross section in Born approximation is modified by the form factor (3.99), again the same for electrons and for positrons. Whereas for electrons relatively good results for low atomic numbers Z have been obtained by multiplying the cross section by the Elwert factor, this Born-Elwert-form-factor approach does not work as well for positrons [139].

If the cross section of positron-nucleus bremsstrahlung is calculated by means of the Sommerfeld-Maue wave function, the results can be easily derived from those of Sec. 3.4.3 by reversing the sign of $a = \alpha Z$. Since the factor F of Eq. (3.176) changes to

$$F^+ = \frac{2\pi a_1}{1 - e^{-2\pi a_1}} \frac{2\pi a_2}{e^{2\pi a_2} - 1}, \tag{7.4}$$

the exponential decrease of the cross section near the short-wavelength limit is ensured.

As in the case of electron-nucleus bremsstrahlung the relativistic distorted partial-wave approximation, describing the process as a positron transition in a self-consistent screened central potential, is the best available approach to calculate the cross section. So far there exist, however, no numerical computations of the triply differential cross section of p-n bremsstrahlung, neither have coincidence experiments been performed. The only numerical results obtained by this method refer to integrated cross sections. Feng et al. [139] presented calculations of photon spectra $d\sigma_+/dk$ both for a point Coulomb potential and for a screened Hartree-Slater central potential.

The data were compared with the corresponding electron bremsstrahlung. The photon spectrum is strongly suppressed for low-energy positrons which cannot penetrate toward the repulsive central potential. For high-energy positrons the photon spectrum approaches the one of electron bremsstrahlung. Tseng [146] calculated photon angular distributions for positron and electron bremsstrahlung in the field of atoms with $Z = 1$, 13, and 79 for kinetic energies of the incident particles $E_0 = 1.0$, 2.0, and 2.5 MeV. In this energy region the Born-approximation predictions agree quite well with the partial-wave results for the shape of the angular distributions of positron bremsstrahlung.

7.3 Electron-positron bremsstrahlung

In the elementary process of electron-positron bremsstrahlung (Fig. 7.1) an electron with four-momentum $\underline{p} = (\epsilon_-, \mathbf{p})$ and a positron with four-momentum $\underline{q} = (\epsilon_+, \mathbf{q})$ collide under the emission of a photon with four-momentum $\underline{k} = (k, \mathbf{k})$. The outgoing particles have the four-momenta $\underline{p}' = (\epsilon'_-, \mathbf{p}')$ and $\underline{q}' = (\epsilon'_+, \mathbf{q}')$. As in electron-electron bremsstrahlung (Chapter 5) eight Feynman diagrams contribute in lowest order to the matrix element, four of them representing scattering graphs and four representing virtual annihilation graphs (Fig 7.2). Therefore the evaluation of the traces is very laborious and the resulting cross-section formula is extremely lengthy. It can, however, be simply derived from the corresponding expression for electron-electron bremsstrahlung (Sec. 5.3) by means of the substitution law.

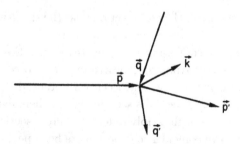

Fig. 7.1 Particle momenta in the elementary process of electron-positron bremsstrahlung.

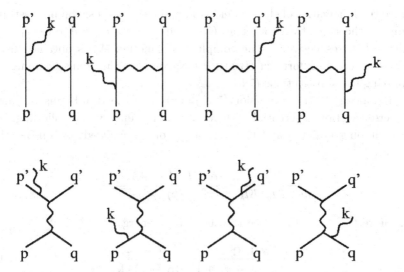

Fig. 7.2 Lowest-order Feynman diagrams for electron-positron bremsstrahlung.

The substitution law [34] states that the four-momenta of electron-electron bremsstrahlung change into the corresponding four-momenta of electron-positron bremsstrahlung as follows:

$$\underline{p_1} \to \underline{p},\ \underline{p_2} \to -\underline{q}',\ \underline{p_1'} \to -\underline{q},\ \underline{p_2'} \to \underline{p}',\ \underline{k} \to \underline{k}. \tag{7.5}$$

Then the cross section of electron-positron bremsstrahlung, summed over the polarizations of the outgoing particles and averaged over the polarizations of the incoming particles, has the form [cf. Eq. (5.70)]

$$\frac{d^3\sigma}{dk\, d\Omega_k\, d\Omega_{q'}} = \frac{\alpha r_0^2}{\pi^2} \frac{k}{R_+ \sqrt{(pq)^2 - 1}} \sum_{\nu=1}^{2} q_\nu'^2 A_+(q_\nu'). \tag{7.6}$$

Here the momenta of the outgoing positrons are given by

$$q_\nu' = \frac{[(p'q') + 1](\mathbf{p} + \mathbf{q} - \mathbf{k}) \cdot \hat{\mathbf{q}}' \pm (\epsilon_+ + \epsilon_- - k) R_+}{(\epsilon_+ + \epsilon_- - k)^2 - [(\mathbf{p} + \mathbf{q} - \mathbf{k}) \cdot \hat{\mathbf{q}}']^2} \tag{7.7}$$

and the square root R_+ by

$$R_+ = \sqrt{[(p'q') + 1]^2 + [(\mathbf{p} + \mathbf{q} - \mathbf{k}) \cdot \hat{\mathbf{q}}']^2 - (\epsilon_+ + \epsilon_- - k)^2}. \tag{7.8}$$

Since the kinematics of electron-electron and electron-positron bremsstrahlung are the same, the expressions (7.7) and (7.8) correspond to Eqs. (5.13) and (5.14), respectively. The complicated function A_+ is obtained from the corresponding formula (B.1) for electron-electron bremsstrahlung by applying the substitutions (7.5).

In analogy to the expression (5.71) of electron-electron bremsstrahlung the cross section differential with respect to the energy and solid angle of the outgoing positron and the solid angle of the emitted photon has the form

$$\frac{d^3\sigma}{d\epsilon'_+\,d\Omega_{q'}\,d\Omega_k} = \frac{\alpha r_0^2}{\pi^2}\frac{k^2 q'}{(q'k)}\frac{A_+}{\sqrt{(pq)^2-1}}. \qquad (7.9)$$

In the rest system of the electron the photon energy

$$k = \frac{\mathbf{q}\cdot\mathbf{q}' - (\epsilon_+ + 1)(\epsilon'_+ - 1)}{\epsilon_+ - \epsilon'_+ + 1 + (\mathbf{q}' - \mathbf{q})\cdot\hat{\mathbf{k}}} \qquad (7.10)$$

[cf. Eq. (5.28)] is angle dependent, and $(p'k)$ is given by

$$\begin{aligned}(p'k) &= (pk) + (qk) - (q'k)\\ &= (\epsilon_+ - \epsilon'_+ + 1)k + (\mathbf{q}' - \mathbf{q})\cdot\mathbf{k}\,.\end{aligned} \qquad (7.11)$$

Figure 7.3 shows an example of the photon angular distribution (7.9) in the laboratory frame (rest system of the electron) for coplanar geometry and incident positrons of energy $E_+ = 300\,\text{keV}$. The scattered positrons at fixed angle $\theta_+ = 28°$ have the energy $E'_+ = 160\,\text{keV}$. These are the same parameters as for the angular distribution of electron-electron bremsstrahlung presented in Fig. 5.3. As in the case of electron-nucleus bremsstrahlung (see Fig. 3.2a) this is a beamed dipole distribution with two maxima, in distinct contrast to the quadrupole distribution of e-e bremsstrahlung. Moreover, the e^--e^+ cross section is larger by an order of magnitude than the e-e cross section [e^--e^+ : $\sigma_{\max} \approx 75.6\,\mu\text{b}/(\text{keV sr}^2)$ at $\theta_k = 19.4°$; e-e: $\sigma_{\max} \approx 6.72\,\mu\text{b}/(\text{keV sr}^2)$ at $\theta_k = 15.5°$]. The main lobe of the e^--e^+ radiation exceeds also that of electron-proton bremsstrahlung by a factor of about 6, whereas the side lobes are of comparable magnitude. Both maxima occur at different angles in the two radiation processes which is due to the different kinematics.

Again these cross sections are exact within lowest-order perturbation theory (Born approximation). Thus correct results are expected if the con-

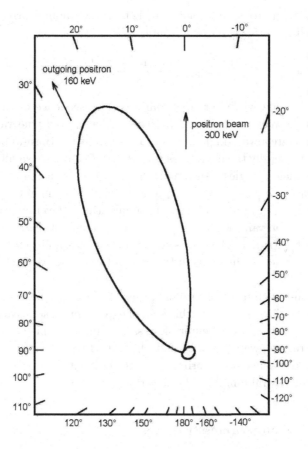

Fig. 7.3 Photon angular distribution in the elementary process of electron-positron bremsstrahlung for incident positrons of 300 keV and coplanar geometry. The scattered positrons at angle $\theta_+ = 28°$ have energy $E'_+ = 160\,\text{keV}$.

ditions (5.72) are satisfied where now

$$\beta_{12} = \frac{\sqrt{(pq)^2 - 1}}{(pq)} \ , \quad \beta'_{12} = \frac{\sqrt{(p'q')^2 - 1}}{(p'q')} \tag{7.12}$$

are the relative velocities (in units of the speed of light) of the electrons and positrons in the initial and final state, respectively. Due to the small value of the fine-structure constant the conditions (5.72) are only violated for low values of the relative velocities (7.12). The Coulomb correction factor corresponding to (5.74) has the form [140]

$$F_{e^-e^+} = \frac{a_2}{a_1} \frac{1 - e^{-2\pi a_1}}{1 - e^{-2\pi a_2}} .\tag{7.13}$$

This factor is always larger than unity as a consequence of the Coulomb attraction between electron and positron. That is, the true values of the cross section are always higher than those given by the Born-approximation formulae. Generally the deviations from unity of (7.13) are small, especially at high particle energies. An important exception is the short-wavelength limit given by $(p'q') = 1$ where a_2 and $F_{e^-e^+}$ tend to infinity. As may be seen from Eq. (7.6) and the form of the quantity A given in Appendix B, the cross section vanishes for $(p'q') \to 1$. By applying the factor (7.13) the square root $\sqrt{(p'q')^2 - 1}$ cancels out resulting in a finite cross section at the short-wavelength limit, as in the case of the attractive electron-nucleus interaction.

Electron-positron bremsstrahlung was detected with colliding beams at ultra-relativistic energies. For this kind of experiments radiative corrections of relative order α in the center-of-mass system, exact in the high-energy, low-angle regime, were given by Jadach et al. [147].

To our knowledge no experiments on the elementary process of electron-positron bremsstrahlung have been performed so far.

7.4 Two-photon bremsstrahlung

Two-photon bremsstrahlung (or double bremsstrahlung) is a process in which two photons are emitted simultaneously in a single collision of an incident electron on an atom. Multi-photon bremsstrahlung is also possible [148] but the emission probability of more than one photon decreases rapidly with the number of photons. Therefore we will confine ourselves to the process of two-photon bremsstrahlung (2PB) which has been investigated in photon-photon coincidence experiments.

2PB is a quantum-electrodynamic process of fourth order. It is a typical quantum effect in the sense that its probability cannot be calculated from classical theory. Two-photon bremsstrahlung is described by six Feynman diagrams, which are depicted in Fig. 7.4. The lower graphs are obtained from the upper ones by interchanging the photon lines k_1 and k_2.

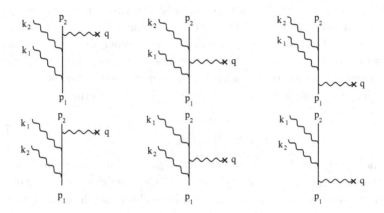

Fig. 7.4 Lowest-order Feynman diagrams describing two-photon bremsstrahlung.

There are only very little experimental data available on 2PB. The experiments performed so far are photon-photon coincidence measurements ignoring the scattered electron. The experimental status was summarized by Quarles and Liu [149]. The authors state that there is reasonable order-of-magnitude agreement with calculations for some ranges of radiated photon energy but that there are significant discrepancies for certain photon energies. Since the work of Quarles and Liu no additional experiments became known to us.

The process of 2PB was first considered in 1934 by Heitler and Nordheim [150] as a radiative correction to the single-bremsstrahlung theory. The calculation was performed by means of the Weizsäcker-Williams method of virtual quanta (see Sec. 2.3) which was established in the same year. Heitler and Nordheim suggested that the 2PB cross section would be reduced from that of one-photon bremsstrahlung by a power of the fine-structure constant $\alpha \approx 1/137$.

The cross section for the elementary process of two-photon bremsstrahlung is five-fold differential,

$$d^5\sigma \equiv \frac{d^5\sigma}{dk_1\,dk_2\,d\Omega_{k_1}\,d\Omega_{k_2}\,d\Omega_{p_2}}\,, \qquad (7.14)$$

where k_1, k_2 denote the energies and $d\Omega_{k_1}$, $d\Omega_{k_2}$ the elements of solid angle of the emitted photons, and $d\Omega_{p_2}$ the element of solid angle of the outgoing electron. To measure $d^5\sigma$, a triple coincidence experiment (the scattered electron in coincidence with the emitted photons) is needed in which the directions of the three outgoing particles and the energies of two of the particles are recorded. Since only experiments with two-photon coincidences have been performed so far, with the scattered electron not observed, the quantity of interest is the four-fold differential cross section

$$d^4\sigma \equiv \frac{d^4\sigma}{dk_1\,dk_2\,d\Omega_{k_1}\,d\Omega_{k_2}} = \int d^5\sigma\,d\Omega_{p_2}\,. \tag{7.15}$$

Smirnov [151] has investigated the 2PB process in relativistic Born approximation ignoring the polarizations of the particles. Designating the electron variables as in the case of one-photon bremsstrahlung, the five-fold differential cross section can be written in the form

$$\frac{d^5\sigma_B}{dk_1\,dk_2\,d\Omega_{k_1}\,d\Omega_{k_2}\,d\Omega_{p_2}} = \frac{\alpha^2 Z^2 r_0^2}{4\pi^4}\frac{p_2}{p_1 q^4}k_1 k_2 X\,, \tag{7.16}$$

where

$$\mathbf{q} = \mathbf{p}_1 - \mathbf{p}_2 - \mathbf{k}_1 - \mathbf{k}_2 \tag{7.17}$$

denotes the recoil momentum imparted to the atom, and $X/\epsilon_1\epsilon_2$ is the trace of the matrix element squared; X was given as a complicated function of the particle energies and angles [151].

In the nonrelativistic limit the cross section in Born approximation is greatly simplified reducing to [152]

$$d^5\sigma_B \approx \frac{\alpha^2 Z^2 r_0^2}{4\pi^4}\frac{p_2}{k_1 k_2 p_1 q^4}(\mathbf{e}_1\cdot\mathbf{q})^2(\mathbf{e}_2\cdot\mathbf{q})^2\,. \tag{7.18}$$

Summing (7.18) over the polarization vectors $\mathbf{e}_1$ and $\mathbf{e}_2$ of the photons by means of the relation (3.79) we get[‡]

$$d^5\sigma_B \approx \frac{\alpha^2 Z^2 r_0^2}{4\pi^4}\frac{p_2}{p_1 k_1 k_2}(\hat{\mathbf{k}}_1\times\hat{\mathbf{q}})^2(\hat{\mathbf{k}}_2\times\hat{\mathbf{q}})^2\,. \tag{7.19}$$

At nonrelativistic energies, where the photon momentum is of order p^2, the recoil momentum (7.17) reduces to $\mathbf{q}\approx\mathbf{p}_1-\mathbf{p}_2$. In order to allow for the atomic screening of the nuclear Coulomb field [for spherically symmetric

---

[‡]The cross section (7.19) was derived by means of the exact nonrelativistic dipole approximation [152]. It disagrees with the nonrelativistic limit of Smirnov's result [151] where the last two angle-dependent factors are missing.

charge densities the form factor (3.99) is only a function of q], it is expedient to transform the integration in (7.15) into one over q^2 and the azimuthal angle φ_q. Using

$$q^2 \approx (\mathbf{p}_1 - \mathbf{p}_2)^2 = p_1^2 + p_2^2 - 2p_1 p_2 \cos \theta_e , \qquad (7.20)$$

we get for fixed $\mathbf{p}_1$

$$d\Omega_{p_2} = -d(\cos\theta_e)\,d\varphi_q = \frac{1}{2p_1 p_2}\,dq^2\,d\varphi_q = \frac{q}{p_1 p_2}\,dq\,d\varphi_q . \qquad (7.21)$$

Then, substituting the cross section (7.19) including form-factor screening, the integral (7.15) takes the form [153]

$$\frac{d^4 \sigma_B}{dk_1\,dk_2\,d\Omega_{k_1}\,d\Omega_{k_2}} \approx \frac{\alpha^2 Z^2 r_0^2}{4\pi^4 p_1^2 k_1 k_2} \int_{p_1-p_2}^{p_1+p_2} dq\, q\,[1 - F(q)]^2$$

$$\cdot \int_0^{2\pi} d\varphi_q\,(\hat{\mathbf{k}}_1 \times \hat{\mathbf{q}})^2 (\hat{\mathbf{k}}_2 \times \hat{\mathbf{q}})^2 . \qquad (7.22)$$

The second integral can be performed analytically resulting in[§]

$$\int_0^{2\pi} (\hat{\mathbf{k}}_1 \times \hat{\mathbf{q}})^2 (\hat{\mathbf{k}}_2 \times \hat{\mathbf{q}})^2\,d\varphi_q = \frac{\pi}{(2qp_1)^4} \Big\{ 4(qp_1)^4 \big[1 + 3(\hat{\mathbf{p}}_1 \cdot \hat{\mathbf{k}}_1)^2$$
$$+ 3(\hat{\mathbf{p}}_1 \cdot \hat{\mathbf{k}}_2)^2 + 3(\hat{\mathbf{p}}_1 \cdot \hat{\mathbf{k}}_1)^2(\hat{\mathbf{p}}_1 \cdot \hat{\mathbf{k}}_2)^2 - 4(\hat{\mathbf{p}}_1 \cdot \hat{\mathbf{k}}_1)(\hat{\mathbf{p}}_1 \cdot \hat{\mathbf{k}}_2)(\hat{\mathbf{k}}_1 \cdot \hat{\mathbf{k}}_2)$$
$$+ 2(\hat{\mathbf{k}}_1 \cdot \hat{\mathbf{k}}_2)^2 \big] + 2q^2 p_1^2 (p_1^2 - p_2^2 + q^2)^2 \big[3 - 3(\hat{\mathbf{p}}_1 \cdot \hat{\mathbf{k}}_1)^2 - 3(\hat{\mathbf{p}}_1 \cdot \hat{\mathbf{k}}_2)^2$$
$$- 15(\hat{\mathbf{p}}_1 \cdot \hat{\mathbf{k}}_1)^2(\hat{\mathbf{p}}_1 \cdot \hat{\mathbf{k}}_2)^2 + 12(\hat{\mathbf{p}}_1 \cdot \hat{\mathbf{k}}_1)(\hat{\mathbf{p}}_1 \cdot \hat{\mathbf{k}}_2)(\hat{\mathbf{k}}_1 \cdot \hat{\mathbf{k}}_2) - 2(\hat{\mathbf{k}}_1 \cdot \hat{\mathbf{k}}_2)^2\big]$$
$$+ (p_1^2 - p_2^2 + q^2)^4 \big[\tfrac{1}{4} - \tfrac{5}{4}(\hat{\mathbf{p}}_1 \cdot \hat{\mathbf{k}}_1)^2 - \tfrac{5}{4}(\hat{\mathbf{p}}_1 \cdot \hat{\mathbf{k}}_2)^2 + \tfrac{1}{2}(\hat{\mathbf{k}}_1 \cdot \hat{\mathbf{k}}_2)^2$$
$$+ \tfrac{35}{4}(\hat{\mathbf{p}}_1 \cdot \hat{\mathbf{k}}_1)^2(\hat{\mathbf{p}}_1 \cdot \hat{\mathbf{k}}_2)^2 - 5(\hat{\mathbf{p}}_1 \cdot \hat{\mathbf{k}}_1)(\hat{\mathbf{p}}_1 \cdot \hat{\mathbf{k}}_2)(\hat{\mathbf{k}}_1 \cdot \hat{\mathbf{k}}_2)\big] \Big\} . \qquad (7.23)$$

If the energy of the incident electron and the photon energies are chosen such that the lower limit of integration in (7.22), $q_{\min} = p_1 - p_2$, satisfies the condition $q_{\min}^{-1} \ll r_{1s}$, where r_{1s} is the radius of the 1s-subshell of the target atom, the screening effect is negligible, and one may put $F(q) = 0$ for all q. In this case the q integration can also be done analytically yielding

[§] Even though the integrand contains only the unit vector $\hat{\mathbf{q}}$ the integral is dependent on the magnitude q. This is a consequence of the change of integration variables since $\cos\theta_e$ is now a function of q according to Eq. (7.20).

[154] (misprints have been corrected in Ref. [155])

$$\frac{d^4\sigma_B}{dk_1\, dk_2\, d\Omega_{k_1}\, d\Omega_{k_2}} \approx \frac{\alpha^2 Z^2 r_0^2}{64\pi^3}\frac{1}{k_1 k_2}\Big\{ax - bx^3 + 2cx^5$$
$$+ (d - cx^2)(1 - x^2)^2 \ln\frac{1+x}{1-x}\Big\}, \qquad (7.24)$$

where $x = p_2/p_1$ and

$$\begin{aligned}
a &= 50 - 42\big[(\hat{\mathbf{p}}_1\cdot\hat{\mathbf{k}}_1)^2 + (\hat{\mathbf{p}}_1\cdot\hat{\mathbf{k}}_2)^2\big] + 54(\hat{\mathbf{p}}_1\cdot\hat{\mathbf{k}}_1)^2(\hat{\mathbf{p}}_1\cdot\hat{\mathbf{k}}_2)^2 \\
&\quad - 8(\hat{\mathbf{p}}_1\cdot\hat{\mathbf{k}}_1)(\hat{\mathbf{p}}_1\cdot\hat{\mathbf{k}}_2)(\hat{\mathbf{k}}_1\cdot\hat{\mathbf{k}}_2) + 4(\hat{\mathbf{k}}_1\cdot\hat{\mathbf{k}}_2)^2, \\
b &= \tfrac{4}{3}\big\{13 - 29\big[(\hat{\mathbf{p}}_1\cdot\hat{\mathbf{k}}_1)^2 + (\hat{\mathbf{p}}_1\cdot\hat{\mathbf{k}}_2)^2\big] + 95(\hat{\mathbf{p}}_1\cdot\hat{\mathbf{k}}_1)^2(\hat{\mathbf{p}}_1\cdot\hat{\mathbf{k}}_2)^2 \\
&\quad - 44(\hat{\mathbf{p}}_1\cdot\hat{\mathbf{k}}_1)(\hat{\mathbf{p}}_1\cdot\hat{\mathbf{k}}_2)(\hat{\mathbf{k}}_1\cdot\hat{\mathbf{k}}_2) + 2(\hat{\mathbf{k}}_1\cdot\hat{\mathbf{k}}_2)^2\big\}, \\
c &= 1 - 5\big[(\hat{\mathbf{p}}_1\cdot\hat{\mathbf{k}}_1)^2 + (\hat{\mathbf{p}}_1\cdot\hat{\mathbf{k}}_2)^2\big] + 35(\hat{\mathbf{p}}_1\cdot\hat{\mathbf{k}}_1)^2(\hat{\mathbf{p}}_1\cdot\hat{\mathbf{k}}_2)^2 \\
&\quad - 20(\hat{\mathbf{p}}_1\cdot\hat{\mathbf{k}}_1)(\hat{\mathbf{p}}_1\cdot\hat{\mathbf{k}}_2)(\hat{\mathbf{k}}_1\cdot\hat{\mathbf{k}}_2) + 2(\hat{\mathbf{k}}_1\cdot\hat{\mathbf{k}}_2)^2, \\
d &= 7 - 11\big[(\hat{\mathbf{p}}_1\cdot\hat{\mathbf{k}}_1)^2 + (\hat{\mathbf{p}}_1\cdot\hat{\mathbf{k}}_2)^2\big] + 5(\hat{\mathbf{p}}_1\cdot\hat{\mathbf{k}}_1)^2(\hat{\mathbf{p}}_1\cdot\hat{\mathbf{k}}_2)^2 \\
&\quad + 4(\hat{\mathbf{p}}_1\cdot\hat{\mathbf{k}}_1)(\hat{\mathbf{p}}_1\cdot\hat{\mathbf{k}}_2)(\hat{\mathbf{k}}_1\cdot\hat{\mathbf{k}}_2) - 2(\hat{\mathbf{k}}_1\cdot\hat{\mathbf{k}}_2)^2.
\end{aligned} \qquad (7.25)$$

Another important limiting case is that in which the photon energy k_2 is low, $k_2 \ll 1$, $k_2 \ll k_1$. Then the cross section (7.16) takes the form [151]

$$d^5\sigma_B \approx \frac{d^3\sigma_B}{dk_1\, d\Omega_{k_1}\, d\Omega_{p_2}}\frac{\alpha}{\pi^2}\left\{\frac{q^2 + b_1 - a_1}{a_2 b_2} - \left(\frac{1}{a_2} - \frac{1}{b_2}\right)^2\right\} k_2, \qquad (7.26)$$

where $d^3\sigma_B/(dk_1\, d\Omega_{k_1}\, d\Omega_{p_2})$ is the Bethe-Heitler cross section (3.81), and

$$\begin{aligned}
a_1 &= 2(\epsilon_1 k_1 - \mathbf{p}_1\cdot\mathbf{k}_1), & a_2 &= 2(\epsilon_1 k_2 - \mathbf{p}_1\cdot\mathbf{k}_2), \\
b_1 &= 2(\epsilon_2 k_1 - \mathbf{p}_2\cdot\mathbf{k}_1), & b_2 &= 2(\epsilon_2 k_2 - \mathbf{p}_2\cdot\mathbf{k}_2).
\end{aligned} \qquad (7.27)$$

The physical meaning of the expression (7.26) is quite clear: The differential cross section for 2PB is equal to the product of the differential cross section for one-photon bremsstrahlung and the probability of emission of a soft photon. This is an extension to the 2PB of the low-energy theorem [156] for one-photon bremsstrahlung. Taking the nonrelativistic limit, the cross section (7.26) transforms to the expression (7.19).

Following the early work of Smirnov [151] theoretical investigations were based mainly on second-order nonrelativistic perturbation theory. In this approximation exact analytic formulae were derived for the 2PB cross section in the Coulomb field [152; 157; 158; 159]. After tedious analytical calculations the amplitudes of the matrix element can be expressed as one-dimensional integrals over Gauss hypergeometric functions representing the analogue of Sommerfeld's cross section [10] for the one-photon case, Eq. (3.203). In contrast to the Born approximation the initial and final electron velocities can be arbitrarily low even for targets with high atomic numbers.

The complicated analytical cross section simplifies considerably in a number of limiting cases. At high (but still nonrelativistic) initial and final electron energies where the conditions (3.89) are satisfied, Véniard et al. [152] got the formula (7.19) corresponding to the Born approximation. Comparisons with the exact nonrelativistic dipole approximation shows that the Born-approximation results are improved by using the Elwert factor (3.90) [155].

The 2PB cross section has also been calculated in the framework of the nonrelativistic distorted partial-wave approximation (DPWA) which permits the use of realistic wave functions for the incident electron with the screening effect of the atomic electrons taken into account [160]. Numerical results were presented for incident electrons of kinetic energy 8.82 keV impinging on targets of He, Ne, Ar, Kr, and Xe atoms [153]. These parameters were chosen to meet the experimental conditions of Hippler [161]. There is good quantitative agreement with the nonrelativistic Born approximation [Eq. (7.24) including the Elwert factor] for helium targets and satisfactory agreement with the exact nonrelativistic point Coulomb results [152] for krypton. In the latter case some discrepancy may be due to the screening effect of the atomic electrons leading to a decrease of the cross section. The nonrelativistic Born approximation fails to provide reliable quantitative results. The theory cannot explain the large values of the experimental cross sections of Hippler [161].

It was noted [154] that the role of retardation effects in the 2PB process is much more significant than in the one-photon bremsstrahlung. Therefore, in order to obtain reliable theoretical results, one has to go beyond the dipole approximation even when considering initial electron energies as low as a few keV. Retardation effects in 2PB have first been investigated in the nonrelativistic Born approximation [155]. When retardation is included, the structure of the four-fold differential cross section is much more complex than in Eq. (7.24) because of the existence of two new vectors, the photon momenta $\mathbf{k}_1$ and $\mathbf{k}_2$. It is demonstrated that, as in one-photon bremsstrahlung, the region of validity of the Born approximation is extended by the Elwert factor (3.90). Since the first Born approximation in many cases essentially underestimates the 2PB cross section [152; 158] it is necessary to take into account the distortion of the electron wave functions by the (point or screened) Coulomb field of the target. The matrix element of 2PB in an exact nonrelativistic treatment including retardation was presented for the Coulomb case by Dondera and Florescu [162]. The exact calculation of the nonrelativistic 2PB matrix element with retardation by means of distorted wave functions is much more laborious than in the dipole case, and this problem has not been solved so far. An approximate procedure, representing an extension of the dipole DPWA, was described by Korol [163]. Numerical results for incident electrons of 70 keV [164] colliding with targets of atomic numbers $Z = 2$ to 92 show clearly that the 2PB cross section increases considerably if the retardation is taken into account and if the angles of emission with respect to the electron beam satisfy $\theta < 90°$. Apart from the difference in magnitude the maxima of the curves with retardation are shifted to emission angles $\theta < 90°$ whereas the dipole curves have maxima at $\theta = 90°$ and are symmetrical about this value. The experimental data of Kahler et al. [165] agree well with the theoretical curves. Results including retardation were found to be in good agreement with those of Smirnov [151] at incident electron energies below 50 keV.

The problem which still remains concerns the relativistic treatment beyond the Born approximation of the 2PB process in the atomic field. The relativistic formalism will be much more complicated than the nonrelativistic approach because the evaluation of the matrix element can be achieved only numerically. A relativistic calculation was presented by Fedorova et al. [166] using an approximate treatment of the matrix element. Numerical results were obtained by means of Sommerfeld-Maue wave functions. The differential cross section $d^4\sigma$ was given as a function of the photon energy k_2 for incident electron energies between 100 and 1000 keV and compared with previous results. There is good agreement with calculations carried out in relativistic Born approximation.

In addition to the case of a static atomic field the 2PB process has been studied taking into account emission due to the virtual excitations of target electrons (polarization bremsstrahlung) [167; 168]. In general, the polarization contributions are less than one percent of the total emission and cannot explain the large discrepancies between theory and experiment.

However, these contributions are expected to be significant at lower impact electron and photon energies, as in the case of one-photon bremsstrahlung. A systematic study of the role of the dynamic atomic structure in the 2PB process is not yet available.

Two-photon bremsstrahlung produced in the collisions of electrons with electrons or positrons has been investigated for high incident energies for arbitrary lepton and photon polarizations [169]. This process is of great interest for experiments on colliding beams. In general, the simultaneously detected photons are emitted at small angles $\theta_i \approx 1/\epsilon_i$ with respect to the beams; the scattering angles of the leptons are small as well. Formulae for the four-fold differential cross section in the center-of-mass system were given. When at least one photon is emitted at angles larger than $1/\epsilon$, the single-bremsstrahlung cross section decreases faster with increasing photon emission angle than the 2PB cross section.

7.5 Polarization bremsstrahlung

Polarization bremsstrahlung arises when charged particles, such as electrons, positrons, protons, or ions, collide with an atom whereby the radiation emitted is due to a polarization of the structured atomic target [170; 171; 172; 173]. During the collision of the incident particle and the atom, the internal structure of the atom is deformed or polarized and an electric dipole moment is induced. Being time-dependent, it becomes a source of continuous electromagnetic radiation, the polarization bremsstrahlung. While the amplitude of ordinary bremsstrahlung is inversely proportional to the mass of the projectile, the amplitude of polarization bremsstrahlung is almost independent of it since the projectile serves mainly as a source of the external field acting on the atomic electrons [172]. We will restrict ourselves to polarization bremsstrahlung produced in the collision of electrons with atoms.

Whereas a variety of calculations of polarization bremsstrahlung is available, there is no convincing experimental evidence for the existence of this radiation mechanism so far. From precise measurements of the doubly differential cross section it is concluded that the data fit very well to ordinary bremsstrahlung without the contribution of polarization bremsstrahlung [174]. However, recent experiments with gaseous targets [138] seem to have provided evidence for a contribution of polarization bremsstrahlung. The effects of polarization, which can be dominant in ion-atom scattering, are generally smaller in electron-atom bremsstrahlung owing to the smaller projectile mass.

The mechanism of polarization bremsstrahlung was suggested in 1975 by Buĭmistrov and Trakhtenberg [175]. It is more complicated than ordinary bremsstrahlung since in addition to the description of the electron-photon coupling one has to consider the dynamic atomic response to the action of two fields created by the incident electron and the emitted photon. Therefore the problem becomes essentially of many-body nature. The only exception is the collision with a hydrogen atom (or a hydrogen-like ion) where simpler approaches can be used [175; 176]. For a nonrelativistic projectile electron (no retardation) the amplitude of polarization bremsstrahlung is proportional to the scalar dipole polarizability α_d which describes the dynamic atomic excitations due to the interaction with the Coulomb field of the projectile. In addition to this the polarization bremsstrahlung of a relativistic projectile (including retardation) contains the second polarizability β_d which describes the dynamic atomic response to the vector part of the virtual photon field.

At least in the soft-photon region of relativistic bremsstrahlung from many-electron atoms and ions and for the entire spectrum of nonrelativistic bremsstrahlung, polarization bremsstrahlung should be taken into account [177]. In contrast to the ordinary bremsstrahlung amplitude, the amplitude of polarization bremsstrahlung is a more rapidly varying function of $h\nu$, E_0, and Z, with maxima in the vicinity of the subshell ionization thresholds. These maxima reflect the resonant response of the electrons of each subshell to the external field of the projectile electron. Taking this process into account therefore leads to a complicated behaviour of the bremsstrahlung cross section near the ionization threshold of atomic subshells.

Two main approximations are used to investigate polarization bremsstrahlung. These are the Born approximation for both nonrelativistic and relativistic collisions, and the formalism based on the distorted partial-wave approximation (DPWA). For high, but still nonrelativistic, electron energies it was shown in Born approximation [178], that polarization bremsstrahlung can be included with ordinary bremsstrahlung in a 'stripped' atom approximation [179], neglecting the specific details of the structure of the bremsstrahlung cross section near each subshell threshold. In this approximation the total bremsstrahlung amplitude, $A_{ob} + A_{pb}$, for emitted photon energies higher than the ionization threshold of a given subshell, is described simply by an amplitude A_{ob} for ordinary bremsstrahlung from the ionic species containing only the atomic electrons with binding energies exceeding the energy of the emitted photon. At high nonrelativistic incident electron energies (much higher than the binding energy of the K-shell, but still much lower than the energy of the incident electron), the spectrum is therefore described by Coulombic bremsstrahlung from the nucleus. The fact that the decrease of ordinary bremsstrahlung due to screening is completely compensated by the additional polarization radiation in the harder photon region,has also been proved without assuming the Born approximation [180], but still for nonrelativistic energies. In this region, where the photon takes a major fraction of the incident electron energy, polarization bremsstrahlung becomes negligible and the ordinary bremsstrahlung itself has this Coulombic behaviour.

For high relativistic electron energies but soft photon energies an expression for A_{pb} was derived [181; 182]. It was shown that in the relativistic regime the contribution of polarization bremsstrahlung to the soft photon region ($h\nu \ll mc^2 Z$) is larger than in the nonrelativistic case, increasing logarithmically with incident electron energy. Since it is presumed that polarization bremsstrahlung does not play a role in the hard photon region, this contribution should decrease with increasing photon energy. In the nonrelativistic case polarization bremsstrahlung decreases with increasing photon energy in the same way as the screening contribution to ordinary bremsstrahlung, leading to the Coulombic behaviour of the cross section. In the relativistic regime there is at present not enough information, neither theoretical nor experimental, about the interplay of the two kinds of bremsstrahlung in the intermediate part of the spectrum, where polarization bremsstrahlung is decreasing but screening in ordinary bremsstrahlung is still significant, to be certain whether the nonrelativistic stripping approximation remains valid.

A completely relativistic formalism was developed by Korol et al. [183] to calculate the production of 'elastic' bremsstrahlung (i.e., without target excitation or ionization) in collisions between charged particles and many-electron targets whose states could be described in terms of the relativistic one-particle self-consistent field approximation. Both bremsstrahlung mechanisms, usual and polarization, and their interference are considered. The motion of the incident particle is described by the formalism of DPWA. Limiting cases of the general formulae are shown to reproduce correctly the results of simpler theories. So far, the method is applied only to collisions of heavy incident particles with atoms [184].

7.6 Crystalline targets: coherent bremsstrahlung

A relativistic electron (or positron) traversing a monocrystalline target produces coherent radiation resulting from two physically different radiation processes. The first one is coherent bremsstrahlung caused by bremsstrahlung of the electron in the crystal field (as discussed in Sec. 4.4). The second process is parametric x-radiation or coherent polarization radiation. It originates from the coherent superposition of the electromagnetic waves emitted due to the dynamical polarization of the crystal atoms periodically distributed along the particle trajectory. The interference of the two radiation processes has been studied as well [185].

The highly monochromatic parametric x-radiation, that can intuitively be understood as Bragg reflection of virtual photons associated with the incoming electron passing through crystal planes, is emitted into a small angular cone well separated from the electron beam and virtually free of bremsstrahlung background [186]. It can be emitted by heavy particles as well as by electrons or positrons. This is because, unlike the bremsstrahlung-type radiation, polarization radiation is emitted by the target atoms and hence the deflection of the trajectory of the incident particles is not necessary.

Calculations have shown that the coherence effect disappears when averaged over crystal orientation, to a high degree of accuracy. The bremsstrahlung theories for single atoms therefore remain valid for polycrystalline material, except at very high energies.

For further information on the properties of coherent bremsstrahlung we refer to [187; 188; 189; 190; 191; 192].

7.7 Bremsstrahlung from heavy particles

Bremsstrahlung is emitted whenever a charge is accelerated and its intensity varies inversely with the square of the mass of the radiating particle. Thus bremsstrahlung from heavier particles, such as muons, pions, protons, or α particles, is much weaker than electron or positron bremsstrahlung. Nucleon-nucleon bremsstrahlung is the most fundamental reaction used in studying effects beyond elastic scattering in the nucleon-nucleon interaction. The investigation of nucleon-nucleon bremsstrahlung has been revived in the past decade with the advent of modern detectors covering a large part of phase space and capable of dealing with high count rates [193; 194; 195]. In particular, the bremsstrahlung produced in proton-proton collisions has been studied as a possible means of gaining greater understanding of the nucleon-nucleon force.

7.8 Bremsstrahlung in nuclear decays

Bremsstrahlung emission occurs in nuclear decays, since they are usually associated with the separation and acceleration of charged particles within a very short time period. This process is called internal bremsstrahlung to distinguish it from the (external) bremsstrahlung emitted by the same particle in passing through matter. We will briefly discuss some examples.

7.8.1 Bremsstrahlung in β decay

In the process of β decay an unstable nucleus transforms spontaneously into another nucleus of higher or lower atomic number while emitting an electron or positron (and a neutrino). The sudden creation of a rapidly moving charged particle will be accompanied by the emission of bremsstrahlung. The radiated energy constitutes a very small fraction of the total energy released in β decay [20]. The heavy nucleus receives a negligible acceleration and so does not appreciably contribute to the radiation. The process of bremsstrahlung in β decay provides useful information for nuclear physicists.

7.8.2 Bremsstrahlung in orbital-electron capture

In orbital-electron capture by an unstable nucleus the sudden disappearance of an electron gives rise to radiation. Calculation by means of a simplified model is given by Jackson [20]. The disappearance of the magnetic moment also gives rise to radiation, but with a spectrum of quite different character.

7.8.3 Bremsstrahlung in α decay

During the α decay of a nucleus the α particle tunnels through the Coulomb barrier and is accelerated beyond the classical turning point to its final energy. In the classical picture, the α particle is accelerated outside the Coulomb barrier and can emit the bremsstrahlung photon there. It is of particular interest to ask whether α particles emit photons only during acceleration in the Coulomb field or also in tunnelling. In a quantum-mechanical analysis of the bremsstrahlung process in α decay of ^{210}Po [196] it is shown that the final bremsstrahlung spectrum results from a subtle interference of the contributions of the tunnelling and the classical regions as well as of further contributions. Semiclassical as well as classical theories do not seem to be reliable for describing these interference effects, though they give some clear understanding of the phenomena.

7.9 Bremsstrahlung in magnetic fields

7.9.1 Electron-nucleus bremsstrahlung in strong magnetic fields

The bremsstrahlung process is the dominant mechanism for producing photons in hot plasmas. The existence of strongly magnetized plasmas in the vicinity of neutron stars gave rise to the investigation of bremsstrahlung in magnetic fields with strengths of up to 10^9 T. In such fields, the motion of the electrons perpendicular to the field is quantized in Landau orbitals, and the motion remains free only parallel to the field, i.e., the electrons behave like particles in one-dimensional space [197]. As a consequence of this completely different structure of the initial and final electron states, the field-free results are no longer valid and a recalculation from first principles is necessary. After a number of approaches in this direction, the peculiarities in the cross section of the elementary bremsstrahlung process caused by strong magnetic fields were studied in nonrelativistic Born approximation by Lauer et al. [198]. The general expression of the cross section was given assuming that the incoming and outgoing electrons occupied the ground-state Landau level $n = 0$¶ with momenta $\mathbf{p}_1$ and $\mathbf{p}_2$, respectively, parallel to the field. Then the cross section of the elementary process is doubly differential with respect to the photon energy and direction. After averaging over the component of the orbital angular momentum parallel to

---

¶For nonrelativistic electrons in a magnetic field of sufficiently high intensity the energy differences between the quantized states of perpendicular motion are much greater than other energies involved.

the field l_z, and summing over the degenerate final states l'_z, it depends on the photon polarization, the initial electron energy, and the magnetic field strength. Besides, one has to distinguish between forward and backward processes, in which the decelerated electron moves in the same direction as the incident one, or opposite to it. These two discrete possibilities of motion along the magnetic field replace the complete solid angle 4π available to the outgoing electron without the magnetic field.

The authors [198] presented detailed numerical results on the dependence of the cross section on the magnetic field strength, the initial electron energy, and the energy, polarization, and direction of the emitted photon. In this context it is convenient to express the magnetic field strength B in terms of the critical field

$$B_{cr} = \frac{m^2 c^2}{e\hbar} \approx 4.414 \cdot 10^9 \text{ T}, \qquad (7.28)$$

corresponding to the cyclotron energy $\hbar\omega_B = mc^2$. For $B = 0.1 B_{cr}$ and electron energies well below $\hbar\omega_B$ the angular dependence of the cross section for polarization in the **k-B** plane is typical of a dipole ($\sim \sin^2 \theta$), while emission with polarization perpendicular to it is almost isotropic, although reduced by at least two orders of magnitude. If the incident electron energy E_0 approaches the Landau threshold $\hbar\omega_B$, for both polarizations the cross section becomes of the same magnitude on the short-wavelength side. For low-energy photons, perpendicular polarization is still strongly suppressed, and parallel polarization exhibits a pronounced forward-backward asymmetry. This asymmetry is also present for $h\nu = 0.9 E_0$, it only disappears when the electron completely comes to rest.

It is instructive to consider the doubly differential cross section, summed over the photon polarizations, as a function of the fraction of energy transferred to the photon, for various emission angles. The curves show a sharp increase near the short-wavelength limit, opposite to the behaviour of the $B = 0$ case (the weak divergence of the cross section for $p_2 \to 0$ is probably due to the use of the Born approximation which breaks down in this limit). In particular, for photon energies near the Landau threshold $\hbar\omega_B$, the influence of the cyclotron resonance is drastically changing the angle dependence and the rise of the cross section in the short-wavelength regime.

7.9.2 Synchrotron radiation (magnetobremsstrahlung)

Electromagnetic radiation emitted by charged particles moving at relativistic energies in curved paths in a magnetic field is called synchrotron radiation, sometimes magnetobremsstrahlung. The radiation patterns of electrons (or positrons) moving in a circular orbit are shown schematically in Fig. 2.2b for nonrelativistic and in Fig. 2.2d for semirelativistic energies. In high-energy ring accelerators and in storage rings the relativistic beaming collimates the radiation into very small solid angles. Quantum-mechanical effects in synchrotron radiation may be neglected if the energy of the emitted photon is small compared to the energy of the electrons.

Synchrotron radiation is of great importance in astrophysics. In the universe there exist many sources of synchrotron radiation reaching from the sun to distant quasars.

For further information on the process of synchrotron radiation we refer to the reviews by Wille [199] and Ternov [200], and the textbook [201].

7.10 Stimulated bremsstrahlung

All bremsstrahlung processes considered in this book are based on spontaneous emission. The enormous progress in laser technology with intensities exceeding 10^{19} W cm^{-2} has strongly enhanced the interest in the subject of stimulated bremsstrahlung. Such intensities correspond to fields much stronger than the intra-atomic field. The second order of perturbation theory is the lowest order in which stimulated bremsstrahlung is possible. One-photon stimulated bremsstrahlung occurs in a weak stimulating light field. Multiphoton stimulated bremsstrahlung requires a strong light field. For the discussion of stimulated bremsstrahlung emitted in electron-atom scattering in the field of an electromagnetic wave we refer the reader, e.g., to the book of Fedorov [202]. Here the topic is treated under the keywords 'field-assisted electron-atom scattering' and 'multiphoton stimulated bremsstrahlung'.

Conclusion

The investigation of the elementary bremsstrahlung process by electron-photon coincidence experiments has proved to be a very sensitive method for testing the bremsstrahlung theory. Such tests are especially important since the general problem of relativistic bremsstrahlung theory is extremely difficult and the numerical results from various authors differ in some cases. Approximate calculations for low atomic numbers of the target nucleus are found to be in fairly good agreement with the experiments.

The few available measurements of the elementary process including polarization variables proved to provide not only a stronger test of the theory but also a more direct physical insight into the mechanism of the radiation process. Especially conspicuous effects of the weak spin-orbit interaction could be observed which are usually masked in non-coincidence experiments. This should motivate us to investigate more of the numerous polarization correlations. However, a complete experiment on the elementary process of bremsstrahlung which includes all polarization variables, is not likely to be done soon.

Considering the role of the electron spin in the radiation process acting via spin-orbit coupling it would be interesting to make comparisons with bremsstrahlung emitted from particles having spin 0 (pions, α particles).

Regarding electron-electron bremsstrahlung from (initially) bound electrons it would be desirable to separate the radiation processes from the individual shells, i.e., combining bremsstrahlung and ionization.

Comparison of positron-nucleus and electron-nucleus bremsstrahlung would offer some further understanding of the elementary bremsstrahlung process as well as the investigation of electron-positron bremsstrahlung.

A measurement of the elementary process of two-photon bremsstrahlung would yield an interesting check of quantum electrodynamic calculations, in particular since there are still considerable discrepancies between experiments and theory.

Electron-photon coincidence experiments on polarization bremsstrahlung would enable more insight into bremsstrahlung from structured particles.

In summary, after more than 100 years of Röntgen's discovery of x-rays, the future holds many challenges for both experiment and theory.

Appendix A
Problems

2.1 In nonrelativistic approximation the angular distribution (power per unit solid angle) of the radiation emitted by a charge e which is accelerated parallel to its velocity $\mathbf{v}$ is given by [20]

$$\frac{dI}{d\Omega} = \frac{e^2}{4\pi\epsilon_0} \frac{\dot{\mathbf{v}}^2}{4\pi c^3} \sin^2\theta \qquad (A.1)$$

(dipole radiation), where θ is the angle between the velocity vector and the direction of emission.

Calculate the relativistic angular distribution and the total radiation power by means of a Lorentz transformation from the instantaneous rest system S' of the charge to the laboratory system S. At which angle is the maximum radiation emitted?

Solution:
Denoting the polar and azimuth angle of the emitted radiation with respect to the velocity vector $\mathbf{v}$ by θ' and φ', respectively, the radiation power in the rest system S' has the form

$$dI' = \frac{e^2}{4\pi\epsilon_0} \frac{\dot{\mathbf{v}}'^2}{4\pi c^3} \sin^2\theta' \, d\Omega' \qquad (A.2)$$

with $d\Omega' = \sin\theta' \, d\theta' \, d\varphi'$; $\dot{\mathbf{v}}' = d\mathbf{v}'/dt'$ is the proper acceleration of the charge. The angles in the systems S' and S are connected by the transformation formulae

$$\cos\theta' = \frac{\cos\theta - \beta}{1 - \beta\cos\theta}, \quad \sin\theta' = \frac{\sin\theta}{\gamma(1 - \beta\cos\theta)},$$
$$d\theta' = \frac{d\theta}{\gamma(1 - \beta\cos\theta)}, \quad \varphi' = \varphi, \qquad (A.3)$$

where $\beta = v/c$ and $\gamma = (1 - \beta^2)^{-1/2}$ is the Lorentz factor.

The energy emitted in the rest system S' during the time interval dt' transforms as

$$dE' = \gamma(dE - \mathbf{v}\cdot d\mathbf{p}). \qquad (A.4)$$

Using the relation between momentum and energy of the photon, $|d\mathbf{p}| = dE/c$, this can be written as

$$dE' = \gamma(1 - \beta\cos\theta)\, dE. \qquad (A.5)$$

The time intervals in the two systems are related by $dt = \gamma\, dt'$. Thus the radiated power in the laboratory system has the form

$$\begin{aligned}
dI &= \frac{dE}{dt} = \frac{1}{\gamma^2(1 - \beta\cos\theta)}\frac{dE'}{dt'} \\
&= \frac{1-\beta^2}{1-\beta\cos\theta}\frac{e^2}{4\pi\epsilon_0}\frac{\dot{\mathbf{v}}'^2}{4\pi c^3}\frac{\sin^2\theta}{\gamma^2(1-\beta\cos\theta)^2}\frac{1-\beta^2}{(1-\beta\cos\theta)^2}d\Omega \\
&= \frac{e^2}{4\pi\epsilon_0}\frac{\dot{\mathbf{v}}'^2}{4\pi c^3}\frac{(1-\beta^2)^3\sin^2\theta}{(1-\beta\cos\theta)^5}d\Omega. \qquad (A.6)
\end{aligned}$$

By means of the transformation formula for the acceleration, $\dot{\mathbf{v}}' = \gamma^3\dot{\mathbf{v}}$, we get finally

$$\frac{dI}{d\Omega} = \frac{e^2}{4\pi\epsilon_0}\frac{\dot{\mathbf{v}}^2}{4\pi c^3}\frac{\sin^2\theta}{(1-\beta\cos\theta)^5}. \qquad (A.7)$$

Here $dI/d\Omega$ refers to the power radiated at the retarded time $t_r = t - R(t_r)/c$ where $R(t_r)$ is the distance between charge and observer. The maximum radiation is emitted at the angle

$$\theta_m = \arccos\left\{\left(\sqrt{1 + 15\beta^2} - 1\right)\big/3\beta\right\}. \qquad (A.8)$$

At relativistic energies $\beta \to 1$ or $\gamma \gg 1$, this results in $\theta_m \approx 1/2\gamma$. Here the angular distribution is strongly beamed in forward direction due to the high power of the denominator of (A.7).

The total instantaneous power radiated is obtained by integrating (A.7) over all solid angle yielding the Liénard result [20]

$$P = \frac{e^2}{4\pi\epsilon_0} \frac{2\dot{\mathbf{v}}^2}{3c^3} \gamma^6, \qquad (A.9)$$

differing from the nonrelativistic Larmor formula (2.1) by the factor γ^6.

3.1 Verify that $\Phi_2(\mathbf{r}) = \int \Phi_2(\mathbf{k})\, e^{i\mathbf{k}\cdot\mathbf{r}} d^3k$ with $\Phi_2(\mathbf{k})$ given by Eq. (3.48) is a solution of the wave equation (3.47),

$$(-i\vec{\alpha}\cdot\nabla + \beta - \epsilon)\Phi_2(\mathbf{r}) = (1/r)\Phi_1(\mathbf{r}). \qquad (A.10)$$

Solution:
Applying the operator $(-i\vec{\alpha}\cdot\nabla + \beta - \epsilon)$ to the wave function $\Phi_2(\mathbf{r})$ yields

$$\begin{aligned}
F(\mathbf{r}) &= \lim_{\lambda\to 0}(2\pi^2)^{-2}\left(-i\vec{\alpha}\cdot\nabla + \beta - \epsilon\right)\int d^3k\, e^{i\mathbf{k}\cdot\mathbf{r}}\frac{\vec{\alpha}\cdot\mathbf{k}+\beta+\epsilon}{k^2-p^2-i\lambda} \\
&\quad \cdot \int d^3k'\frac{\vec{\alpha}\cdot\mathbf{k}'+\beta+\epsilon}{(\mathbf{k}'-\mathbf{p})^2(\mathbf{k}'-\mathbf{k})^2(k'^2-p^2-i\lambda)} u(\mathbf{p}) \\
&= \lim_{\lambda\to 0}(2\pi^2)^{-2}\int d^3k\, e^{i\mathbf{k}\cdot\mathbf{r}}\underbrace{\frac{(\vec{\alpha}\cdot\mathbf{k}+\beta-\epsilon)(\vec{\alpha}\cdot\mathbf{k}+\beta+\epsilon)}{k^2-p^2-i\lambda}}_{=1} \\
&\quad \cdot \int d^3k'\frac{\vec{\alpha}\cdot\mathbf{k}'+\beta+\epsilon}{(\mathbf{k}'-\mathbf{p})^2(\mathbf{k}'-\mathbf{k})^2(k'^2-p^2-i\lambda)} u(\mathbf{p}) \\
&= \lim_{\lambda\to 0}(2\pi^2)^{-2}\int d^3k'\frac{\vec{\alpha}\cdot\mathbf{k}'+\beta+\epsilon}{(\mathbf{k}'-\mathbf{p})^2(k'^2-p^2-i\lambda)} \\
&\quad \cdot \int d^3k\frac{e^{i\mathbf{k}\cdot\mathbf{r}}}{(\mathbf{k}-\mathbf{k}')^2} u(\mathbf{p}).
\end{aligned} \qquad (A.11)$$

The second integral can be readily solved,

$$\begin{aligned}
\int \frac{e^{i\mathbf{k}\cdot\mathbf{r}}}{(\mathbf{k}-\mathbf{k}')^2} d^3k &= e^{i\mathbf{k}'\cdot\mathbf{r}}\int \frac{e^{i(\mathbf{k}-\mathbf{k}')\cdot\mathbf{r}}}{(\mathbf{k}-\mathbf{k}')^2} d^3k = e^{i\mathbf{k}'\cdot\mathbf{r}}\int \frac{e^{i\mathbf{q}\cdot\mathbf{r}}}{q^2} d^3q \\
&= 2\pi\, e^{i\mathbf{k}'\cdot\mathbf{r}}\int_0^\infty dq \int_{-1}^{+1} dy\, e^{iqry} \\
&= \frac{4\pi}{r} e^{i\mathbf{k}'\cdot\mathbf{r}}\int_0^\infty \frac{\sin(qr)}{q} dq = \frac{2\pi^2}{r} e^{i\mathbf{k}'\cdot\mathbf{r}},
\end{aligned} \qquad (A.12)$$

resulting in

$$F(\mathbf{r}) = \lim_{\lambda \to 0} \frac{1}{2\pi^2 r} \int d^3k' \frac{\vec{\alpha} \cdot \mathbf{k}' + \beta + \epsilon}{(\mathbf{k}' - \mathbf{p})^2(k'^2 - p^2 - i\lambda)} e^{i\mathbf{k}' \cdot \mathbf{r}} u(\mathbf{p})$$

$$= \frac{1}{r} \int \Phi_1(\mathbf{k}') e^{i\mathbf{k}' \cdot \mathbf{r}} d^3k' = \frac{\Phi_1(\mathbf{r})}{r} \quad (A.13)$$

with $\Phi_1(\mathbf{k})$ given by (3.44).

3.2 Show that in the bremsstrahlung production process the recoil momentum of the atom, $\mathbf{q} = \mathbf{p}_1 - \mathbf{p}_2 - \mathbf{k}$, can never vanish. This is equivalent to the fact that the emission of a photon by a free electron is not possible which is easily proved by means of the four-vector formalism.

Solution:
The minimum value of $q = |\mathbf{q}|$ is

$$q_{\min} = p_1 - p_2 - k = p_1 - p_2 - (\epsilon_1 - \epsilon_2) = (\epsilon_2 - p_2) - (\epsilon_1 - p_1). \quad (A.14)$$

From

$$\frac{d}{d\epsilon}(\epsilon - p) = \frac{d}{d\epsilon}\left[\epsilon - \sqrt{\epsilon^2 - 1}\right] = 1 - \frac{\epsilon}{p} < 0 \quad (A.15)$$

it follows that $\epsilon - p$ is a monotonically decreasing function of ϵ. Since $\epsilon_2 < \epsilon_1$ we have $q_{\min} > 0$.

In the four-vector formalism the emission of a photon by a free electron is described by the energy-momentum conservation $\underline{k} = \underline{p}_1 - \underline{p}_2$. Squaring this equation yields

$$\underline{k}^2 = (\underline{p}_1 - \underline{p}_2)^2 = 2 - 2(p_1 p_2) = 2(1 - \epsilon_1 \epsilon_2 + \mathbf{p}_1 \cdot \mathbf{p}_2) < 0. \quad (A.16)$$

Because $\underline{k}^2 = 0$, this leads to a contradiction.

3.3 Use Green's theorem and the Poisson equation $\nabla^2 \varphi(\mathbf{r}) = -\rho(\mathbf{r})/\epsilon_0$ to derive the Fourier transform $\varphi(\mathbf{q})$ of an arbitrary potential, vanishing at infinity, from the source charge distribution $\rho(\mathbf{r})$. Apply the result to get Eq. (3.92) for a point charge.

Solution:
The Fourier transform of the electrostatic potential $\varphi(\mathbf{r})$ is given by

$$\varphi(\mathbf{q}) = \int \varphi(\mathbf{r})\, e^{i\mathbf{q}\cdot\mathbf{r}}\, d^3r\,. \tag{A.17}$$

Writing

$$e^{i\mathbf{q}\cdot\mathbf{r}} = -\frac{1}{q^2}\nabla^2 e^{i\mathbf{q}\cdot\mathbf{r}} \tag{A.18}$$

we can apply Green's theorem and obtain

$$\int \varphi(\mathbf{r})\, e^{i\mathbf{q}\cdot\mathbf{r}}\, d^3r = -\frac{1}{q^2}\int \varphi(\mathbf{r})\nabla^2 e^{i\mathbf{q}\cdot\mathbf{r}}\, d^3r = -\frac{1}{q^2}\int e^{i\mathbf{q}\cdot\mathbf{r}}\nabla^2\varphi(\mathbf{r})\, d^3r\,, \tag{A.19}$$

because the surface integrals vanish at infinity. Using Poisson's equation this yields

$$\varphi(\mathbf{q}) = \frac{1}{\epsilon_0 q^2}\int \rho(\mathbf{r})\, e^{i\mathbf{q}\cdot\mathbf{r}}\, d^3r\,. \tag{A.20}$$

For a point nucleus, $\rho(\mathbf{r}) = Ze\,\delta(\mathbf{r})$, resulting in $\varphi(\mathbf{q}) = Ze/(\epsilon_0 q^2)$. If an electron is moving in the field of the point nucleus, the potential energy in momentum space is ($\hbar = c = 1$)

$$V(q) = -e\varphi(q) = -(4\pi/q^2)\,\alpha Z\,, \tag{A.21}$$

where $\alpha = e^2/(4\pi\epsilon_0\hbar c)$ is the fine-structure constant.

3.4 Show that the differential equation for the confluent hypergeometric function $F(a;c;z)$ (primes denote differentiation with respect to the argument z),

$$z\,F''(a;c;z) + (c-z)\,F'(a;c;z) - a\,F(a;c;z) = 0 \tag{A.22}$$

follows from the relation (3.119),

$$\left(\nabla^2 + 2i\mathbf{p}\cdot\nabla + 2\epsilon a/r\right) F(ia\epsilon/p;1;ipr - i\mathbf{p}\cdot\mathbf{r}) = 0\,. \tag{A.23}$$

Solution:
Inserting the expressions

$$\nabla F = i(p\hat{\mathbf{r}} - \mathbf{p})F'\,, \quad \nabla^2 F = -(p\hat{\mathbf{r}} - \mathbf{p})^2 F'' + 2i(p/r)F' \tag{A.24}$$

into the differential equation (A.23) yields

$$-2p^2(1 - \hat{\mathbf{p}} \cdot \hat{\mathbf{r}})F'' + 2i(p/r)F' - 2p^2(\hat{\mathbf{p}} \cdot \hat{\mathbf{r}} - 1)F' + 2\epsilon(a/r)F = 0 . \quad (A.25)$$

If this equation is multiplied by $(-ir/2p)$, we get

$$i(pr - \mathbf{p} \cdot \mathbf{r})F'' + [1 - i(pr - \mathbf{p} \cdot \mathbf{r})]F' - ia(\epsilon/p)F = 0 , \quad (A.26)$$

in agreement with (A.22).

3.5 The confluent hypergeometric function $F(a;c;z)$ behaves asymptotically for $|z| \gg 1$ and $-\pi < \arg z < \pi$ as [53]

$$F(a;c;z) = \frac{\Gamma(c)}{\Gamma(c-a)} e^{\pm i\pi a} z^{-a} \sum_{s=0}^{N} \frac{(a)_s(a-c+1)_s}{s!} (-z)^{-s} + O(|z|^{-N-a-1})$$

$$+ \frac{\Gamma(c)}{\Gamma(a)} e^z z^{a-c} \sum_{s=0}^{M} \frac{(1-a)_s(c-a)_s}{s!} z^{-s} + O(|e^z z^{-M+a-c-1}|) ,$$

$$(A.27)$$

where $(a)_0 = 1$, $(a)_s = a(a+1)\ldots(a+s-1)$; the signs $\pm$ in the first exponential hold for $\operatorname{Im} z > 0$ and $\operatorname{Im} z < 0$, respectively.

Show that the Sommerfeld-Maue functions (3.129) and (3.134) have the asymptotic form of a plane wave plus outgoing and ingoing spherical waves, respectively.

Solution:
Taking into account that Eq. (3.134) represents the adjoint of $\psi^-(\mathbf{r})$ the SM function for the incident and outgoing electrons can be written as

$$\psi^{\pm}_{\text{SM}}(\mathbf{r}) = \Gamma(1 \mp ia_1) e^{\pi a_1/2} e^{i\mathbf{p}\cdot\mathbf{r}} \left(1 - \frac{i}{2\epsilon}\vec{\alpha}\cdot\nabla\right) F(\pm ia_1; 1; \pm ipr - i\mathbf{p}\cdot\mathbf{r}) u(\mathbf{p}) ,$$

$$(A.28)$$

where $a_1 = a\epsilon/p$. Using $F'(a;c;z) = (a/c)F(a+1;c+1;z)$ we get

$$\psi^{\pm}_{\text{SM}}(\mathbf{r}) = \Gamma(1 \mp ia_1)e^{\pi a_1/2}e^{i\mathbf{p}\cdot\mathbf{r}}\{F(\pm ia_1; 1; \pm ipr - i\mathbf{p}\cdot\mathbf{r})$$
$$+ \tfrac{1}{2}ia\,\vec{\alpha}\cdot(\hat{\mathbf{r}} \mp \hat{\mathbf{p}})\,F(1 \pm ia_1; 2; \pm ipr - i\mathbf{p}\cdot\mathbf{r})\} u(\mathbf{p}) . \quad (A.29)$$

If we consider only the leading terms in the asymptotic expansion, the second term in curly brackets can be neglected resulting in

$$\psi_{\text{SM}}^{\pm}(\mathbf{r}) \sim \Gamma(1 \mp ia_1) e^{\pi a_1/2} e^{i\mathbf{p}\cdot\mathbf{r}} \left\{ \frac{e^{-\pi a_1}}{\Gamma(1 \mp ia_1)} [\pm i(pr \mp \mathbf{p}\cdot\mathbf{r})]^{\mp ia_1} \right.$$
$$\left. + \frac{e^{i(\pm pr - \mathbf{p}\cdot\mathbf{r})}}{\Gamma(\pm ia_1)} [\pm i(pr \mp \mathbf{p}\cdot\mathbf{r})]^{\pm ia_1 - 1} \right\} u(\mathbf{p}) . \quad (A.30)$$

Exploiting the well-known properties of the gamma function and using

$$(\pm i)^{\mp ia_1} = e^{\pi a_1/2}, \quad (\pm i)^{\pm ia_1} = e^{-\pi a_1/2}, \quad (A.31)$$

we obtain

$$\psi_{\text{SM}}^{\pm}(\mathbf{r}) \sim \left\{ (pr \mp \mathbf{p}\cdot\mathbf{r})^{\mp ia_1} e^{i\mathbf{p}\cdot\mathbf{r}} \right.$$
$$\left. - \frac{a_1 \Gamma(\mp ia_1)}{\Gamma(\pm ia_1)} (p \mp \mathbf{p}\cdot\hat{\mathbf{r}})^{-1} (pr \mp \mathbf{p}\cdot\mathbf{r})^{\pm ia_1} \frac{e^{\pm ipr}}{r} \right\} u(\mathbf{p})$$
$$= \left\{ (pr \mp \mathbf{p}\cdot\mathbf{r})^{\mp ia_1} e^{i\mathbf{p}\cdot\mathbf{r}} \right.$$
$$\left. - a_1 e^{\mp 2i \arg \Gamma(ia_1)} (p \mp \mathbf{p}\cdot\hat{\mathbf{r}})^{-1} (pr \mp \mathbf{p}\cdot\mathbf{r})^{\pm ia_1} \frac{e^{\pm ipr}}{r} \right\} u(\mathbf{p})$$
$$= \left\{ e^{i\mathbf{p}\cdot\mathbf{r} \mp ia_1 \ln(pr - \mathbf{p}\cdot\mathbf{r})} \right.$$
$$\left. - a_1 e^{\mp 2i \arg \Gamma(ia_1)} (pr \mp \mathbf{p}\cdot\mathbf{r})^{-1} e^{\pm ipr \pm ia_1 \ln(pr \mp \mathbf{p}\cdot\mathbf{r})} \right\} u(\mathbf{p}) .$$
$$(A.32)$$

As was pointed out in Section 3.4.1, the logarithmic terms in the phase of both the plane and the spherical wave indicate the distortion of the plane wave by the Coulomb field even at large distances from the scattering center.

3.6 Show that the residual of the wave equation adjoint to Eq. (3.115), when applied to the wave function $\psi_2^\dagger(\mathbf{r})$ of Eq. (3.134), is equal to

$$N_2^* \frac{ia}{2\epsilon_2 r} e^{-i\mathbf{p}_2 \cdot \mathbf{r}} u^\dagger(\mathbf{p}_2) \vec{\alpha} \cdot \nabla F(ia\epsilon_2/p_2; 1; ip_2 r + i\mathbf{p}_2 \cdot \mathbf{r}) . \quad (A.33)$$

Hint: Note that the ∇ operator of the wave equation acts also on $e^{-i\mathbf{p}_2 \cdot \mathbf{r}}$ whereas the ∇ operator of the wave function does not.

Appendix A

Solution:
In the expression
$$\mathcal{F}(\mathbf{r}) = \psi_2^\dagger(\mathbf{r})(-i\vec{\alpha}\cdot\nabla - \beta + \epsilon_2 + a/r) \tag{A.34}$$
the operator ∇ is understood to act backwards on $\psi_2^\dagger(\mathbf{r})$. Taking this into account, inserting (3.134), and applying
$$(\vec{\alpha}\cdot\mathbf{a})(\vec{\alpha}\cdot\mathbf{b}) + (\vec{\alpha}\cdot\mathbf{b})(\vec{\alpha}\cdot\mathbf{a}) = 2\,\mathbf{a}\cdot\mathbf{b} \tag{A.35}$$
we get
$$\begin{aligned}
\mathcal{F}(\mathbf{r}) &= N_2^* e^{-i\mathbf{p}_2\cdot\mathbf{r}} u^\dagger(\mathbf{p}_2)\Big\{\Big(1+\frac{i}{2\epsilon_2}\vec{\alpha}\cdot\nabla\Big) F(ia\epsilon_2/p_2;1;ip_2 r + i\mathbf{p}_2\cdot\mathbf{r})\Big\} \\
&\quad \cdot\big(-i\vec{\alpha}\cdot\nabla - \vec{\alpha}\cdot\mathbf{p}_2 - \beta + \epsilon_2 + a/r\big) \\
&= N_2^* e^{-i\mathbf{p}_2\cdot\mathbf{r}} u^\dagger(\mathbf{p}_2)\Big\{-i\vec{\alpha}\cdot\nabla - \vec{\alpha}\cdot\mathbf{p}_2 - \beta + \epsilon_2 + \frac{a}{r} + \frac{1}{2\epsilon_2}\nabla^2 \\
&\quad - \frac{i}{\epsilon_2}\mathbf{p}_2\cdot\nabla + \frac{i}{2\epsilon_2}(\vec{\alpha}\cdot\mathbf{p}_2)(\vec{\alpha}\cdot\nabla) + \frac{i}{2\epsilon_2}\beta(\vec{\alpha}\cdot\nabla) \\
&\quad + \frac{i}{2}\vec{\alpha}\cdot\nabla + \frac{ia}{2\epsilon_2 r}\vec{\alpha}\cdot\nabla\Big\} F(ia\epsilon_2/p_2;1;ip_2 r + i\mathbf{p}_2\cdot\mathbf{r})\,. \tag{A.36}
\end{aligned}$$
Using $u^\dagger(\mathbf{p}_2)(\vec{\alpha}\cdot\mathbf{p}_2 + \beta - \epsilon_2) = 0$ and $(\nabla^2 - 2i\mathbf{p}_2\cdot\nabla + 2\epsilon_2 a/r)F = 0$ this reduces to the residual
$$\mathcal{F}(\mathbf{r}) = N_2^* \frac{ia}{2\epsilon_2 r} e^{-i\mathbf{p}_2\cdot\mathbf{r}} u^\dagger(\mathbf{p}_2)\,\vec{\alpha}\cdot\nabla F(ia\epsilon_2/p_2;1;ip_2 r + i\mathbf{p}_2\cdot\mathbf{r})\,. \tag{A.37}$$

3.7 Show that the two-component spinor
$$v = \frac{1}{\sqrt{2(1+\zeta_z)}}\begin{pmatrix} 1+\zeta_z \\ \zeta_x + i\zeta_y \end{pmatrix} \tag{A.38}$$
satisfies the equations $\vec{\zeta}\cdot\vec{\sigma}\,v = v$ and $vv^\dagger = \frac{1}{2}(1+\vec{\zeta}\cdot\vec{\sigma})$, where $\vec{\zeta} = \{\zeta_x,\zeta_y,\zeta_z\}$ is the spin unit vector and $\vec{\sigma}$ is the Pauli spin matrix.

Solution:
$$\begin{aligned}
\vec{\zeta}\cdot\vec{\sigma}\,v &= \frac{1}{\sqrt{2(1+\zeta_z)}}\begin{pmatrix} \zeta_z & \zeta_x - i\zeta_y \\ \zeta_x + i\zeta_y & -\zeta_z \end{pmatrix}\begin{pmatrix} 1+\zeta_z \\ \zeta_x + i\zeta_y \end{pmatrix} \\
&= \frac{1}{\sqrt{2(1+\zeta_z)}}\begin{pmatrix} \zeta_z + \zeta_z^2 + \zeta_x^2 + \zeta_y^2 \\ \zeta_x + \zeta_x\zeta_z + i\zeta_y + i\zeta_y\zeta_z - \zeta_z\zeta_x - i\zeta_z\zeta_y \end{pmatrix}
\end{aligned}$$

$$= \frac{1}{\sqrt{2(1+\zeta_z)}} \begin{pmatrix} 1+\zeta_z \\ \zeta_x + i\zeta_y \end{pmatrix} = v \ . \tag{A.39}$$

$$vv^\dagger = \frac{1}{2(1+\zeta_z)} \begin{pmatrix} 1+\zeta_z \\ \zeta_x + i\zeta_y \end{pmatrix} \begin{pmatrix} 1+\zeta_z & \zeta_x - i\zeta_y \end{pmatrix}$$

$$= \frac{1}{2(1+\zeta_z)} \begin{pmatrix} (1+\zeta_z)^2 & (1+\zeta_z)(\zeta_x - i\zeta_y) \\ (1+\zeta_z)(\zeta_x + i\zeta_y) & \zeta_x^2 + \zeta_y^2 \end{pmatrix}$$

$$= \frac{1}{2(1+\zeta_z)} \begin{pmatrix} (1+\zeta_z)^2 & (1+\zeta_z)(\zeta_x - i\zeta_y) \\ (1+\zeta_z)(\zeta_x + i\zeta_y) & 1 - \zeta_z^2 \end{pmatrix}$$

$$= \frac{1}{2} \begin{pmatrix} 1+\zeta_z & \zeta_x - i\zeta_y \\ \zeta_x + i\zeta_y & 1-\zeta_z \end{pmatrix}$$

$$= \frac{1}{2}(1 + \vec{\zeta} \cdot \vec{\sigma}) \tag{A.40}$$

3.8 Calculate the extreme values for the argument x [Eq. (3.174)] of the hypergeometric functions V and W. Show that $x \leq 0$ and that the minimum value of x is $x_m = -p_1 p_2 \mu / k^2$ for all angles θ_1 of the incident electrons with respect to the photon direction.

Solution:
Using the coordinate system (3.86) the argument x has the form

$$x = 1 - \frac{\mu q^2}{D_1 D_2}$$

$$= 1 - \frac{\mu}{4k^2(\epsilon_1 - p_1 \cos\theta_1)(\epsilon_2 - p_2 \cos\theta_2)} [p_1^2 + p_2^2 + k^2 - 2kp_1 \cos\theta_1 + 2kp_2 \cos\theta_2 - 2p_1 p_2 (\cos\theta_1 \cos\theta_2 + \sin\theta_1 \sin\theta_2 \cos\varphi)] \ , \tag{A.41}$$

where $\mu = 2(\epsilon_1 \epsilon_2 + p_1 p_2 - 1)$.

Obviously the extreme values occur for $\cos\varphi = \pm 1$, the two signs corresponding to maximum and minimum, respectively. In order to get the angles θ_2 where x has its extreme values, we set

$$\frac{\partial}{\partial \theta_2} \frac{p_1^2 + p_2^2 + k^2 - 2k(p_1 \cos\theta_1 - p_2 \cos\theta_2) - 2p_1 p_2 \cos(\theta_2 \mp \theta_1)}{\epsilon_2 - p_2 \cos\theta_2} = 0 \tag{A.42}$$

yielding

$$\sin\theta_2 (p_1 - \epsilon_1 \cos\theta_1) = \pm \sin\theta_1 (p_2 - \epsilon_2 \cos\theta_2) \tag{A.43}$$

with the solutions

$$\sin\theta_2 = \frac{\epsilon_2(\epsilon_1 - p_1\cos\theta_1) \mp p_2(\epsilon_1\cos\theta_1 - p_1)}{(\epsilon_1 - p_1\cos\theta_1)^2 + p_2^2\sin^2\theta_1}\sin\theta_1, \quad (A.44)$$

$$\cos\theta_2 = \frac{\epsilon_2 p_2 \sin^2\theta_1 \pm (\epsilon_1 - p_1\cos\theta_1)(\epsilon_1\cos\theta_1 - p_1)}{(\epsilon_1 - p_1\cos\theta_1)^2 + p_2^2\sin^2\theta_1}. \quad (A.45)$$

At these angles θ_2 and $\cos\varphi = \pm 1$ we have $D_2 = (\sin\theta_2/\sin\theta_1)D_1$ and

$$\begin{aligned}q^2 &= 2(\epsilon_1\epsilon_2 - \mathbf{p}_1\cdot\mathbf{p}_2 - 1) + D_1 - D_2 \\ &= 2(\epsilon_1 - p_1\cos\theta_1)^2 \frac{k(\epsilon_1 - p_1\cos\theta_1) + p_2(p_2 \mp p_1 \pm k\cos\theta_1)}{(\epsilon_1 - p_1\cos\theta_1)^2 + p_2^2\sin^2\theta_1}.\end{aligned}$$
(A.46)

Inserting these expressions into the above formula for x results in

$$x = 1 - \frac{\epsilon_1\epsilon_2 + p_1 p_2 - 1}{k^2}\frac{k(\epsilon_1 - p_1\cos\theta_1) + p_2(p_2 \mp p_1 \pm k\cos\theta_1)}{\epsilon_2(\epsilon_1 - p_1\cos\theta_1) \mp p_2(\epsilon_1\cos\theta_1 - p_1)}. \quad (A.47)$$

If both the numerator and the denominator of the last fraction is multiplied by $\epsilon_2(\epsilon_1 - p_1\cos\theta_1) \pm p_2(\epsilon_1\cos\theta_1 - p_1)$, it turns out that it is independent of the angle θ_1,

$$\frac{k(\epsilon_1 - p_1\cos\theta_1) + p_2(p_2 \mp p_1 \pm k\cos\theta_1)}{\epsilon_2(\epsilon_1 - p_1\cos\theta_1) \mp p_2(\epsilon_1\cos\theta_1 - p_1)} = \epsilon_1\epsilon_2 \mp p_1 p_2 - 1. \quad (A.48)$$

Thus, the argument x has the extreme values

$$x = \begin{cases} 0 \\ 1 - (\mu/2k)^2 = -(p_1 p_2/k^2)\mu \end{cases} \quad (A.49)$$

independent of the angle θ_1. Since x has the maximum value $x = 0$, it can never be positive.

3.9 If the Schrödinger equation with potential $U(r)$ is solved by means of the partial wave expansion, the solution is given by [56]

$$\psi(\mathbf{r}) = \sum_{l=0}^{\infty} A_l L_l(r) P_l(\hat{\mathbf{p}}\cdot\hat{\mathbf{r}}), \quad (A.50)$$

where the radial wave function $G_l(r) = rL_l(r)$ satisfies the differential equation

$$G_l'' + \left[p^2 - \frac{l(l+1)}{r^2} - U(r)\right]G_l = 0. \qquad (A.51)$$

The function $\bar{g}_\kappa(r) = rg_\kappa(r)$ which is r times the Dirac radial wave function, satisfies the second-order differential equation (3.267). Show that the substitution $\bar{g}_\kappa(r) = \sqrt{h(r)}G_\kappa(r)$, where $h(r) = \epsilon + 1 - V(r)$, brings (3.267) into the form

$$G_\kappa'' + \left[p^2 - \frac{l(l+1)}{r^2} - U_\kappa(r)\right]G_\kappa = 0 \qquad (A.52)$$

and that the effective potential $U_\kappa(r)$ has the form

$$U_\kappa(r) = 2\epsilon V - V^2 - \frac{\kappa}{rh}\frac{dV}{dr} + \frac{3}{4h^2}\left(\frac{dV}{dr}\right)^2 + \frac{1}{2h}\frac{d^2V}{dr^2}. \qquad (A.53)$$

Solution:
Inserting

$$\bar{g}_\kappa = \sqrt{h}\,G_\kappa\,,\quad \bar{g}_\kappa' = \sqrt{h}\,G_\kappa' + \frac{h'}{2\sqrt{h}}G_\kappa\,,$$

$$\bar{g}_\kappa'' = \sqrt{h}\,G_\kappa'' + \frac{h'}{\sqrt{h}}G_\kappa' - \frac{h'^2}{4h^{3/2}}G_\kappa + \frac{h''}{2\sqrt{h}}G_\kappa \qquad (A.54)$$

into (3.267) yields

$$\sqrt{h}\,G_\kappa'' + \frac{h'}{\sqrt{h}}G_\kappa' - \frac{h'^2}{4h^{3/2}}G_\kappa + \frac{h''}{2\sqrt{h}}G_\kappa$$
$$+ \left[p^2 - 2\epsilon V + V^2 - \frac{l(l+1)}{r^2}\right]\sqrt{h}\,G_\kappa$$
$$- \frac{h'}{h}\left(\frac{h'}{2\sqrt{h}}G_\kappa + \sqrt{h}\,G_\kappa' + \frac{\kappa}{r}\sqrt{h}\,G_\kappa\right) = 0. \qquad (A.55)$$

Dividing by $\sqrt{h}$ and noting that $h' = -dV/dr$, one gets

$$G_\kappa'' + \left[p^2 - \frac{l(l+1)}{r^2} - 2\epsilon V + V^2 + \frac{\kappa}{rh}\frac{dV}{dr} - \frac{3}{4h^2}\left(\frac{dV}{dr}\right)^2 - \frac{1}{2h}\frac{d^2V}{dr^2}\right]G_\kappa = 0. \qquad (A.56)$$

The first two terms of (A.53) are independent of the electron spin and are typical of the Klein-Gordon equation for particles without spin. The remaining terms are a consequence of the spin-orbit interaction and depend not only on the potential but also on its radial derivatives.

3.10 The three real Stokes parameters ξ_i satisfying $\sum_i \xi_i^2 = 1$ can be represented by

$$\xi_i = e_s^\dagger \sigma_i e_s, \qquad (A.57)$$

where σ_i are the Pauli spin matrices, the polarization vector is given by $\mathbf{e} = a_1 \mathbf{e}_1 + a_2 \mathbf{e}_2$ with $|a_1|^2 + |a_2|^2 = 1$,

$$e_s = \begin{pmatrix} a_1 \\ a_2 \end{pmatrix}, \qquad (A.58)$$

and $\mathbf{e}_1$, $\mathbf{e}_2$ are orthonormal vectors orthogonal to the photon direction $\hat{\mathbf{k}}$, i.e., $\mathbf{e}_1 \cdot \hat{\mathbf{k}} = \mathbf{e}_2 \cdot \hat{\mathbf{k}} = \mathbf{e}_1 \cdot \mathbf{e}_2 = 0$. Express the ξ_i by the components a_1 and a_2 and calculate the Stokes parameters for linear and circular photon polarization.

Hint: In order to specify the photon polarization completely, the determination of the three Stokes parameters may be achieved by polarization measurements with respect to three linearly independent bases. For these bases one can choose (1) two states of linear polarization with plane of polarization perpendicular and parallel to the production plane [see Eq. (3.77)], (2) two other states of linear polarization making angles $\pi/4$ with the polarization planes in (1), and (3) the two states of left and right circularly polarized waves.

Solution:
With the representation of the Pauli spin matrix σ_1 we get

$$\xi_1 = \begin{pmatrix} a_1^* & a_2^* \end{pmatrix} \begin{pmatrix} 0 & 1 \\ 1 & 0 \end{pmatrix} \begin{pmatrix} a_1 \\ a_2 \end{pmatrix} = a_1 a_2^* + a_1^* a_2 = 2\,\mathrm{Re}(a_1 a_2^*). \qquad (A.59)$$

Correspondingly, the two other Stokes parameters have the form

$$\xi_2 = \begin{pmatrix} a_1^* & a_2^* \end{pmatrix} \begin{pmatrix} 0 & -i \\ i & 0 \end{pmatrix} \begin{pmatrix} a_1 \\ a_2 \end{pmatrix}$$
$$= i(a_1^* a_2 - a_1 a_2^*) = 2\,\mathrm{Im}(a_1 a_2^*), \qquad (A.60)$$
$$\xi_3 = \begin{pmatrix} a_1^* & a_2^* \end{pmatrix} \begin{pmatrix} 1 & 0 \\ 0 & -1 \end{pmatrix} \begin{pmatrix} a_1 \\ a_2 \end{pmatrix} = |a_1|^2 - |a_2|^2. \qquad (A.61)$$

For linear photon polarization with $\mathbf{e} = \mathbf{e}_\perp$ we have $a_1 = 1$, $a_2 = 0$, resulting in $\xi_1 = \xi_2 = 0$, $\xi_3 = 1$.
$\mathbf{e} = \mathbf{e}_\parallel$: $a_1 = 0$, $a_2 = 1$ $\Rightarrow$ $\xi_1 = \xi_2 = 0$, $\xi_3 = -1$.
$\mathbf{e} = (\mathbf{e}_\perp \pm \mathbf{e}_\parallel)/\sqrt{2}$: $a_1 = 1/\sqrt{2}$, $a_2 = \pm 1/\sqrt{2}$ $\Rightarrow$ $\xi_1 = \pm 1$, $\xi_2 = \xi_3 = 0$.
For circular photon polarization we have $a_1 = 1/\sqrt{2}$, $a_2 = \pm i/\sqrt{2}$ yielding $\xi_1 = \xi_3 = 0$, $\xi_2 = \pm 1$.

3.11 Derive the solution (3.299) (up to a real normalization factor) of the radial wave equation for a Coulomb potential $V(r) = -a/r$.
Hint: Use the ansatz

$$rg_\kappa = \bar{g}_\kappa(r) = \sqrt{\epsilon+1}\big[\Phi_1(x) + \Phi_2(x)\big], \tag{A.62}$$

$$rf_\kappa = \bar{f}_\kappa(r) = i\sqrt{\epsilon-1}\big[\Phi_1(x) - \Phi_2(x)\big] \tag{A.63}$$

with $x = 2ipr$ to solve the wave equations (3.265) and (3.266). Then show that $\Phi_2 = \Phi_1^*$. The solution (3.299), up to a real normalization factor N, is obtained by setting

$$\Phi_1(x) = (\gamma + ia_1) e^{i\eta} x^\gamma e^{-x/2} F(x) \tag{A.64}$$

with $\gamma = \sqrt{\kappa^2 - a^2}$, $a_1 = (\epsilon/p)a$. Show that $F(x)$ is a confluent hypergeometric function satisfying the differential equation (A.22) and employ the condition $\Phi_2 = \Phi_1^*$ to evaluate the phase η. This is achieved by applying Kummer's formula

$$F(a; c; x) = e^x F(c - a; c; -x) \tag{A.65}$$

and the contiguous relation of the confluent hypergeometric function [53],

$$x F(a+1; c+1; x) = c\big[F(a+1; c; x) - F(a; c; x)\big]. \tag{A.66}$$

Solution:
We first show that near the origin $r = 0$ the radial wave function $g_\kappa(r)$ behaves as

$$g_\kappa \propto r^{\gamma-1}. \tag{A.67}$$

For small values of r the radial wave equation (3.254) with $V(r) = -a/r$

reduces to [keep in mind that $l(l+1) = \kappa(\kappa+1)$]

$$g_\kappa'' + \frac{3}{r}g_\kappa' + \frac{a^2 - \kappa^2 + 1}{r^2}g_\kappa = 0 , \qquad (A.68)$$

where $g_\kappa' = dg_\kappa/dr$. Setting $g_\kappa \propto r^{\gamma-1}$ this leads to $\gamma^2 = \kappa^2 - a^2$. Thus, for wave functions which are regular at $r = 0$, $\gamma = +\sqrt{\kappa^2 - a^2}$.

Now we introduce the relations (A.62) and (A.63) into the coupled differential equations (3.265) and (3.266). Denoting derivatives with respect to x by primes, this results in

$$2p\sqrt{\epsilon - 1}(\Phi_1' - \Phi_2') = \frac{2\kappa p}{x}\sqrt{\epsilon - 1}(\Phi_1 - \Phi_2)$$
$$+ (\epsilon - 1 + 2ipa/x)\sqrt{\epsilon + 1}(\Phi_1 + \Phi_2), \quad (A.69)$$

$$2p\sqrt{\epsilon + 1}(\Phi_1' + \Phi_2') = -\frac{2\kappa p}{x}\sqrt{\epsilon + 1}(\Phi_1 + \Phi_2)$$
$$+ (\epsilon + 1 + 2ipa/x)\sqrt{\epsilon - 1}(\Phi_1 - \Phi_2). \quad (A.70)$$

If we multiply the first of these equations by $\sqrt{\epsilon + 1}$ and the second one by $\sqrt{\epsilon - 1}$, we get

$$2p^2(\Phi_1' - \Phi_2') = \frac{2\kappa p^2}{x}(\Phi_1 - \Phi_2) + [p^2 + (\epsilon + 1)2ipa/x](\Phi_1 + \Phi_2) ,$$
$$(A.71)$$

$$2p^2(\Phi_1' + \Phi_2') = -\frac{2\kappa p^2}{x}(\Phi_1 + \Phi_2) + [p^2 + (\epsilon - 1)2ipa/x](\Phi_1 - \Phi_2) .$$
$$(A.72)$$

By adding and subtracting these equations we arrive at

$$\Phi_1' = \frac{d\Phi_1}{dx} = \left(\frac{1}{2} + \frac{ia_1}{x}\right)\Phi_1 - \left(\frac{\kappa}{x} - \frac{ia}{px}\right)\Phi_2 \qquad (A.73)$$

and

$$\Phi_2' = \frac{d\Phi_2}{dx} = -\left(\frac{\kappa}{x} + \frac{ia}{px}\right)\Phi_1 - \left(\frac{1}{2} + \frac{ia_1}{x}\right)\Phi_2 , \qquad (A.74)$$

where $a_1 = a\epsilon/p$. The complex conjugate of Eqs. (A.73) and (A.74) has the form (remember that x is purely imaginary)

$$\frac{d\Phi_1^*}{dx} = -\left(\frac{1}{2} + \frac{ia_1}{x}\right)\Phi_1^* - \left(\frac{\kappa}{x} + \frac{ia}{px}\right)\Phi_2^* , \qquad (A.75)$$

$$\frac{d\Phi_2^*}{dx} = -\left(\frac{\kappa}{x} - \frac{ia}{px}\right)\Phi_1^* + \left(\frac{1}{2} + \frac{ia_1}{x}\right)\Phi_2^* . \qquad (A.76)$$

If we set $\Phi_2 = \Phi_1^*$, the latter equations are identical with (A.73) and (A.74). Therefore, $\bar{f}_\kappa$ and $\bar{g}_\kappa$ can be chosen real, as is obvious from the original system of radial equations (3.265) and (3.266). Eliminating Φ_2 from (A.73) and (A.74), the second-order differential equation for $\Phi_1(x)$ has the form

$$\Phi_1'' + \frac{1}{x}\Phi_1' - \left[\tfrac{1}{4} + (\tfrac{1}{2} + ia_1)/x + \gamma^2/x^2\right]\Phi_1 = 0 \ . \tag{A.77}$$

Now we use the result (A.67), i.e., $\Phi_1 \propto r^\gamma \propto x^\gamma$ for small values of x, and introduce the ansatz (C is a constant)

$$\Phi_1(x) = Cx^\gamma e^{-x/2} F(x) \tag{A.78}$$

into (A.77) yielding

$$xF'' + (2\gamma + 1 - x)F' - (\gamma + 1 + ia_1)F = 0 \ . \tag{A.79}$$

This differential equation is satisfied by the confluent hypergeometric function $F(\gamma + 1 + ia_1; 2\gamma + 1; x)$ [see Eq. (A.22)]. We write the regular solution of (A.77) in the form

$$\Phi_1(r) = N(\gamma + ia_1)e^{i\eta}(2pr)^\gamma e^{-ipr} F(\gamma + 1 + ia_1; 2\gamma + 1; 2ipr) \ , \tag{A.80}$$

where N is a real normalization factor. The phase η has to be determined such that Φ_2, evaluated from (A.73), obeys the condition $\Phi_2 = \Phi_1^*$. This requires that

$$\Phi_2 = \frac{x}{\kappa - ia/p}\left[\left(\frac{1}{2} + \frac{ia_1}{x}\right)\Phi_1 - \frac{d\Phi_1}{dx}\right] = \Phi_1^* \ . \tag{A.81}$$

Denoting the r-dependent terms of Φ_1 by $\Phi(r)$, we write Eq. (A.80) in the form

$$\Phi_1(r) = N(\gamma + ia_1)e^{i\eta}(2p)^\gamma \Phi(r) \ . \tag{A.82}$$

Then (A.81) results in

$$e^{-2i\eta} = \frac{\gamma + ia_1}{\gamma - ia_1}\frac{r}{\kappa - ia/p}\left[i(p + a/r)\Phi - \frac{d\Phi}{dr}\right]\frac{1}{\Phi^*} \ . \tag{A.83}$$

In order to evaluate the right-hand side of (A.83) we use Kummer's formula (A.65) in the form

$$e^{-x/2}F(\gamma + 1 + ia_1; 2\gamma + 1; x) = e^{x/2}F(\gamma - ia_1; 2\gamma + 1; -x) \ , \tag{A.84}$$

and apply the derivative

$$\frac{dF(a;c;x)}{dx} = \frac{a}{c}F(a+1;c+1;x) \tag{A.85}$$

and the contiguous relation (A.66) of the confluent hypergeometric function. This yields

$$\begin{aligned}\Phi(r) &= r^\gamma e^{-ipr} F(\gamma+1+ia_1; 2\gamma+1; 2ipr) \\ &= r^\gamma e^{ipr} F(\gamma-ia_1; 2\gamma+1; -2ipr)\end{aligned} \tag{A.86}$$

and

$$\begin{aligned}\frac{d\Phi}{dr} &= r^{\gamma-1} e^{ipr}\{(\gamma+ipr)F(\gamma-ia_1; 2\gamma+1; -2ipr) \\ &\quad - 2ipr\frac{\gamma-ia_1}{2\gamma+1}F(\gamma+1-ia_1; 2\gamma+2; -2ipr)\} \\ &= r^{\gamma-1} e^{ipr}\{i(a_1+pr)F(\gamma-ia_1; 2\gamma+1; -2ipr) \\ &\quad + (\gamma-ia_1)F(\gamma+1-ia_1; 2\gamma+2; -2ipr)\} \,.\end{aligned} \tag{A.87}$$

From this we get

$$\begin{aligned}i(pr+a_1)\Phi - r\frac{d\Phi}{dr} &= r^\gamma e^{ipr}(ia_1-\gamma)F(\gamma+1-ia_1; 2\gamma+1; -2ipr) \\ &= (ia_1-\gamma)\Phi^*(r) \,,\end{aligned} \tag{A.88}$$

so that Eq. (A.83) reduces to

$$e^{2i\eta} = -\frac{\kappa - ia/p}{\gamma + ia_1} = -\frac{\gamma - ia_1}{\kappa + ia/p}, \tag{A.89}$$

which is given in (3.302).

The radial wave functions (A.62) and (A.63) take the form

$$\begin{aligned}g_\kappa(r) &= 2(\epsilon+1)\sqrt{\epsilon-1}\,N(2pr)^{\gamma-1}\{(\gamma+ia_1)e^{-i(pr-\eta)} \\ &\quad \cdot F(\gamma+1+ia_1; 2\gamma+1; 2ipr) \\ &\quad + (\gamma-ia_1)e^{i(pr-\eta)}F(\gamma+1-ia_1; 2\gamma+1; -2ipr)\}, \quad (A.90) \\ f_\kappa(r) &= 2i(\epsilon-1)\sqrt{\epsilon+1}\,N(2pr)^{\gamma-1}\{(\gamma+ia_1)e^{-i(pr-\eta)} \\ &\quad \cdot F(\gamma+1+ia_1; 2\gamma+1; 2ipr) \\ &\quad - (\gamma-ia_1)e^{i(pr-\eta)}F(\gamma+1-ia_1; 2\gamma+1; -2ipr)\} \,. \quad (A.91)\end{aligned}$$

The choice of the real normalization factor N depends on the normalization prescription.[||] Since the phase η is defined in (A.89) only to within an additive multiple of π, there is a sign ambiguity in f_κ and g_κ, but the ratio f_κ/g_κ is unambiguous. The only factor in Eqs. (A.90) and (A.91) which depends on the sign of κ is $\exp(\pm i\eta)$.

3.12 Use the asymptotic behaviour of the confluent hypergeometric function [Eq. (3.131)] to derive that the radial wave function $g_\kappa(r)$ in Eq. (3.299) behaves for $r \to \infty$ as $g_\kappa \sim \cos(pr + \delta_\kappa)/(pr)$ with

$$\delta_\kappa = a_1 \ln(2pr) - \arg\Gamma(\gamma + ia_1) + \eta - \tfrac{1}{2}\pi\gamma. \tag{A.92}$$

The occurrence of the r-dependent logarithmic term is characteristic of the Coulomb potential and arises from the long range of the field. It would not appear for screened Coulomb potentials with $\lim_{r\to\infty}(rV) = 0$.

Solution:
For $|x| \gg 1$ the asymptotic form of the confluent hypergeometric function $F(a; c; x)$ is given by

$$F(a; c; x) \sim \frac{\Gamma(c)}{\Gamma(c-a)} e^{i\pi a} x^{-a} + \frac{\Gamma(c)}{\Gamma(a)} e^x x^{a-c}. \tag{A.93}$$

Because the argument $x = 2ipr$ of the confluent hypergeometric function F in (3.300) is purely imaginary, the decrease of F is determined by the power factors and not by the exponential e^x of the second term. Since $\operatorname{Re} a = \gamma+1$, $\operatorname{Re} c = 2\gamma + 1$, the first term decreases more rapidly than the second one. Hence we have to consider only the second term, and Eq. (3.300) gives

$$\Phi(r) \sim e^{-ipr+i\eta}(\gamma + ia_1)\frac{\Gamma(2\gamma+1)}{\Gamma(\gamma+ia_1+1)}(2ipr)^{ia_1-\gamma}e^{2ipr}. \tag{A.94}$$

If we write

$$\frac{\gamma + ia_1}{\Gamma(\gamma+ia_1+1)} = \frac{1}{\Gamma(\gamma+ia_1)} = \frac{e^{-i\arg\Gamma(\gamma+ia_1)}}{|\Gamma(\gamma+ia_1)|}, \tag{A.95}$$

$$(2ipr)^{ia_1} = e^{-\pi a_1/2}e^{ia_1\ln(2pr)}, \quad i^{-\gamma} = e^{-i\pi\gamma/2}, \tag{A.96}$$

[||] The normalization of the function (3.299) is such that $g_\kappa(r) \sim \cos(pr + \delta)/(pr)$.

Eq. (A.94) takes the form

$$\Phi(r) \sim \frac{\Gamma(2\gamma+1)}{|\Gamma(\gamma+ia_1)|} (2pr)^{-\gamma} e^{i(pr+\delta_\kappa)} e^{-\pi a_1/2} \qquad (A.97)$$

with the phase δ_κ given by (A.92). Then we have

$$\Phi(r) + \Phi^*(r) \sim 2 \frac{\Gamma(2\gamma+1)}{|\Gamma(\gamma+ia_1)|} e^{-\pi a_1/2} (2pr)^{-\gamma} \cos(pr+\delta_\kappa) \qquad (A.98)$$

and

$$g_\kappa(r) \sim \frac{\cos(pr+\delta_\kappa)}{pr}. \qquad (A.99)$$

Correspondingly

$$f_\kappa(r) \sim -\sqrt{\frac{\epsilon-1}{\epsilon+1}} \frac{\sin(pr+\delta_\kappa)}{pr} = -\frac{\sin(pr+\delta_\kappa)}{(\epsilon+1)r}. \qquad (A.100)$$

3.13 The radial wave functions $g_\kappa(r)$ and $f_\kappa(r)$ which are exact solutions of the Dirac equation with Coulomb potential are given by Eqs. (3.299) and (3.304). Calculate the limit of these functions for vanishing momentum p of the electrons [see Eqs. (3.314) and (3.315)] corresponding to the short-wavelength limit of the bremsstrahlung spectrum. Show that the resulting functions satisfy the wave equation (3.262).
Hint: Since the factor $(\gamma+ia_1)$ of the function $\Phi(r)$ [Eq. (3.300)] tends to infinity for $p \to 0$, all the factors of $g_\kappa(r)$ and $f_\kappa(r)$ have to be expanded in terms of the momentum p. Use the formulae [53]

$$\lim_{y\to\infty} |\Gamma(x+iy)| = \sqrt{2\pi} e^{-\pi y/2} y^{x-1/2}, \qquad (A.101)$$

$$F(a;c;z) = \Gamma(c) \Big\{ (-az)^{-(c-1)/2} J_{c-1}(2\sqrt{-az}) + \tfrac{1}{2} az^2 (-az)^{-(c+1)/2} J_{c+1}(2\sqrt{-az}) + \ldots \Big\}, \qquad (A.102)$$

and the relations between the Bessel functions of different order and their derivatives.

Solution:
The radial wave functions (3.299) and (3.304) are

$$g_\kappa(r) = (2pr)^{\gamma-1} e^{\pi a_1/2} \frac{|\Gamma(\gamma + ia_1)|}{\Gamma(2\gamma + 1)} [\Phi(r) + \Phi^*(r)] \qquad \text{(A.103)}$$

and

$$f_\kappa(r) = i\frac{p}{\epsilon + 1}(2pr)^{\gamma-1} e^{\pi a_1/2} \frac{|\Gamma(\gamma + ia_1)|}{\Gamma(2\gamma + 1)} [\Phi(r) - \Phi^*(r)] \qquad \text{(A.104)}$$

with

$$\Phi(r) = e^{-ipr+i\eta}(\gamma + ia_1) F(\gamma + 1 + ia_1; 2\gamma + 1; 2ipr) \qquad \text{(A.105)}$$

and

$$e^{2i\eta} = \frac{ia_1/\epsilon - \kappa}{ia_1 + \gamma}, \quad \gamma = \sqrt{\kappa^2 - a^2}, \quad a_1 = (\epsilon/p)a. \qquad \text{(A.106)}$$

For $p \to 0$ we have $\epsilon \to 1$ and hence $a_1 \to a/p$. Then the relation (A.101) yields

$$\lim_{p \to 0} e^{\pi a/2p} |\Gamma(\gamma + ia/p)| = \sqrt{2\pi}(a/p)^{\gamma-1/2}. \qquad \text{(A.107)}$$

Up to first order in p we get

$$e^{2i\eta} = \frac{\epsilon a^2 - \gamma\kappa p^2 + iap(\gamma + \epsilon\kappa)}{a^2 + p^2\kappa^2} \approx 1 + i(\kappa + \gamma)(p/a). \qquad \text{(A.108)}$$

Thus the exponential has the approximate form

$$e^{i(\eta - pr)} \approx 1 + i(\kappa + \gamma)(p/2a) - ipr. \qquad \text{(A.109)}$$

Using (A.102) and the notation $\rho = 2\sqrt{2ar}$ the confluent hypergeometric function can be expanded into

$$F(\gamma + 1 + ia_1; 2\gamma + 1; 2ipr) \approx \Gamma(2\gamma + 1) \Big\{ (2ar)^{-\gamma} [1 + i\gamma(\gamma + 1)(p/a)]$$
$$\cdot \big[J_{2\gamma}(\rho) - i(\gamma + 1)(p/a)\sqrt{2ar} J'_{2\gamma}(\rho) \big]$$
$$- i(a/2p)(2pr)^2 (2ar)^{-(\gamma+1)} J_{2\gamma+2}(\rho) \Big\}$$
$$\approx \Gamma(2\gamma + 1)(2ar)^{-\gamma} \Big\{ [1 + i\gamma(\gamma + 1)(p/a)] J_{2\gamma}(\rho)$$
$$- i(\gamma + 1)(p/a) \big[\sqrt{2ar} J_{2\gamma-1}(\rho) - \gamma J_{2\gamma}(\rho) \big]$$
$$- ipr \big[(2\gamma + 1) J_{2\gamma+1}(\rho)/\sqrt{2ar} - J_{2\gamma}(\rho) \big] \Big\}$$

$$\begin{aligned}
&= \Gamma(2\gamma+1)(2ar)^{-\gamma}\Big\{\big[1+2i\gamma(\gamma+1)(p/a)+ipr\big]J_{2\gamma}(\rho)\\
&\quad - i(\gamma+1)(p/a)\sqrt{2ar}\,J_{2\gamma-1}(\rho)\\
&\quad - ipr(2\gamma+1)\big[(\gamma/ar)J_{2\gamma}(\rho)-J_{2\gamma-1}(\rho)/\sqrt{2ar}\big]\Big\}\\
&= \Gamma(2\gamma+1)(2ar)^{-\gamma}\Big\{(1+i\gamma p/a+ipr)J_{2\gamma}(\rho)-ipr\,J_{2\gamma-1}(\rho)/\sqrt{2ar}\Big\}.
\end{aligned}$$
(A.110)

Applying the above expansions one obtains

$$\begin{aligned}
\Phi(r)+\Phi^*(r) &= 2\,\text{Re}\,\Phi(r)\\
&\approx 2\text{Re}\Big\{\big[1+i(\gamma+\kappa)(p/2a)-ipr\big](\gamma+ia/p)\Gamma(2\gamma+1)(2ar)^{-\gamma}\\
&\quad \cdot \big[(1+i\gamma p/a+ipr)J_{2\gamma}(\rho)-ipr\,J_{2\gamma-1}(\rho)/\sqrt{2ar}\big]\Big\}\\
&\approx \Gamma(2\gamma+1)(2ar)^{-\gamma}\big\{\tfrac{1}{2}\rho J_{2\gamma-1}(\rho)-(\gamma+\kappa)J_{2\gamma}(\rho)\big\}.
\end{aligned}$$
(A.111)

If we utilize the relations (A.107) and (A.111) the radial wave function $\bar{g}_\kappa(r)=rg_\kappa(r)$ at the short-wavelength limit takes the form

$$\bar{g}_\kappa(r)=\sqrt{\frac{\pi}{2ap}}\,\big\{\tfrac{1}{2}\rho J_{2\gamma-1}(\rho)-(\kappa+\gamma)J_{2\gamma}(\rho)\big\}.$$
(A.112)

The calculation of the component $f_\kappa(r)$ is more simple due to its additional factor p. For $p\to 0$, $\epsilon\to 1$ we get

$$ip\big[\Phi(r)-\Phi^*(r)\big]=-2p\,\text{Im}\,\Phi(r)=-2\Gamma(2\gamma+1)a(2ar)^{-\gamma}J_{2\gamma}(\rho)$$
(A.113)

which results in

$$\begin{aligned}
\bar{f}_\kappa(r) &= rf_\kappa(r)\\
&= -r(2pr)^{\gamma-1}a(2ar)^{-\gamma}(a/p)^\gamma\sqrt{2\pi p/a}\,J_{2\gamma}(\rho)\\
&= -\sqrt{\pi a/2p}\,J_{2\gamma}(\rho).
\end{aligned}$$
(A.114)

Except for a constant factor the two radial wave functions for $p\to 0$ can be written as

$$g_\kappa(r)=\frac{\rho}{2r}J_{2\gamma-1}(\rho)-\frac{\kappa+\gamma}{r}J_{2\gamma}(\rho),$$
(A.115)

$$f_\kappa(r)=-\frac{a}{r}J_{2\gamma}(\rho).$$
(A.116)

Inserting these expressions into the wave equation (3.262) yields (primes denote derivatives with respect to ρ)

$$-\sqrt{\frac{a}{2r^3}} J_{2\gamma-1}(\rho) + \frac{2a}{r} J'_{2\gamma-1}(\rho) + \frac{\kappa+\gamma}{r^2} J_{2\gamma}(\rho)$$

$$-\frac{\kappa+\gamma}{r}\sqrt{\frac{2a}{r}} J'_{2\gamma}(\rho) + (\kappa+1)\frac{\rho}{2r^2} J_{2\gamma-1}(\rho)$$

$$-\frac{1}{r^2}(\kappa+\gamma)(\kappa+1) J_{2\gamma}(\rho) + (2+a/r)\frac{a}{r} J_{2\gamma}(\rho) = 0 . \quad \text{(A.117)}$$

The derivatives of the Bessel functions can be eliminated by applying the relation [53]

$$J'_\nu(z) = J_{\nu-1}(z) - \frac{\nu}{z} J_\nu(z) = \frac{\nu}{z} J_\nu(z) - J_{\nu+1}(z) \quad \text{(A.118)}$$

which results in

$$J_{2\gamma-1}(\rho)\Big[(\kappa+1)\sqrt{2a/r^3} - \sqrt{a/2r^3}\Big] + \frac{2a}{r}\left[\frac{2\gamma-1}{2\sqrt{2ar}} J_{2\gamma-1}(\rho) - J_{2\gamma}(\rho)\right]$$

$$-\frac{\kappa}{r^2}(\kappa+\gamma) J_{2\gamma}(\rho) + \left(a^2/r^2 + 2a/r\right) J_{2\gamma}(\rho)$$

$$-\frac{\kappa+\gamma}{r}\sqrt{2a/r}\left[J_{2\gamma-1}(\rho) - \frac{\gamma}{\sqrt{2ar}} J_{2\gamma}(\rho)\right] = 0 . \quad \text{(A.119)}$$

Thus the wave equation (3.262) is satisfied by the functions $g_\kappa(\rho)$ and $f_\kappa(\rho)$. Clearly, the second wave equation (3.264) is satisfied as well.

3.14 Show that the function L_1 in Eq. (3.308) is of order a^2 by proving that $L_1 = 0$ for $\gamma = n$.
Hint: Use the recursion relation for confluent hypergeometric functions,

$$\frac{c-1}{x}[F(a;c;x) - F(a;c-1;x)] = -\frac{a}{c} F(a+1;c+1;x) . \quad \text{(A.120)}$$

Solution:
According to Eq. (3.309) we have $\gamma = (n^2 - a^2)^{1/2}$. Expanding L_1 into a Taylor series gives

$$L_1 = (L_1)_{\gamma=n} + \left(\frac{\partial L_1}{\partial \gamma}\frac{\partial \gamma}{\partial a^2}\right)_{\gamma=n} a^2 + \ldots . \quad \text{(A.121)}$$

If we verify that $(L_1)_{\gamma=n} = 0$ we have proved that L_1 is of order a^2. Setting $\gamma = n$ in (3.308) results in

$$(L_1)_{\gamma=n} = \sum_{n=1}^{\infty}(-1)^n \frac{\Gamma(n-ia_1)}{\Gamma(2n+1)} e^{\pi a_1/2}(-2ip_1 r)^{n-1} e^{ip_1 r}$$
$$\cdot \left\{-2n\, F(n-ia_1; 2n; -2ip_1 r) P'_{n-1} + 2n\left[F(n-ia_1; 2n+1; -2ip_1 r)\right.\right.$$
$$\left.\left. - F(n-ia_1; 2n; -2ip_1 r)\right] P'_n \right\}. \tag{A.122}$$

The first term in curly brackets vanishes for $n = 1$ since $P'_0 = 0$. Therefore, by means of the substitution $n \to n+1$, it can be rewritten in the form

$$\sum_{n=1}^{\infty}(-1)^{n+1} \frac{\Gamma(n+1-ia_1)}{\Gamma(2n+3)} e^{\pi a_1/2}(-2ip_1 r)^n e^{ip_1 r}\left[-2(n+1)\right]$$
$$\cdot F(n+1-ia_1; 2n+2; -2ip_1 r)\, P'_n$$
$$= \sum_{n=1}^{\infty}(-1)^n \frac{n-ia_1}{2n+1} \frac{\Gamma(n-ia_1)}{\Gamma(2n+1)} e^{\pi a_1/2}(-2ip_1 r)^n$$
$$\cdot e^{ip_1 r} F(n+1-ia_1; 2n+2; -2ip_1 r)\, P'_n\,. \tag{A.123}$$

If the terms in (A.122) proportional to P'_n are transformed by means of the relation (A.120) one sees that they are exactly equal to (A.123) with opposite sign so that the coefficient of P'_n is identically zero; whence $L_1 = 0$ for $\gamma = n$.

3.15 At the short-wavelength limit ($p_2 = 0$) the quantities **I** and **J** [see Eqs. (3.322) and (3.323)] reduce to

$$\mathbf{I}_0 = \frac{\mathbf{p}_1}{\epsilon_1 + 1} \mathbf{I}_{10} + \mathbf{I}_{20} + \mathbf{I}_{30}$$
$$= \left(\frac{\mathbf{p}_1}{\epsilon_1 + 1} + \frac{\mathbf{q}}{2}\right) \mathbf{I}_{10} + (\epsilon_1 + 1)\, \mathbf{I}_{20} \tag{A.124}$$

and

$$\mathbf{J}_0 = \frac{\mathbf{p}_1}{\epsilon_1 + 1} \mathbf{I}_{10} + \mathbf{I}_{20} - \mathbf{I}_{30}$$
$$= \left(\frac{\mathbf{p}_1}{\epsilon_1 + 1} - \frac{\mathbf{q}}{2}\right) \mathbf{I}_{10} - (\epsilon_1 - 1)\, \mathbf{I}_{20}\,, \tag{A.125}$$

where the integrals $\mathbf{I}_{10}$, $\mathbf{I}_{20}$, and $\mathbf{I}_{30}$ are given by (3.208) to (3.210); besides $R = 0$ and $\mathbf{T} = 0$. Use the identity (3.243) and Eq. (3.329) to derive from Eq. (3.326) an expression for the summed matrix element squared, $\sum_{\vec{\zeta}_2} |M_0|^2$, and evaluate this expression for linear photon polarization (**e** real) and circular polarization $[\mathbf{e} = (\mathbf{e}_\perp + i\delta \mathbf{e}_\parallel)/\sqrt{2}]$. Give the degree of circular photon polarization for longitudinally polarized electrons ($\vec{\zeta}_1 = \hat{\mathbf{p}}_1$) and the degree of linear photon polarization for transversely polarized electrons ($\vec{\zeta} = \mathbf{e}_\perp$).

Solution:
For $p_2 = 0$ we get from (3.326) the matrix element

$$M_0 = N_1 N_2^* \sqrt{\frac{\epsilon_1 + 1}{2\epsilon_1}} \left(v_2^\dagger \{ \mathbf{e}^* \cdot \mathbf{I}_0 + i\vec{\sigma} \cdot (\mathbf{e}^* \times \mathbf{J}_0) \} v_1 \right). \quad (A.126)$$

Squaring M_0 and summing over the spins of the outgoing electrons by means of (3.329) leads to

$$\sum_{\vec{\zeta}_2} |M_0|^2 = |N_1 N_2|^2 \frac{\epsilon_1 + 1}{2\epsilon_1} \left(v_1^\dagger \{ \mathbf{e} \cdot \mathbf{I}_0^* - i\vec{\sigma} \cdot (\mathbf{e} \times \mathbf{J}_0^*) \} \{ \mathbf{e}^* \cdot \mathbf{I}_0 \right.$$

$$\left. + i\vec{\sigma} \cdot (\mathbf{e}^* \times \mathbf{J}_0) \} v_1 \right)$$

$$= |N_1 N_2|^2 \frac{\epsilon_1 + 1}{2\epsilon_1} \left(v_1^\dagger \{ |\mathbf{e}^* \cdot \mathbf{I}_0|^2 + i (\mathbf{e} \cdot \mathbf{I}_0^*) \vec{\sigma} \cdot (\mathbf{e}^* \times \mathbf{J}_0) \right.$$

$$- i(\mathbf{e}^* \cdot \mathbf{I}_0) \vec{\sigma} \cdot (\mathbf{e} \times \mathbf{J}_0^*) + |\mathbf{e}^* \times \mathbf{J}_0|^2$$

$$\left. + i\vec{\sigma} \cdot [(\mathbf{e} \times \mathbf{J}_0^*) \times (\mathbf{e}^* \times \mathbf{J}_0)] \} v_1 \right). \quad (A.127)$$

Using the diagonal matrix element (3.149) results in

$$\sum_{\vec{\zeta}_2} |M_0|^2 = |N_1 N_2|^2 \frac{\epsilon_1 + 1}{2\epsilon_1} \left\{ |\mathbf{e}^* \cdot \mathbf{I}_0|^2 + |\mathbf{e}^* \times \mathbf{J}_0|^2 \right.$$

$$\left. + 2 \operatorname{Im}[(\mathbf{e}^* \cdot \mathbf{I}_0)(\mathbf{e} \times \mathbf{J}_0^*) \cdot \vec{\zeta}_1] + i\vec{\zeta}_1 \cdot [(\mathbf{e} \times \mathbf{J}_0^*) \times (\mathbf{e}^* \times \mathbf{J}_0)] \right\}. \quad (A.128)$$

For real values of **e** this has the form

$$\sum_{\vec{\zeta}_2} |M_0|^2 = |N_1 N_2|^2 \frac{\epsilon_1 + 1}{2\epsilon_1} \left\{ |\mathbf{e} \cdot \mathbf{I}_0|^2 + |\mathbf{e} \times \mathbf{J}_0|^2 \right.$$

$$\left. + 2 \operatorname{Im}[(\mathbf{e} \cdot \mathbf{I}_0)(\mathbf{e} \times \mathbf{J}_0^*) \cdot \vec{\zeta}_1] + i[\mathbf{J}_0^* \times \mathbf{J}_0) \cdot \mathbf{e}](\vec{\zeta}_1 \cdot \mathbf{e}) \right\}. \quad (A.129)$$

For $\mathbf{e} = (\mathbf{e}_\perp + i\,\delta \mathbf{e}_\parallel)/\sqrt{2}$, with $\delta = \pm 1$, we obtain from (A.128)

$$\sum_{\vec{\zeta}_2} |M_0|^2 = |N_1 N_2|^2 \frac{\epsilon_1 + 1}{4\epsilon_1} \Big\{ |\mathbf{e}_\perp \cdot \mathbf{I}_0|^2 + |\mathbf{e}_\parallel \cdot \mathbf{I}_0|^2 + i\delta\big[(\mathbf{e}_\perp \cdot \mathbf{I}_0)(\mathbf{e}_\parallel \cdot \mathbf{I}_0^*)$$
$$- (\mathbf{e}_\parallel \cdot \mathbf{I}_0)(\mathbf{e}_\perp \cdot \mathbf{I}_0^*)\big] + 2|\mathbf{J}_0|^2 - |\mathbf{e}_\perp \cdot \mathbf{J}_0|^2 - |\mathbf{e}_\parallel \cdot \mathbf{J}_0|^2$$
$$- i\delta\big[(\mathbf{e}_\perp \cdot \mathbf{J}_0)(\mathbf{e}_\parallel \cdot \mathbf{J}_0^*) - (\mathbf{e}_\parallel \cdot \mathbf{J}_0)(\mathbf{e}_\perp \cdot \mathbf{J}_0^*)\big]$$
$$+ 2\,\mathrm{Im}\big[(\mathbf{e}_\perp \cdot \mathbf{I}_0)\,\mathbf{e}_\perp \cdot (\mathbf{J}_0^* \times \vec{\zeta}_1) + (\mathbf{e}_\parallel \cdot \mathbf{I}_0)\,\mathbf{e}_\parallel \cdot (\mathbf{J}_0^* \times \vec{\zeta}_1)\big]$$
$$+ 2\delta\,\mathrm{Re}\big[(\mathbf{e}_\perp \cdot \mathbf{I}_0)\,\mathbf{e}_\parallel \cdot (\mathbf{J}_0^* \times \vec{\zeta}_1) - (\mathbf{e}_\parallel \cdot \mathbf{I}_0)\,\mathbf{e}_\perp \cdot (\mathbf{J}_0^* \times \vec{\zeta}_1)\big]$$
$$+ i\big[\mathbf{e}_\perp \cdot (\mathbf{J}_0^* \times \mathbf{J}_0)(\mathbf{e}_\perp \cdot \vec{\zeta}_1) + \mathbf{e}_\parallel \cdot (\mathbf{J}_0^* \times \mathbf{J}_0)(\mathbf{e}_\parallel \cdot \vec{\zeta}_1)\big]$$
$$+ \delta\big[\mathbf{e}_\perp \cdot (\mathbf{J}_0^* \times \mathbf{J}_0)(\mathbf{e}_\parallel \cdot \vec{\zeta}_1) + 2(\mathbf{J}_0 \cdot \hat{\mathbf{k}})(\mathbf{J}_0 \cdot \vec{\zeta}_1)\big] \Big\}. \qquad (A.130)$$

Using the relations

$$|\mathbf{e}_\perp \cdot \mathbf{I}_0|^2 + |\mathbf{e}_\parallel \cdot \mathbf{I}_0|^2 = (\mathbf{I}_0 \times \hat{\mathbf{k}})^2 \qquad (A.131)$$

[see Eq. (3.79] and

$$i\,(\mathbf{e}_\perp \cdot \mathbf{I}_0)(\mathbf{e}_\parallel \cdot \mathbf{I}_0^*) - i\,(\mathbf{e}_\parallel \cdot \mathbf{I}_0)(\mathbf{e}_\perp \cdot \mathbf{I}_0^*) = i\,(\mathbf{e}_\perp \times \mathbf{e}_\parallel) \cdot (\mathbf{I}_0 \times \mathbf{I}_0^*)$$
$$= i\,\hat{\mathbf{k}} \cdot (\mathbf{I}_0 \times \mathbf{I}_0^*), \qquad (A.132)$$

this can be written as

$$\sum_{\vec{\zeta}_2} |M_0|^2 = |N_1 N_2|^2 \frac{\epsilon_1 + 1}{4\epsilon_1} \Big\{ |\mathbf{I}_0|^2 - |\mathbf{I}_0 \cdot \hat{\mathbf{k}}|^2 + |\mathbf{J}_0|^2 + |\mathbf{J}_0 \cdot \hat{\mathbf{k}}|^2$$
$$+ i\delta\big[(\mathbf{I}_0 \times \mathbf{I}_0^*) \cdot \hat{\mathbf{k}} - (\mathbf{J}_0 \times \mathbf{J}_0^*) \cdot \hat{\mathbf{k}}\big]$$
$$+ 2\,\mathrm{Im}\big[\mathbf{I}_0 \cdot (\mathbf{J}_0^* \times \vec{\zeta}_1) - (\mathbf{I}_0 \cdot \hat{\mathbf{k}})(\mathbf{J}_0^* \times \vec{\zeta}_1) \cdot \hat{\mathbf{k}}\big]$$
$$+ 2\delta\,\mathrm{Re}\big[(\mathbf{I}_0 + \mathbf{J}_0) \cdot \vec{\zeta}_1(\mathbf{J}_0^* \cdot \hat{\mathbf{k}}) - (\mathbf{I}_0 \cdot \mathbf{J}_0^*)(\vec{\zeta}_1 \cdot \hat{\mathbf{k}})\big]$$
$$+ i\big[(\mathbf{J}_0^* \times \mathbf{J}_0) \cdot \vec{\zeta}_1 - (\mathbf{J}_0^* \times \mathbf{J}_0 \cdot \hat{\mathbf{k}}(\vec{\zeta}_1 \cdot \hat{\mathbf{k}}))\big] \Big\}. \qquad (A.133)$$

The summation of (A.129) over the photon polarizations yields, using (3.79),

$$\sum_{\vec{\zeta}_2,\mathbf{e}} |M_0|^2 = |N_1 N_2|^2 \frac{\epsilon_1 + 1}{2\epsilon_1} \Big\{ |\mathbf{I}_0 \times \hat{\mathbf{k}}|^2 + |\mathbf{J}_0|^2 + |\mathbf{J}_0 \cdot \hat{\mathbf{k}}|^2$$
$$+ 2\,\mathrm{Im}\big[(\mathbf{I}_0 \times \hat{\mathbf{k}}) \cdot \{(\mathbf{J}_0^* \times \vec{\zeta}_1) \times \hat{\mathbf{k}}\}\big] + i\big[(\mathbf{J}_0^* \times \mathbf{J}_0) \times \hat{\mathbf{k}}\big] \cdot (\vec{\zeta}_1 \times \hat{\mathbf{k}}) \Big\}$$
$$= |N_1 N_2|^2 \frac{\epsilon_1 + 1}{2\epsilon_1} \Big\{ |\mathbf{I}_0|^2 - |\mathbf{I}_0 \cdot \hat{\mathbf{k}}|^2 + |\mathbf{J}_0|^2 + |\mathbf{J}_0 \cdot \hat{\mathbf{k}}|^2$$

$$+ 2 \operatorname{Im} \left[(\mathbf{I}_0 \times \mathbf{J}_0^*) \cdot \vec{\zeta}_1 + (\mathbf{I}_0 \cdot \hat{\mathbf{k}})(\mathbf{J}_0^* \times \hat{\mathbf{k}}) \cdot \vec{\zeta}_1 \right]$$
$$+ 2 \left[(\operatorname{Im} \mathbf{J}_0 \times \operatorname{Re} \mathbf{J}_0) \cdot \vec{\zeta}_1 - (\operatorname{Im} \mathbf{J}_0 \times \operatorname{Re} \mathbf{J}_0) \cdot \hat{\mathbf{k}} (\vec{\zeta}_1 \cdot \hat{\mathbf{k}}) \right] \Big\} . \quad (A.134)$$

Applying these formulae the linear and circular photon polarization is, respectively, given by

$$P_l(\vec{\zeta}_1) = \frac{1}{\mathcal{N}_1} \Big\{ |\mathbf{e}_\perp \mathbf{I}_0|^2 - |\mathbf{e}_\parallel \mathbf{I}_0|^2 - |\mathbf{e}_\perp \mathbf{J}_0|^2 + |\mathbf{e}_\parallel \mathbf{J}_0|^2$$
$$+ 2 \operatorname{Im} \left[(\mathbf{e}_\perp \mathbf{I}_0)(\mathbf{e}_\perp \times \mathbf{J}_0^*) - (\mathbf{e}_\parallel \mathbf{I}_0)(\mathbf{e}_\parallel \times \mathbf{J}_0^*) \right] \cdot \vec{\zeta}_1$$
$$+ 2 \left[\mathbf{e}_\perp \cdot (\operatorname{Im} \mathbf{J}_0 \times \operatorname{Re} \mathbf{J}_0)(\mathbf{e}_\perp \cdot \vec{\zeta}_1) - \mathbf{e}_\parallel \cdot (\operatorname{Im} \mathbf{J}_0 \times \operatorname{Re} \mathbf{J}_0)(\mathbf{e}_\parallel \cdot \vec{\zeta}_1) \right] \Big\}$$
$$(A.135)$$

and

$$P_c(\vec{\zeta}_1) = \frac{1}{\mathcal{N}_1} \Big\{ i \left[(\mathbf{I}_0 \times \mathbf{I}_0^*) - (\mathbf{J}_0 \times \mathbf{J}_0^*) \right] \cdot \hat{\mathbf{k}}$$
$$+ 2 \operatorname{Re} \left[(\mathbf{I}_0 + \mathbf{J}_0) \cdot \vec{\zeta}_1 (\mathbf{J}_0^* \cdot \hat{\mathbf{k}}) - (\mathbf{I}_0 \cdot \mathbf{J}_0^*)(\vec{\zeta}_1 \cdot \hat{\mathbf{k}}) \right] \Big\}, \quad (A.136)$$

where

$$\mathcal{N}_1 = |\mathbf{I}_0|^2 - |\mathbf{I}_0 \cdot \hat{\mathbf{k}}|^2 + |\mathbf{J}_0|^2 + |\mathbf{J}_0 \cdot \hat{\mathbf{k}}|^2$$
$$+ 2 \operatorname{Im} \left[\mathbf{I}_0 \times \mathbf{J}_0^* + (\mathbf{I}_0 \cdot \hat{\mathbf{k}})(\mathbf{J}_0^* \times \hat{\mathbf{k}}) \right] \cdot \vec{\zeta}_1$$
$$+ 2 \left[(\operatorname{Im} \mathbf{J}_0 \times \operatorname{Re} \mathbf{J}_0) \cdot \vec{\zeta}_1 - (\operatorname{Im} \mathbf{J}_0 \times \operatorname{Re} \mathbf{J}_0) \cdot \hat{\mathbf{k}} (\vec{\zeta}_1 \cdot \hat{\mathbf{k}}) \right] . \quad (A.137)$$

For transversely polarized incident electrons ($\vec{\zeta}_1 = \mathbf{e}_\perp$) the degree of linear polarization reduces to

$$P_l(\mathbf{e}_\perp) = \frac{1}{\mathcal{N}_2} \Big\{ |\mathbf{e}_\perp \cdot \mathbf{I}_0|^2 - |\mathbf{e}_\parallel \cdot \mathbf{I}_0|^2 - |\mathbf{e}_\perp \cdot \mathbf{J}_0|^2 + |\mathbf{e}_\parallel \cdot \mathbf{J}_0|^2$$
$$- 2 \operatorname{Im}\{(\mathbf{e}_\parallel \cdot \mathbf{I}_0)(\hat{\mathbf{k}} \cdot \mathbf{J}_0^*)\} + 2 (\operatorname{Im} \mathbf{J}_0 \times \operatorname{Re} \mathbf{J}_0) \cdot \mathbf{e}_\perp \Big\}, \quad (A.138)$$

where

$$\mathcal{N}_2 = |\mathbf{I}_0|^2 - |\mathbf{I}_0 \cdot \hat{\mathbf{k}}|^2 + |\mathbf{J}_0|^2 + |\mathbf{J}_0 \cdot \hat{\mathbf{k}}|^2$$
$$+ 2 \operatorname{Im} \left[(\mathbf{I}_0 \times \mathbf{J}_0^*) \cdot \mathbf{e}_\perp + (\mathbf{I}_0 \cdot \hat{\mathbf{k}})(\mathbf{J}_0^* \cdot \mathbf{e}_\parallel) \right] + 2 (\operatorname{Im} \mathbf{J}_0 \times \operatorname{Re} \mathbf{J}_0) \cdot \mathbf{e}_\perp .$$
$$(A.139)$$

If the incident electrons are longitudinally polarized ($\vec{\zeta}_1 = \hat{\mathbf{p}}_1$), the degree of circular photon polarization is given by

$$P_c(\hat{\mathbf{p}}_1) = \frac{1}{\mathcal{N}_3}\left\{i(\mathbf{I}_0 \times \mathbf{I}_0^* - \mathbf{J}_0 \times \mathbf{J}_0^*) \cdot \hat{\mathbf{k}}\right.$$
$$\left. + 2\operatorname{Re}\left[(\mathbf{I}_0 + \mathbf{J}_0) \cdot \hat{\mathbf{p}}_1 (\mathbf{J}_0^* \cdot \hat{\mathbf{k}}) - (\mathbf{I}_0 \cdot \mathbf{J}_0^*)(\hat{\mathbf{p}}_1 \cdot \hat{\mathbf{k}})\right]\right\}, \quad (A.140)$$

where

$$\mathcal{N}_3 = |\mathbf{I}_0|^2 - |\mathbf{I}_0 \cdot \hat{\mathbf{k}}|^2 + |\mathbf{J}_0|^2 + |\mathbf{J}_0 \cdot \hat{\mathbf{k}}|^2$$
$$+ 2\operatorname{Im}\left[\mathbf{I}_0 \times \mathbf{J}_0^* + (\mathbf{I}_0 \cdot \hat{\mathbf{k}})(\mathbf{J}_0^* \times \hat{\mathbf{k}})\right] \cdot \hat{\mathbf{p}}_1$$
$$+ 2(\operatorname{Im}\mathbf{J}_0 \times \operatorname{Re}\mathbf{J}_0) \cdot \left[\hat{\mathbf{p}}_1 - (\hat{\mathbf{p}}_1 \cdot \hat{\mathbf{k}})\hat{\mathbf{k}}\right]. \quad (A.141)$$

4.1 Use the system of polar coordinates (3.86) to calculate the angle $\theta_{2\min}$ where the term $A = |\mathbf{A}|$ of the bremsstrahlung cross section in Born approximation [Eq. (3.83)] vanishes, for given energies ϵ_1, ϵ_2, and angles θ_1, $\varphi = 0$ (coplanar geometry). Evaluate the terms A, $B = |\mathbf{B}|$, and $C = |\mathbf{C}|$ around $\theta_2 = \theta_{2\min}$ for the parameters $E_0 = 140$ MeV, $h\nu = 95$ MeV, $\theta_1 = 1°$, and show that there is a sharp dip in the cross section (3.82) (cf. Fig. 4.21). What is the behaviour of the photon linear polarization P_l near $\theta_2 = \theta_{2\min}$?

Solution
For coplanar geometry ($\varphi = 0$) the quantities A, B, and C [Eqs. (3.83) to (3.85)] have the form

$$A = \frac{\epsilon_2 p_1 \sin\theta_1}{\epsilon_1 - p_1 \cos\theta_1} - \frac{\epsilon_1 p_2 \sin\theta_2}{\epsilon_2 - p_2 \cos\theta_2}, \quad (A.142)$$

$$B = \frac{q}{2}\left(\frac{p_1 \sin\theta_1}{\epsilon_1 - p_1 \cos\theta_1} - \frac{p_2 \sin\theta_2}{\epsilon_2 - p_2 \cos\theta_2}\right), \quad (A.143)$$

$$C = \frac{k(p_1 \sin\theta_1 - p_2 \sin\theta_2)}{2\sqrt{(\epsilon_1 - p_1 \cos\theta_1)(\epsilon_2 - p_2 \cos\theta_2)}}. \quad (A.144)$$

A has a zero at the angle $\theta_{2\min}$ given by

$$\cos\theta_{2\min} = \frac{\epsilon_2^3 p_1^2 \sin^2\theta_1 - \epsilon_1(\epsilon_1 - p_1 \cos\theta_1)S}{p_2[\epsilon_1^2(\epsilon_1 - p_1 \cos\theta_1)^2 + \epsilon_2^2 p_1^2 \sin^2\theta_1]}, \quad (A.145)$$

where
$$S = \sqrt{\epsilon_1^2 p_2^2 (\epsilon_1 - p_1 \cos\theta_1)^2 - \epsilon_2^2 p_1^2 \sin^2\theta_1}. \quad (A.146)$$

For the parameters considered we get $\theta_{2\min} \approx 3.0866°$.

The cross section (3.82) is proportional to

$$X = A^2 - B^2 + 2C^2. \quad (A.147)$$

In the following table are given some values of X for angles θ_2 around $\theta_{2\min}$, illustrating the sharp dip of the cross section.

θ_2	2.9°	3.0°	3.05°	$\theta_{2\min}$	3.0877°	3.1°	3.2°	3.3°
A	−573	−258	−107.6	0	3.091	38.7	319	585
B	10.9	5.92	3.94	3.385	3.387	3.49	7.53	13.2
C	377	172	74.0	4.295	2.295	−20.7	−202	−372
X	6.12 10⁵	1.26 10⁵	22514	25.44	8.610	2345	1.83 10⁵	6.19 10⁵
P_l	−0.536	−0.530	−0.513	+0.450	+0.223	−0.633	−0.556	−0.553

The minimum of the dip is reached at $\theta_2 \approx 3.0877°$, that is very close to $\theta_{2\min}$. The full width of the dip, where $X = 2X_{\min}$, is $\Delta\theta_2 \approx 0.0015°$.

For coplanar geometry the respective cross sections (3.350) for perpendicular and parallel photon polarization are given by

$$\frac{d^3\sigma_\perp}{dk\, d\Omega_k\, d\Omega_{p_2}} = \frac{\alpha Z^2 r_0^2}{\pi^2} \frac{p_2}{k p_1 q^4} C^2 \quad (A.148)$$

and

$$\frac{d^3\sigma_\parallel}{dk\, d\Omega_k\, d\Omega_{p_2}} = \frac{\alpha Z^2 r_0^2}{\pi^2} \frac{p_2}{k p_1 q^4} \left(A^2 - B^2 + C^2\right). \quad (A.149)$$

The degree of linear polarization (3.344) has the form

$$P_l = \frac{B^2 - A^2}{A^2 - B^2 + 2C^2} = \frac{B^2 - A^2}{X}. \quad (A.150)$$

At most angles θ_2 the term A^2 is larger than B^2; hence $P_l < 0$, i.e., the radiation is polarized parallel to the reaction plane. However, near $\theta_2 = \theta_{2\min}$, where $A = 0$, the degree of linear polarization takes positive values within a narrow angular range. As stated in Sec. 3.7.2, the linear polarization is complete, $P_l = -1$, for $C = 0$ or $\sin\theta_2 = (p_1/p_2)\sin\theta_1$. For the present parameters this occurs at $\theta_2 \approx 3.0889°$, again very close to $\theta_{\min}$.

5.1 The maximum photon energy of electron-electron bremsstrahlung in the center-of-mass system (cms) is $k_{\max} = p^2/\epsilon$, independent of the photon direction [see Eq. (5.18)]. Derive the corresponding formula for the maximum energy of the photon in the laboratory system (rest system of electron 2), emitted at angle θ_k with respect to the incoming electron, by means of a Lorentz transformation from the cms.

Solution:
For arbitrary momenta $\mathbf{p}_1$ and $\mathbf{p}_2$ of the initial electrons the velocity (in units of the speed of light, c) of the center of mass is $\vec{\beta}_c = (\mathbf{p}_1+\mathbf{p}_2)/(\epsilon_1+\epsilon_2)$. In the laboratory system we have $\mathbf{p}_2 = 0$, $\epsilon_2 = 1$, hence $\vec{\beta}_c = \mathbf{p}_1/(\epsilon_1 + 1)$. The corresponding Lorentz factor is

$$\gamma = (1 - \beta_c^2)^{-1/2} = \sqrt{\tfrac{1}{2}(\epsilon_1 + 1)}\,. \qquad (A.151)$$

Labelling the quantities in the cms by a bar, the energy-momentum transformation formulae yield

$$\bar{k} = \gamma(k - \vec{\beta}_c \cdot \mathbf{k}) = \frac{(\epsilon_1 + 1)k - \mathbf{p}_1 \cdot \mathbf{k}}{\sqrt{2(\epsilon_1 + 1)}}\,,$$

$$\bar{\epsilon}_1 = \gamma(\epsilon_1 - \vec{\beta}_c \cdot \mathbf{p}_1) = \gamma = \sqrt{\tfrac{1}{2}(\epsilon_1 + 1)}\,, \qquad (A.152)$$

$$\bar{p}_1 = \sqrt{\gamma^2 - 1} = \sqrt{\tfrac{1}{2}(\epsilon_1 - 1)}\,.$$

Inserting these expressions into the formula for the maximum photon energy in the cms, $\bar{k}_{\max} = \bar{p}_1^2/\bar{\epsilon}_1$, we get

$$\frac{k_{\max}}{\sqrt{2(\epsilon_1 + 1)}}(\epsilon_1 + 1 - \mathbf{p}_1 \cdot \hat{\mathbf{k}}) = \frac{\epsilon_1 - 1}{\sqrt{2(\epsilon_1 + 1)}} \qquad (A.153)$$

or

$$k_{\max}(\epsilon_1, \theta_k) = \frac{\epsilon_1 - 1}{\epsilon_1 + 1 - p_1 \cos \theta_k} \qquad (A.154)$$

which is identical to formula (5.34) in Sec. 5.2.2.

Appendix B

Squared matrix element of electron-electron bremsstrahlung

The squared matrix element of electron-electron bremsstrahlung averaged over the spins of the incoming electrons and summed over the spins of the outgoing electrons and the polarizations of the emitted photon is given by

$$
\begin{aligned}
8A = 2\sum |M|^2 \\
= \frac{1}{2(kp_1)}\Big(\frac{1}{N_1}+\frac{1}{N_3}\Big) + \frac{1}{2(kp_2)}\Big(\frac{1}{N_2}+\frac{1}{N_4}\Big) \\
- \frac{1}{2(kp'_1)}\Big(\frac{1}{N_1}+\frac{1}{N_4}\Big) - \frac{1}{2(kp'_2)}\Big(\frac{1}{N_2}+\frac{1}{N_3}\Big) \\
- 3\big[(p_1p_2)-1\big]\Big\{\frac{1}{(kp_1)}\Big(\frac{1}{N_2}+\frac{1}{N_4}\Big)+\frac{1}{(kp_2)}\Big(\frac{1}{N_1}+\frac{1}{N_3}\Big)\Big\} \\
+ 3\big[(p'_1p'_2)-1\big]\Big\{\frac{1}{(kp'_1)}\Big(\frac{1}{N_2}+\frac{1}{N_3}\Big)+\frac{1}{(kp'_2)}\Big(\frac{1}{N_1}+\frac{1}{N_4}\Big)\Big\} \\
- 4\Big(\frac{1}{N_1N_2}+\frac{1}{N_3N_4}\Big)\big[(p_1p_2)+(p'_1p'_2)\big] - \frac{2}{N_1N_3}\big[(kp_1)+1\big] \\
- \frac{2}{N_2N_4}\big[(kp_2)+1\big] + \frac{2}{N_1N_4}\big[(kp'_1)-1\big] + \frac{2}{N_2N_3}\big[(kp'_2)-1\big] \\
+ \frac{2}{(kp_1)^2}\Big\{\big[2-(p'_1p'_2)\big]\Big(\frac{1}{N_2}+\frac{1}{N_4}\Big) - 1 + \frac{1}{N_2N_4} \\
- (p'_1p'_2)\Big[\frac{(p'_1p_2)}{N_2^2}+\frac{(p_2p'_2)}{N_4^2}\Big]\Big\} \\
+ \frac{2}{(kp_2)^2}\Big\{\big[2-(p'_1p'_2)\big]\Big(\frac{1}{N_1}+\frac{1}{N_3}\Big) - 1 + \frac{1}{N_1N_3}
\end{aligned}
$$

$$-(p_1'p_2')\left[\frac{(p_1p_2')}{N_1^2}+\frac{(p_1p_1')}{N_3^2}\right]\bigg\}$$

$$+\frac{2}{(kp_1')^2}\bigg\{[2-(p_1p_2)]\left(\frac{1}{N_2}+\frac{1}{N_3}\right)-1+\frac{1}{N_2N_3}$$

$$-(p_1p_2)\left[\frac{(p_1p_2')}{N_2^2}+\frac{(p_2p_2')}{N_3^2}\right]\bigg\}$$

$$+\frac{2}{(kp_2')^2}\bigg\{[2-(p_1p_2)]\left(\frac{1}{N_1}+\frac{1}{N_4}\right)-1+\frac{1}{N_1N_4}$$

$$-(p_1p_2)\left[\frac{(p_1'p_2)}{N_1^2}+\frac{(p_1p_1')}{N_4^2}\right]\bigg\}$$

$$+2\bigg\{\frac{(p_1p_2)}{(kp_1)(kp_2)}+\frac{(p_1'p_2')}{(kp_1')(kp_2')}\bigg\}\left(\frac{1}{N_1}+\frac{1}{N_2}+\frac{1}{N_3}+\frac{1}{N_4}\right)$$

$$-\frac{3(p_1p_1')}{(kp_1)(kp_1')}\left(\frac{2}{N_2}+\frac{1}{N_3}+\frac{1}{N_4}\right)-\frac{3(p_1p_2')}{(kp_1)(kp_2')}\left(\frac{1}{N_1}+\frac{1}{N_2}+\frac{2}{N_4}\right)$$

$$-\frac{3(p_1'p_2)}{(kp_1')(kp_2)}\left(\frac{1}{N_1}+\frac{1}{N_2}+\frac{2}{N_3}\right)-\frac{3(p_2p_2')}{(kp_2)(kp_2')}\left(\frac{2}{N_1}+\frac{1}{N_3}+\frac{1}{N_4}\right)$$

$$+\frac{1}{2(kp_1)(kp_1')}\bigg\{\frac{(kp_2)}{N_4}-\frac{(kp_2')}{N_3}\bigg\}+\frac{1}{2(kp_1')(kp_2)}\bigg\{\frac{(kp_1)}{N_1}-\frac{(kp_2')}{N_2}\bigg\}$$

$$+\frac{1}{2(kp_1)(kp_2')}\bigg\{\frac{(kp_2)}{N_2}-\frac{(kp_1')}{N_1}\bigg\}+\frac{1}{2(kp_2)(kp_2')}\bigg\{\frac{(kp_1)}{N_3}-\frac{(kp_1')}{N_4}\bigg\}$$

$$-\frac{1}{(kp_1)N_2^2}\bigg\{(p_1p_2)(p_1'p_2')+(p_1p_2')(p_1'p_2)+2(p_1p_2)(p_1p_2')$$

$$-(p_1'p_2')(kp_2)-(p_1'p_2)(kp_2')+(kp_1')\bigg\}$$

$$-\frac{1}{(kp_1)N_4^2}\bigg\{(p_1p_2)(p_1'p_2')+(p_1p_1')(p_2p_2')+2(p_1p_2)(p_1p_1')$$

$$-(p_1'p_2')(kp_2)-(p_2p_2')(kp_1')+(kp_2')\bigg\}$$

$$-\frac{1}{(kp_2)N_1^2}\bigg\{(p_1p_2)(p_1'p_2')+(p_1p_2')(p_1'p_2)+2(p_1p_2)(p_1'p_2)$$

$$-(p_1'p_2')(kp_1)-(p_1p_2')(kp_1')+(kp_2')\bigg\}$$

$$-\frac{1}{(kp_2)N_3^2}\bigg\{(p_1p_2)(p_1'p_2')+(p_1p_1')(p_2p_2')+2(p_1p_2)(p_2p_2')$$

$$-(p_1'p_2')(kp_1)-(p_1p_1')(kp_2')+(kp_1')\bigg\}$$

$$+\frac{1}{(kp_1')N_2^2}\bigg\{(p_1p_2)(p_1'p_2')+(p_1p_2')(p_1'p_2)+2(p_1'p_2')(p_1'p_2)$$

$$+(p_1p_2)(kp_2')+(p_1p_2')(kp_2)-(kp_1)\bigg\}$$

$$+ \frac{1}{(kp'_1)N_3^2}\{(p_1p_2)(p'_1p'_2) + (p_1p'_1)(p_2p'_2) + 2(p'_1p'_2)(p_1p'_1)$$
$$+ (p_1p_2)(kp'_2) + (p_2p'_2)(kp_1) - (kp_2)\}$$
$$+ \frac{1}{(kp'_2)N_1^2}\{(p_1p_2)(p'_1p'_2) + (p_1p'_2)(p'_1p_2) + 2(p'_1p'_2)(p_1p'_2)$$
$$+ (p_1p_2)(kp'_1) + (p'_1p_2)(kp_1) - (kp_2)\}$$
$$+ \frac{1}{(kp'_2)N_4^2}\{(p_1p_2)(p'_1p'_2) + (p_1p'_1)(p_2p'_2) + 2(p'_1p'_2)(p_2p'_2)$$
$$+ (p_1p_2)(kp'_1) + (p_1p'_1)(kp_2) - (kp_1)\}$$
$$+ \frac{1}{(kp_1)N_1N_2}\{(p_1p_2)(p'_1p'_2) - (p_1p'_2)(p'_1p_2) - 2(p'_1p'_2) - 2(p'_1p_2)\}$$
$$+ \frac{1}{(kp_2)N_1N_2}\{(p_1p_2)(p'_1p'_2) - (p_1p'_2)(p'_1p_2) - 2(p'_1p'_2) - 2(p_1p'_2)\}$$
$$+ \frac{1}{(kp_1)N_3N_4}\{(p_1p_2)(p'_1p'_2) - (p_1p'_1)(p_2p'_2) - 2(p'_1p'_2) - 2(p_2p'_2)\}$$
$$+ \frac{1}{(kp_2)N_3N_4}\{(p_1p_2)(p'_1p'_2) - (p_1p'_1)(p_2p'_2) - 2(p'_1p'_2) - 2(p_1p'_1)\}$$
$$+ \frac{1}{(kp'_1)N_1N_2}\{(p_1p'_2)(p'_1p_2) - (p_1p_2)(p'_1p'_2) + 2(p_1p_2) + 2(p_1p'_2)\}$$
$$+ \frac{1}{(kp'_2)N_1N_2}\{(p_1p'_2)(p'_1p_2) - (p_1p_2)(p'_1p'_2) + 2(p_1p_2) + 2(p'_1p_2)\}$$
$$+ \frac{1}{(kp'_1)N_3N_4}\{(p_1p'_1)(p_2p'_2) - (p_1p_2)(p'_1p'_2) + 2(p_1p_2) + 2(p_2p'_2)\}$$
$$+ \frac{1}{(kp'_2)N_3N_4}\{(p_1p'_1)(p_2p'_2) - (p_1p_2)(p'_1p'_2) + 2(p_1p_2) + 2(p_1p'_1)\}$$
$$+ \frac{1}{(kp_1)N_2N_4}\{2(p'_1p'_2)(kp_2) - 2(p_1p_2)(p'_1p'_2) - (p_1p_2)(p_1p'_1)$$
$$- (p_1p_2)(p_1p'_2) + 4(p_1p_2) - 3(kp_1) + (kp_2) + 4\}$$
$$+ \frac{1}{(kp_2)N_1N_3}\{2(p'_1p'_2)(kp_1) - 2(p_1p_2)(p'_1p'_2) - (p_1p_2)(p'_1p_2)$$
$$- (p_1p_2)(p_2p'_2) + 4(p_1p_2) - 3(kp_2) + (kp_1) + 4\}$$
$$+ \frac{1}{(kp'_1)N_2N_3}\{2(p_1p_2)(kp'_2) + 2(p_1p_2)(p'_1p'_2) + (p_1p'_1)(p'_1p'_2)$$
$$+ (p'_1p'_2)(p'_1p_2) - 4(p_1p_2) + (kp'_1) + 5(kp'_2) - 4\}$$
$$+ \frac{1}{(kp'_2)N_1N_4}\{2(p_1p_2)(kp'_1) + 2(p_1p_2)(p'_1p'_2) + (p_1p'_2)(p'_1p'_2)$$

$$+ (p'_1p'_2)(p_2p'_2) - 4(p_1p_2) + (kp'_2) + 5(kp'_1) - 4\}$$

$$+ \frac{1}{(kp_1)N_2N_3}\{(p_1p_2)(p'_1p'_2) - (p_1p_2) - (p'_1p'_2) + 1\}$$

$$+ \frac{1}{(kp_2)N_2N_3}\{(p_1p_2)(p'_1p'_2) - (p_1p_2) - (p_1p'_1) + 1\}$$

$$+ \frac{1}{(kp_1)N_1N_4}\{(p_1p_2)(p'_1p'_2) - (p_1p_2) - (p_2p'_2) + 1\}$$

$$+ \frac{1}{(kp_2)N_1N_4}\{(p_1p_2)(p'_1p'_2) - (p_1p_2) - (p_1p'_2) + 1\}$$

$$- \frac{1}{(kp'_1)N_1N_3}\{(p_1p_2)(p'_1p'_2) - (p'_1p'_2) - (p_2p'_2) + 1\}$$

$$- \frac{1}{(kp'_2)N_1N_3}\{(p_1p_2)(p'_1p'_2) - (p'_1p'_2) - (p'_1p_2) + 1\}$$

$$- \frac{1}{(kp'_1)N_2N_4}\{(p_1p_2)(p'_1p'_2) - (p'_1p'_2) - (p_1p'_2) + 1\}$$

$$- \frac{1}{(kp'_2)N_2N_4}\{(p_1p_2)(p'_1p'_2) - (p'_1p'_2) - (p_1p'_1) + 1\}$$

$$+ \frac{(p_1p_2)}{(kp_1)(kp_2)N_1N_2}\{(kp'_1)(p_1p'_2) + (kp'_2)(p'_1p_2)$$
$$- 2(p_1p_2)(p'_1p'_2) - 2(p_1p'_2)(p'_1p_2)\}$$

$$+ \frac{(p_1p_2)}{(kp_1)(kp_2)N_3N_4}\{(kp'_1)(p_2p'_2) + (kp'_2)(p_1p'_1)$$
$$- 2(p_1p_2)(p'_1p'_2) - 2(p_1p'_1)(p_2p'_2)\}$$

$$- \frac{(p'_1p'_2)}{(kp'_1)(kp'_2)N_1N_2}\{(kp_1)(p'_1p_2) + (kp_2)(p_1p'_2)$$
$$+ 2(p_1p_2)(p'_1p'_2) + 2(p_1p'_2)(p'_1p_2)\}$$

$$- \frac{(p'_1p'_2)}{(kp'_1)(kp'_2)N_3N_4}\{(kp_1)(p_2p'_2) + (kp_2)(p_1p'_1)$$
$$+ 2(p_1p_2)(p'_1p'_2) + 2(p_1p'_1)(p_2p'_2)\}$$

$$+ \frac{2}{(kp_1)(kp'_1)N_2^2}\{(p_1p'_1)[(p_1p_2)(p'_1p'_2) + (p_1p'_2)(p'_1p_2)$$
$$+ (kp_2)(kp'_2)] - (kp_2)(kp'_2)\}$$

$$+ \frac{2}{(kp_2)(kp'_2)N_1^2}\{(p_2p'_2)[(p_1p_2)(p'_1p'_2) + (p_1p'_2)(p'_1p_2)$$
$$+ (kp_1)(kp'_1)] - (kp_1)(kp'_1)\}$$

$$+ \frac{2}{(kp_1)(kp'_2)N_4^2}\{(p_1p'_2)[(p_1p_2)(p'_1p'_2) + (p_1p'_1)(p_2p'_2)$$
$$+ (kp'_1)(kp_2)] - (kp'_1)(kp_2)\}$$
$$+ \frac{2}{(kp'_1)(kp_2)N_3^2}\{(p'_1p_2)[(p_1p_2)(p'_1p'_2) + (p_1p'_1)(p_2p'_2)$$
$$+ (kp_1)(kp'_2)] - (kp_1)(kp'_2)\}$$
$$+ \frac{1}{(kp_1)(kp_2)N_2N_3}\{(p_1p_2)[(p_1p_2) + 3(p'_1p'_2) - 2(p_1p_2)(p'_1p'_2)]$$
$$+ (kp'_1)[(kp'_1) - (p'_1p'_2) + 3] - 2(p'_1p'_2)(kp'_2)\}$$
$$+ \frac{1}{(kp_1)(kp_2)N_1N_4}\{(p_1p_2)[(p_1p_2) + 3(p'_1p'_2) - 2(p_1p_2)(p'_1p'_2)]$$
$$+ (kp'_2)[(kp'_2) - (p'_1p'_2) + 3] - 2(p'_1p'_2)(kp'_1)\}$$
$$+ \frac{1}{(kp'_1)(kp'_2)N_1N_3}\{(p'_1p'_2)[(p'_1p'_2) + 3(p_1p_2) - 2(p_1p_2)(p'_1p'_2)]$$
$$+ (kp_2)[(kp_2) + (p_1p_2) - 3] + 2(p_1p_2)(kp_1)\}$$
$$+ \frac{1}{(kp'_1)(kp'_2)N_2N_4}\{(p'_1p'_2)[(p'_1p'_2) + 3(p_1p_2) - 2(p_1p_2)(p'_1p'_2)]$$
$$+ (kp_1)[(kp_1) + (p_1p_2) - 3] + 2(p_1p_2)(kp_2)\}$$
$$+ \frac{(p_1p'_1)}{(kp_1)(kp'_1)N_3N_4}\{2(p_1p_2)(p'_1p'_2) + 2(p_1p'_1)(p_2p'_2)$$
$$- (p'_1p'_2)(kp_2) + (p_1p_2)(kp'_2)\}$$
$$+ \frac{(p_2p'_2)}{(kp_2)(kp'_2)N_3N_4}\{2(p_1p_2)(p'_1p'_2) + 2(p_1p'_1)(p_2p'_2)$$
$$- (p'_1p'_2)(kp_1) + (p_1p_2)(kp'_1)\}$$
$$+ \frac{(p_1p'_2)}{(kp_1)(kp'_2)N_1N_2}\{2(p_1p_2)(p'_1p'_2) + 2(p_1p'_2)(p'_1p_2)$$
$$- (p'_1p'_2)(kp_2) + (p_1p_2)(kp'_1)\}$$
$$+ \frac{(p'_1p_2)}{(kp'_1)(kp_2)N_1N_2}\{2(p_1p_2)(p'_1p'_2) + 2(p_1p'_2)(p'_1p_2)$$
$$- (p'_1p'_2)(kp_1) + (p_1p_2)(kp'_2)\}$$
$$+ \frac{1}{(kp_1)(kp'_1)N_2N_4}\{(p_1p'_1)[(p_1p_2)(kp'_2) - (p'_1p'_2)(kp_2) + 2(p_1p_2)(p'_1p'_2)$$
$$- 2(p_1p_2) - (p'_1p'_2) - 1] + \tfrac{1}{2}(kp'_2)[(p_1p_2) - 3(kp_2) - 3] - (kp_2)\}$$

$$+ \frac{1}{(kp_1)(kp_1')N_2N_3}\{(p_1p_1')[(p_1p_2)(kp_2') - (p_1'p_2')(kp_2) + 2(p_1p_2)(p_1'p_2')$$
$$- 2(p_1'p_2') - (p_1p_2) - 1] + \tfrac{1}{2}(kp_2)[3 - 3(kp_2') - (p_1'p_2')] + (kp_2')\}$$
$$+ \frac{1}{(kp_1)(kp_2')N_1N_4}\{(p_1p_2')[(p_1p_2)(kp_1') - (p_1'p_2')(kp_2) + 2(p_1p_2)(p_1'p_2')$$
$$- 2(p_1'p_2') - (p_1p_2) - 1] + \tfrac{1}{2}(kp_2)[3 - 3(kp_1') - (p_1'p_2')] + (kp_1')\}$$
$$+ \frac{1}{(kp_1)(kp_2')N_2N_4}\{(p_1p_2')[(p_1p_2)(kp_1') - (p_1'p_2')(kp_2) + 2(p_1p_2)(p_1'p_2')$$
$$- 2(p_1p_2) - (p_1'p_2') - 1] + \tfrac{1}{2}(kp_1')[(p_1p_2) - 3(kp_2) - 3] - (kp_2)\}$$
$$+ \frac{1}{(kp_1')(kp_2)N_1N_3}\{(p_1'p_2)[(p_1p_2)(kp_2') - (p_1'p_2')(kp_1) + 2(p_1p_2)(p_1'p_2')$$
$$- 2(p_1p_2) - (p_1'p_2') - 1] + \tfrac{1}{2}(kp_2')[(p_1p_2) - 3(kp_1) - 3] - (kp_1)\}$$
$$+ \frac{1}{(kp_1')(kp_2)N_2N_3}\{(p_1'p_2)[(p_1p_2)(kp_2') - (p_1'p_2')(kp_1) + 2(p_1p_2)(p_1'p_2')$$
$$- 2(p_1'p_2') - (p_1p_2) - 1] + \tfrac{1}{2}(kp_1)[3 - 3(kp_2') - (p_1'p_2')] + (kp_2')\}$$
$$+ \frac{1}{(kp_2)(kp_2')N_1N_4}\{(p_2p_2')[(p_1p_2)(kp_1') - (p_1'p_2')(kp_1) + 2(p_1p_2)(p_1'p_2')$$
$$- 2(p_1'p_2') - (p_1p_1) - 1] + \tfrac{1}{2}(kp_1)[3 - 3(kp_1') - (p_1'p_2')] + (kp_1')\}$$
$$+ \frac{1}{(kp_2)(kp_2')N_1N_3}\{(p_2p_2')[(p_1p_2)(kp_1') - (p_1'p_2')(kp_1) + 2(p_1p_2)(p_1'p_2')$$
$$- 2(p_1p_2) - (p_1'p_2') - 1] + \tfrac{1}{2}(kp_1')[(p_1p_2) - 3(kp_1) - 3] - (kp_1)\},$$

(B.1)

where

$$N_1 = (p_1p_1') - 1, \quad N_2 = (p_2p_2') - 1, \quad \text{(B.2)}$$
$$N_3 = (p_1p_2') - 1, \quad N_4 = (p_1'p_2) - 1. \quad \text{(B.3)}$$

A is invariant under the transformations

$$\begin{aligned} &\underline{p_1} \leftrightarrow \underline{p_2}, \; \underline{p_1'} \leftrightarrow \underline{p_2'}; \\ &\underline{p_1} \to \underline{p_1}, \; \underline{p_2} \to \underline{p_2}, \; \underline{p_1'} \leftrightarrow \underline{p_2'}; \\ &\underline{p_1} \leftrightarrow \underline{p_2}, \; \underline{p_1'} \to \underline{p_1'}, \; \underline{p_2'} \to \underline{p_2'}. \end{aligned} \quad \text{(B.4)}$$

The above formula for A was obtained by using the traces evaluated by Anders [114]. Even though the expression is very lengthy it is no problem to evaluate it with a computer code. Numerical values of the cross section

(5.70) evaluated by means of the formula (B.1) agree excellently with the results derived independently by Mack and Mitter [130] using computer codes for formula manipulations.

Five of the invariant four-products are independent; if we choose $(p_1 p_2)$, (kp_1), (kp_2), $(p_1 p_1')$ and (kp_1'), the other 5 products are given by

$$(p_1 p_2') = 1 + (p_1 p_2) - (p_1 p_1') - (kp_1) , \tag{B.5}$$

$$(p_1' p_2) = 1 + (p_1 p_2) - (p_1 p_1') - (kp_1) - (kp_2) + (kp_1') , \tag{B.6}$$

$$(p_1' p_2') = (p_1 p_2) - (kp_1) - (kp_2) , \tag{B.7}$$

$$(p_2 p_2') = (p_1 p_1') + (kp_1) - (kp_1') , \tag{B.8}$$

$$(kp_2') = (kp_1) + (kp_2) - (kp_1') . \tag{B.9}$$

As was stated in Section 7.3, the quantity A gives also the cross section of electron-positron bremsstrahlung if the substitutions (7.5) are applied. Thereby the virtual annihilation graphs in electron-positron collisions (Fig. 7.2) are correctly described.

Bibliography

[1] O. Scherzer, Ann. Phys. (Leipzig) **13** (1932) 137
[2] W. Nakel, Phys. Lett. **22** (1966) 614
[3] W. Nakel, Phys. Lett. **25A** (1967) 569
[4] W. Nakel, Z. Phys. **214** (1968) 168
[5] W. Nakel and E. Pankau, Phys. Lett. **38A** (1972) 307
[6] W. Nakel and E. Pankau, Phys. Lett. **44A** (1973) 65
[7] W. Nakel and E. Pankau, Z. Phys. **264** (1973) 139
[8] H.A. Kramers, Phil. Mag. **46** (1923) 836
[9] G. Wentzel, Z. Physik **27** (1924) 257
[10] A. Sommerfeld, Ann. Physik. (Leipzig) **11** (1931) 257
[11] H. Bethe and W. Heitler, Proc. Roy. Soc. London A **146** (1934) 83
[12] H.A. Bethe and L.C. Maximon, Phys. Rev. **93** (1954) 768
[13] G. Elwert and E. Haug, Phys. Rev. **183** (1969) 90.
[14] H.K. Tseng and R.H. Pratt, Phys. Rev. A **3** (1971) 100.
[15] C.D. Shaffer, X.-M. Tong and R.H. Pratt, Phys. Rev. A **53** (1996) 4158
[16] C.D. Shaffer and R.H. Pratt, Phys. Rev. A **56** (1997) 3653
[17] S. Keller and R.M Dreizler, J. Phys. B: At. Mol. Opt. Phys. **30** (1997) 3257 [Erratum: J. Phys. B: At. Mol. Opt. Phys. 36 (2003) 799]
[18] H.K. Tseng, J. Phys. B: At. Mol. Opt. Phys. **35** (2002) 1129
[19] R.H. Pratt and I.J. Feng, in Atomic Inner-Shell Physics, edited by Bernd Crasemann (Plenum Publishing Corporation, 1985), p. 533
[20] J.D. Jackson, Classical Electrodynamics (John Wiley & Sons, New York, 3nd edition, 1999)
[21] L.D. Landau and E.M. Lifshitz, The Classical Theory of Fields (Pergamon Press, Oxford, 4th edition, 1975)
[22] L. Kim and R.H. Pratt, Phys. Rev. A**36** (1987) 45
[23] E.M. Purcell, Electricity and Magnetism, Berkeley Physics Course, Vol. 2 (McGraw-Hill Book Company, New York, 1965)
[24] E. Haug, Astrophys. Lett. **11** (1972) 225
[25] E. Haug, Astron. Astrophys. **406** (2003) 31

[26] G. Elwert, in Physics of the One- and Two-Electron Atoms, edited by F. Bopp and H. Kleinpoppen (North-Holland Publishing Company, Amsterdam, 1969) p. 700
[27] R.L. Gluckstern and M.H. Hull, Phys. Rev. **90** (1953) 1030
[28] J.W. Motz and R.C. Placious, Nuovo Cimento **15** (1960) 571
[29] E.S. Sobolak and P. Stehle, Phys. Rev. **129** (1963) 403
[30] C.F. v. Weizsäcker, Z. Phys. **88** (1934) 612
[31] E.J. Williams, Kgl. Danske Videnskab. Selskab. Mat.-fys. Medd. **13** (1935), No. 4
[32] W. Heitler, The Quantum Theory of Radiation (Oxford University Press, 3rd edition, 1954)
[33] A.I. Achieser and W.B. Berestezki, Quantenelektrodynamik (B.G. Teubner Verlagsgesellschaft, Leipzig, 1962), p. 664
[34] J.M. Jauch and F. Rohrlich, The Theory of Photons and Electrons (Springer, Berlin, 2nd Edition, 1976)
[35] L.I. Schiff, Quantum Mechanics (McGraw-Hill Book Company, New York, 3rd edition, 1968)
[36] W.H. Furry, Phys. Rev. **81** (1951) 115
[37] S.S. Schweber, An Introduction to Relativistic Quantum Field Theory (Harper & Row, New York, 1961)
[38] C. Kacser, Proc. Roy. Soc. **253** (1959) 103
[39] G. Elwert, Ann. Physik (Leipzig) **34** (1939) 178
[40] R.H. Pratt and H.K. Tseng, Phys. Rev. A **11** (1975) 1797
[41] G. Molière, Z. Naturforsch. **2a** (1947) 133
[42] F. Salvat, J.D. Martinez, R. Mayol, and J. Parellada, Phys. Rev. A **36** (1987) 467
[43] S.D. Drell, Phys. Rev. **87** (1952) 753
[44] R.G. Newton, Phys. Rev. **103** (1956) 385
[45] S. Sarkar, Nuovo Cimento **15** (1960) 686
[46] E.E. Ginsberg and R.H. Pratt, Phys. Rev. **134** (1964) B773; Phys. Rev. **137** (1965) B1500
[47] R. Hofstadter, Rev. Mod. Phys. **28** (1956) 214; Ann. Rev. Nucl. Sci. **7** (1957) 231
[48] L.C. Maximon and D.B. Isabelle, Phys. Rev. **136** (1964) B674
[49] D.F. Hubbard and M.E. Rose, Nucl. Phys. **84** (1966) 337
[50] C.G. Darwin, Proc. Roy. Soc. (London) Ser. A **118** (1928) 654
[51] A. Sommerfeld and A.W. Maue, Ann. Physik (Leipzig) **22** (1935) 629
[52] W.H. Furry, Phys. Rev. **46** (1934) 391
[53] M. Abramowitz and I.A. Stegun (eds.), Handbook of Mathematical Functions (Dover Publications, New York, 1972)
[54] A. Sommerfeld, Atombau und Spektrallinien (Vieweg und Sohn, Braunschweig, Germany, 1951), Vol. II.
[55] G. Breit and H.A. Bethe, Phys. Rev. **93** (1954) 888
[56] N.F. Mott and H.S.W. Massey, The Theory of Atomic Collisions (Oxford

University Press, London, 3rd edition, 1965)
- [57] H. Olsen, Phys. Rev. **99** (1955) 1335 (L)
- [58] H. Olsen and L.C. Maximon, Phys. Rev. **114** (1959) 887
- [59] A. Nordsieck, Phys. Rev. **93** (1954) 785
- [60] J.K. Fink and R.H. Pratt, Phys. Rev. A **7** (1973) 392
- [61] W.R. Johnson and C.J. Mullin, Phys. Rev. **119** (1960) 1270
- [62] R.T. Deck, D.S. Moroi and W.R. Alling, Nucl. Phys. A**133** (1969) 321
- [63] A.R. Edmonds, Angular Momentum in Quantum Mechanics (Princeton University Press, Princeton, 2nd edition, 1960)
- [64] M.E. Rose, Relativistic Electron Theory (John Wiley & Sons, New York, 1961)
- [65] H.A. Olsen, Springer Tracts Mod. Phys. **44** (1968) 83
- [66] H. Brysk, C.D. Zerby and S.K. Penny, Phys. Rev. **180** (1969) 104
- [67] H.K. Tseng, Phys. Rev. A **40** (1989) 6826
- [68] W. Kohn and L.S. Sham, Phys. Rev. **140** (1965) A1133; D.A. Liberman, D.J. Cromer, and J.T. Waber, Comput. Phys. Commun. **2** (1971) 107
- [69] C.M. Lee, L. Kissel, R.H. Pratt, and H.K. Tseng, Phys. Rev. A **13** (1976) 1714 [Erratum: Phys. Rev. A **24** (1981) 2866]
- [70] W.R. Johnson and R.T. Deck, J. Math. Phys. **3** (1962) 319
- [71] J.J. Dugne and J. Proriol, Phys. Rev. A **12** (1975) 842
- [72] R.Y. Yin, O.V. Gabriel and R.H. Pratt, Phys. Rev. A **36** (1987) 1207
- [73] J. McEnnan and M. Gavrila, Phys. Rev. A **15** (1977) 1557
- [74] W.R. Johnson and J.D. Rozics, Phys. Rev. **128** (1962) 192
- [75] E. Haug, Z. Physik **37** (1996) 9
- [76] R.L. Gluckstern, M.H. Hull and G. Breit, Phys. Rev. **90** (1953) 1026
- [77] I.B. Zel'dovich, Dokl. Akad. Nauk SSSR **83** (1952) 63
- [78] C. Fronsdal and H.Überall, Phys. Rev. **111** (1958) 580
- [79] H. Frauenfelder and R.M. Steffen, in Alpha-, Beta-, and Gamma-Ray Spectroscopy, ed. by K. Siegbahn, Vol. 2 (North-Holland Publishing Company, Amsterdam, 1974) p. 1431
- [80] L.C. Maximon, Rev. Mod. Phys. **41** (1969) 193
- [81] P.I. Fomin, Zh. Eksp. Teor. Fiz. **35** (1958) 707 [Sov. Phys. JETP **8** (1959) 629]
- [82] A.N. Mitra, P.Narayanaswamy and L.K. Pande, Nucl. Phys. **10** (1959) 629
- [83] K. Kreuzer and W. Nakel, Phys. Lett. **34**A (1971) 407
- [84] W. Nakel and U. Sailer, Phys. Lett. **31**A (1970) 181
- [85] A. Aehlig and M. Scheer, Z. Phys. **250** (1972) 235
- [86] W. Nakel, Phys. Rep. **243** (1994) 317
- [87] W. Nakel and E. Pankau, Z. Phys. A**274** (1975) 319
- [88] R. Hub and W. Nakel, Phys. Lett. **24**A (1967) 601
- [89] J.D. Faulk and C.A. Quarles, Phys. Lett. **44**A (1973) 317; Phys. Rev. A **9** (1974) 732
- [90] A. Aehlig, L. Metzger, and M. Scheer, Z. Phys. A**281** (1977) 205
- [91] M. Komma and W. Nakel, J. Phys. B: At. Mol. Phys. **15** (1982) 1433

[92] C. Bernardini, F. Felicetti, R. Querzoli, V. Silvestrini, G. Vignola, L. Meneghetti, S. Vitale, and G. Penso, Lett. Nuovo Cimento **1** (1969) 15
[93] R.H. Siemann, W.W. Ash, K. Berkelman, D.L. Hartill, C.A. Lichtenstein, and R.M. Littauer, Phys. Rev. Lett. **22** (1969) 421
[94] C.A. Lichtenstein, W.W. Ash, K. Berkelman, D.L. Hartill, R.M. Littauer, and R.H. Siemann, Phys. Rev. D **1** (1970) 825
[95] W. Lichtenberg, A. Przybylski, and M. Scheer, Phys. Rev. A **11** (1975) 480
[96] H.-H. Behncke and W. Nakel, Phys. Lett. **47A** (1974) 149
[97] H.-H. Behncke and W. Nakel, Phys. Rev. A **17** (1978) 1679
[98] W. Bleier and W. Nakel, Phys. Rev. A **30** (1984) 607
[99] U. Fano, K.W. McVoy, and J.R. Albers, Phys. Rev. **116** (1959) 1159
[100] H.K. Tseng, J. Phys. B: At. Mol. Opt. Phys. **30** (1997) L 317 [Erratum: J. Phys. B **33** (2000) 1471]
[101] K. Güthner, Z. Phys. **182** (1965) 278; P.E. Pencynski and H.L. Wehner, Z. Phys. **237** (1970) 75; A. Aehlig, Z. Phys. **A294** (1980) 291
[102] H.R. Schaefer, W. von Drachenfels, and W. Paul, Z. Phys. **A305** (1982) 213
[103] E. Mergl and W. Nakel, Z. Phys. D**17** (1990) 271
[104] E. Mergl, H.-Th. Prinz, C.D. Schröter, and W. Nakel, Phys. Rev. Lett. **69** (1992) 901
[105] E. Mergl, E. Geisenhofer and W. Nakel, Rev. Sci. Instrum. **62** (1991) 2318
[106] E. Geisenhofer and W. Nakel, Z. Phys. D **37** (1996) 123
[107] S.J. Hall, G.J. Miller, R. Beck, and P. Jennewein, Nucl. Instrum. Meth. Phys. Res. A **368** (1996) 698
[108] J. Naumann, G. Anton, A. Bock, P. Grabmayr, K. Helbing, B. Kiel, D. Menze, T. Michel, S. Proff, M. Sauer, M. Schumacher, T. Speckner, W. Weihofen, and G. Zeitler, Nucl. Instrum. Meth. Phys. Res. A **498** (2003) 211
[109] F. Rambo et al., Phys. Rev. C **58** (1998) 489
[110] F.A. Natter, P. Grabmayr, T. Hehl, R.O. Owens, and S. Wunderlich, Nucl. Instrum. Meth. Phys. Res. B **211** (2003) 465
[111] L.C. Maximon, A. Miniac, Th. Aniel, and E. Ganz, Phys. Rep. **147** (1987) 189
[112] R.M. Laszewski, P. Rullhusen, S.D. Hoblit, and S.F. Lebrun, Nucl. Instrum. Meth. Phys. Res. A **228** (1985) 334
[113] I. Hodes, Ph.D. Thesis, University of Chicago (1953) (unpublished)
[114] T.B. Anders, Dissertation, Universität Freiburg, Germany (1961); Nucl. Phys. **59** (1964) 127
[115] N.M. Shumeiko and J.G. Suarez, J. Phys. G: Nucl. Part. Phys. **26** (2000) 113
[116] M.S. Maxon and E.G. Corman, Phys. Rev. **163** (1967) 156
[117] E. Haug, Z. Naturforsch. **30a** (1975) 1099
[118] M.A. Coplan, J.H. Moore, and J.P. Doering, Rev. Mod. Phys. **66** (1994)

985 [Erratum: Rev. Mod. Phys. **66** (1994) 1517]
[119] P. Eisenberger and P.M. Platzman, Phys. Rev. A**2** (1970) 415
[120] Compton Scattering, ed. by B.G. Williams (McGraw-Hill Book Company, New York, 1977)
[121] E. Haug and M. Keppler, J. Phys. B: Atom. Mol. Phys. **17** (1984) 2075
[122] E. Clementi and C. Roetti, At. Data Nucl. Data Tables **14** (1974) 177
[123] A.D. McLean and R.S. McLean, At. Data Nucl. Data Tables **26** (1981) 197
[124] B.G. Williams, G.M. Parkinson, C.J. Eckhardt, J.M. Thomas, and T. Sparrow, Chem. Phys. Lett. **78** (1981) 434
[125] F. Bell, H. Böckl, M.Z. Wu, and H.-D. Betz, J. Phys. B: Atom. Molec. Phys. **16** (1983) 187
[126] D.H. Rester, Nucl. Phys. A**118** (1968) 129
[127] E. Haug, Phys. Lett. **54A** (1975) 339
[128] P. Hackl, Dissertation, University of Vienna, 1970 (unpublished); H. Aiginger and E. Unfried, Acta Phys. Austriaca **35** (1972) 331
[129] W. Nakel and C.T. Whelan, Phys. Rep. **315** (1999) 409
[130] D. Mack and H. Mitter, Phys. Lett. **44A** (1973) 71
[131] W. Bleier and W. Nakel, Phys. Rev. A **30** (1984) 661
[132] H.W. Koch and J.W. Motz, Rev. Mod. Phys. **31** (1959) 920
[133] R.H. Pratt, H.K. Tseng, C.M. Lee, L. Kissel, C. MacCallum, and M. Riley, Atomic Data Nucl. Data Tables **20** (1977) 175 [Erratum: Atomic Data Nucl. Data Tables **26** (1981) 477]
[134] L. Kissel, C.A. Quarles, and R.H. Pratt, Atomic Data Nucl. Data Tables **28** (1983) 381
[135] K. Bernhardi, E. Haug, and K. Wiesemann, Atomic Data Nucl. Data Tables **28** (1983) 461
[136] S.M. Seltzer and M.J. Berger, Atomic Data Nucl. Data Tables **35** (1986) 345
[137] R. Ambrose, J.C. Altman, and C.A. Quarles, Phys. Rev. A **35** (1987) 529
[138] S. Portillo and C.A. Quarles, Phys. Rev. Lett. **91** (2003) 173201
[139] I.J. Feng, R.H. Pratt and H.K. Tseng, Phys. Rev. A **24** (1981) 1358
[140] E. Haug, Phys. Rev. D **31** (1985) 2120 [Erratum: Phys. Rev. D **32** (1985) 1594]
[141] E. Haug, Solar Phys. **178** (1998) 341
[142] E. Haug, Astron. Astrophys. **218** (1989) 330
[143] E. Haug, Eur. Phys. J. C **31** (2003) 365
[144] J.W. Motz, Phys. Rev. **100** (1955) 1560
[145] N. Starfelt and H.W. Koch, Phys. Rev. **102** (1956) 1598
[146] H.K. Tseng, Chin. J. Phys. **40** (2002) 168
[147] S. Jadach, B.F.L. Ward, and S.A. Yost, Phys. Rev. D **47** (1993) 2682; S Jadach, M. Melles, B.F.L. Ward, and S.A. Yost, Phys. Lett. B **377** (1996) 168
[148] L. Rosenberg, Phys. Rev. A **44** (1991) 2949
[149] C.A. Quarles and J. Liu, Nucl. Instrum. Meth. Phys. Res. B **79** (1993) 142

[150] W. Heitler and L. Nordheim, Physica **1** (1934) 1059
[151] A.I. Smirnov, Yad. Fiz. **25** (1977) 1030 [Sov. J. Nucl. Phys. **25** (1977) 548]
[152] V. Véniard, M. Gavrila, and A. Maquet, Phys. Rev. A **35** (1987) 448
[153] A.V. Korol, J. Phys. B: At. Mol. Opt. Phys. **27** (1994) 155
[154] M. Dondera and V. Florescu, Phys. Rev. A **48** (1993) 4267
[155] M. Dondera, V. Florescu, and R.H. Pratt, Phys. Rev. A **53** (1996) 1492
[156] F.E. Low, Phys. Rev. **110** (1958) 974
[157] M. Gavrila, A. Maquet, and V. Véniard, Phys. Rev. A **32** (1985) 2537 [Erratum: Phys. Rev. A **33** (1986) 2826]
[158] V. Florescu and V. Djamo, Phys. Lett. A **119** (1986) 73
[159] M. Gavrila, A. Maquet, and V. Véniard, Phys. Rev. A **42** (1990) 236
[160] A.V. Korol, J. Phys. B: At. Mol. Opt. Phys. **26** (1993) 3137
[161] R. Hippler, Phys. Rev. Lett. **66** (1991) 2197
[162] M. Dondera and V. Florescu, Phys. Rev. A **58** (1998) 2016
[163] A.V. Korol, J. Phys. B: At. Mol. Opt. Phys. **29** (1996) 3257
[164] A.V. Korol, J. Phys. B: At. Mol. Opt. Phys. **30** (1997) 413
[165] D.L. Kahler, J. Liu, and C.A. Quarles, Phys. Rev. Lett. **68** (1992) 1690
[166] T.A. Fedorova, A.V. Korol, and I.A. Solovjev, Surface Rev. Lett. **9** (2002) 1185
[167] G. Kracke, G. Alber, J.S. Briggs, and A. Maquet, J. Phys. B: At. Mol. Opt. Phys. **26** (1993) L561
[168] G. Kracke, J.S. Briggs, A. Dubois, A. Maquet, and V. Véniard, J. Phys. B: At. Mol. Opt. Phys. **27** (1994) 3241
[169] E.A. Kuraev, A. Schiller, and V.G. Serbo, Z. Physik C **30** (1986) 237
[170] M.Ya. Amusia, Phys. Rep. **162** (1988) 249
[171] V.N. Tsytovich and I.M. Ojringel, Polarization Bremsstrahlung (Plenum, New York, 1992)
[172] A.V. Korol and A.V. Solov'yov, J. Phys. B: At. Mol. Opt. Phys. **30** (1997) 1105
[173] V.A. Astapenko, L.A. Bureeva, and V.S. Lisita, Uspekhi Fiz. Nauk **172** (2002) 155 [Phys. – Uspekhi **45** (2002) 149]
[174] C.A. Quarles, Rad. Phys. Chem. **59** (2000) 159
[175] V.M. Buĭmistrov and L.I. Trakhtenberg, Zh. Eksp. Teor. Fiz. **69** (1975) 108 [Sov. Phys. JETP **42** (1975) 54]
[176] A. Dubois and A. Maquet, Phys. Rev. A **40** (1989) 4288
[177] N.B. Avdonina and R.H. Pratt, J. Phys. B: At. Mol. Opt. Phys. **32** (1999) 4261
[178] M.Ya. Amusia, N.B. Avdonina, L.V. Chernysheva, and M.Yu. Kuchiev, J. Phys. B: At. Mol. Opt. Phys. **18** (1985) L791
[179] A.V. Korol, A.G. Lyalin, A.V. Solov'yov, N.B. Avdonina, and R.H. Pratt, J. Phys. B: At. Mol. Phys. **35** (2002) 1197
[180] M.Ya. Amusia and A.V. Korol, J. Phys. B: At. Mol. Phys. **24** (1991) 3251
[181] M.Ya. Amus'ya, M.Yu. Kuchiev, A.V. Korol', and A.V. Solov'ev, Zh. Eksp. Teor. Fiz. **88** (1985) 383 [Sov. Phys. JETP **61** (1985) 224]

[182] V.A. Astapenko, V.M. Buĭmistrov, Yu.A. Krotov, L.K. Mikhaĭlov, and L.I. Trakhtenberg, Zh. Eksp. Teor. Fiz. **88** (1985) 1560 [Sov. Phys. JETP **61** (1985) 930]
[183] A.V. Korol, O.I. Obolensky, A.V. Solov'yov, and I.A. Solov'yov, J. Phys. B: At. Mol. Opt. Phys. **34** (2001) 1589
[184] A.V. Korol', A.G. Lyalin, O.I. Obolenskiĭ, A.V. Solov'ev, and I.A. Solov'ev, Zh. Eksp. Teor. Fiz. **121** (2002) 819 [JETP **94** (2002) 704]
[185] V.B. Gavrikov, V.P. Likhachev, and V.A. Romanov, Nucl. Instrum. Meth. Phys. Res. A **457** (2001) 411
[186] J. Freudenberger et al., Phys. Rev. Lett. **74** (1995) 2487
[187] G. Diambrini Palazzi, Rev. Mod. Phys. **40** (1968) 611
[188] U. Timm, Fortschr. Phys. **17** (1969) 765
[189] A.W. Sáenz and H. Überall (eds.), Coherent Radiation Sources, Topics in Current Physics, Vol. 38 (Springer-Verlag, Berlin, Heidelberg, 1985)
[190] D. Lohmann et al., Nucl. Instrum. Meth. Phys. Res. A **343** (1994) 494
[191] V.N. Baier, V.M. Katkov, and V.M. Strakhovenko, Electromagnetic Processes at High Energies in Oriented Single Crystals (World Scientific Publishing Company, Singapore, 1998)
[192] A. Kubankin, N. Nasanov, and P. Zhukova, Phys. Lett. A **317** (2003) 495
[193] H. Huisman, J.C.S. Bacelar, M.J. van Goethem et al., Phys. Lett. B **476** (2000) 9
[194] H. Huisman, J.C.S. Bacelar, M.J. van Goethem et al., Phys. Rev. C **65** (2002) 031001
[195] M. Volkerts, J.C.S. Bacelar, M.J. van Goethem et al., Phys. Rev. Lett. **90** (2003) 062301
[196] N. Takigawa, Y. Nozawa, K. Hagino, A. Ono, and D.M. Brink, Phys. Rev. C **59** (1999) R593 [Erratum: Phys. Rev. C **60** (1999) 069901]
[197] P. Mészarós, High-Energy Radiation from Magnetized Neutron Stars (The University of Chicago Press, Chicago & London, 1992)
[198] J. Lauer, H. Herold, H. Ruder, and G. Wunner, J. Phys. B: At. Mol. Phys. **16** (1983) 3673
[199] K. Wille, Rep. Prog. Phys. **54** (1991) 1005
[200] I.M. Ternov, Phys. Uspekhi **38** (1995) 409
[201] Synchrotron Radiation Theory and Its Development, ed. by V.A. Bordovitsyn (World Scientific Publishing Company, Singapore, 1999)
[202] M.V. Fedorov, Atomic and Free Electrons in a Strong Light Field (World Scientific Publishing Company, Singapore, 1997)

Index

beaming, relativistic, 11, 15, 70, 74, 162, 210
Bethe, H., 4, 59, 60, 63
Bethe-Heitler formula, 32, 44, 108
Born approximation, 32, 160, 194, 198, 201
   condition for validity of, 46, 160
   cross section, 44, 160, 194, 198
   first, 33, 36, 136
   matrix element, 37, 158
   second, 33, 36, 76, 98, 121, 190
bremsstrahlung
   angular distribution
      electron, 123, 125
      photon, 16, 121, 122, 163, 178
   asymmetry, 97, 104, 136
   atomic-field, 187, 190
   classical calculation, 9, 17
   coherent, 143, 206
   elastic, 27, 79, 205
   electron-atom, 5, 89, 153, 203
   electron-electron, 23, 147, 157, 175, 188
      photon angular distribution, 25, 163 – 165, 178
      from bound electrons, 165
      kinematics, 148
   electron-nucleus, 8, 9, 55, 119
   electron-positron, 188, 192, 195, 247

   energy distribution, 125 – 127, 178
   from heavy particles, 206
   in α decay, 207
   in β decay, 207
   in magnetic fields, 208
   in nuclear decays, 207
   in orbital-electron capture, 207
   polarization, 18, 104, 128, 182
   positron-nucleus, 190
   spontaneous, 210
   stimulated, 210
   two-photon, 196

center-of-mass system, 151, 162
characteristic line spectrum, 4
Clebsch-Gordan coefficients, 81, 85
Coulomb field, 27, 55
Coulomb potential, 27, 32, 45, 55, 93
   screened, 36, 47
cross section, 28, 63, 76, 91, 157, 194, 197
   doubly differential, 74, 187
   fivefold differential, 197
   fourfold differential, 167, 198
   integrated, 187
   spin dependent, 97, 103
   triply differential, 31, 68, 78, 91, 103, 159, 193

crystalline targets, 206

Darwin's solution, 88
dipole approximation, 4, 11, 72, 198, 201
dipole radiation, 10, 20, 74
Dirac equation, 28, 32, 79
Duffin-Kemmer equation, 133

Elwert factor, 46, 74, 160, 191, 201
experimental devices, 119

Feynman diagrams
   electron-electron bremsstrahlung, 148
   electron-nucleus bremsstrahlung, 33
   electron-positron bremsstrahlung, 193
   two-photon bremsstrahlung, 197
form factor
   atomic, 48
   magnetic, 52
   nuclear, 52
Furry picture, 33, 55, 89

Hartree-Slater central potential, 188
high-frequency limit, 13, 18, 24, 96, 124, 161, 175
hypergeometric function, 64
   confluent, 56, 94
      asymptotic behaviour of, 59
      transformations of, 75

independent-particle approximation, 89

laboratory system, 24, 152, 161, 194
longitudinally polarized electrons, 107, 114, 141, 235, 238
long-wavelength limit, 46, 49, 75

matrix element, 38, 63, 158

nonrelativistic approximation, 71, 198, 201, 208
nuclear recoil, 50
nuclear structure, 51

partial-wave expansion, 4, 79
   for Coulomb potential, 93
   for relativistic self-consistent field, 89
   numerical results, 121, 125, 138 – 141
photon emission asymmetry, 97, 104, 136 – 141
photons
   real, 115
   tagged, 143
   virtual, 26, 115
photon spectrum, 167, 171, 177
polarization, 18, 91, 128, 182
   circular, 107, 112
   coefficients, 91, 104, 107, 115, 130, 137
   correlations, 91, 102, 104, 185
   elliptic, 142
   linear, 18, 108, 128, 182
   spin, 61, 136

summation over, 43
polarization bremsstrahlung, 188, 203

quadrupole radiation, 23, 148, 175

radiative corrections, 96, 115
relativistic beaming, 11, 15, 70, 74, 162, 210
relativistic self-consistent field, 89, 191, 205
Röntgen, W.C., 4, 212

Scherzer, O., 4
screening, 16, 47, 76, 92
short-wavelength limit, 46, 69, 72, 78, 93, 96, 115, 150, 161, 191, 196, 209
Sommerfeld, A., 4, 28, 55, 60, 72
Sommerfeld-Maue wave function, 55, 58, 95, 98, 104
spin formalism, 61
spin summation, 39
split spinor representation, 61, 79
Stokes parameters, 91, 107
synchrotron radiation, 210

tagged photons, 143
transversely polarized electrons, 97, 136

Weizsäcker-Williams method, 26, 197